Advances in Autonomous Mini Robots

Ulrich Rückert, Joaquin Sitte,
and Felix Werner (Eds.)

Advances in Autonomous Mini Robots

Proceedings of the 6th AMiRE Symposium

Springer

Editors
Prof. Dr. Ulrich Rückert
Cognitive Interaction Technology Centre of Excellence
Cognitronics and Sensor Systems
Bielefeld
Germany

Dr. Felix Werner
Cognitive Interaction Technology Centre of Excellence
Cognitronics and Sensor Systems
Bielefeld
Germany

Dr. Joaquin Sitte
Queensland University of Technology
Faculty of Science and Engineering
Brisbane
Queensland
Australia

ISBN 978-3-662-52160-1 ISBN 978-3-642-27482-4 (eBook)
DOI 10.1007/978-3-642-27482-4
Springer Heidelberg New York Dordrecht London

Printed on acid-free paper

Springer is part of Springer Science+Business Media (www.springer.com)

Preface
Ten Years of AMiRE Symposia

Introduction

With this volume, containing the extended contributions to the 6-th AMiRE Symposium we celebrate the 10-th anniversary of the AMiRE symposium series (amiresymposia.org). All contributed papers have been rigorously peer reviewed. We thank the authors and the reviewers for their work that made this volume possible.

Before introducing the articles in this volume we wish to share with you some reflections on the motivations for the AMiRE Symposia and the contributions they may have made to the advancement of the field of autonomous robotics.

The first International Symposium on Autonomous Mini robots for Research and Edutainment (AMiRE) was held in 2001 at the Heinz Nixdorf Institute of the University of Paderborn, Germany. Our motivation to call this symposium was to promote the use of mini robots as a tool for research and development of autonomous machine technology. We wanted to assist in giving recognition and legitimacy to researchers that were designing and using small size and inexpensive robots to develop ingenious technical solutions to make substantive research contributions. In 2001 robot soccer tournaments were only a few years old and the Khepera robot had just grabbed a foothold in the market as the quintessential mini robot. The impact on the initial objective of the rapid technology development was reported in the AMiRE Symposia that followed in 2003 (Brisbane), 2005 (Fukui), 2007 (Buenos Aires) and 2009 Incheon.

The question that inevitably comes up is: How small does a robot have to be to be called a mini robot. Obviously, it does not make sense to give a strict size limit. Instead, we have gradually arrived at the following conditions that a robot should meet to qualify as mini. The first criterion is that it must be suitable for *desktop* experimentation. Here, the word desktop is given a broad meaning as it may extend to a large table, the floor of an office, a living room or a small laboratory. This also means that a mini robot can be carried around easily. The next criterion is cost. The robots must be affordable to schools and university departments. Ideally, students and hobbyist should be able to afford them. Taking well known robots as examples, the Sony Aibo robot is near the top of the *mini* scale and the Alice robot is near the bottom of the scale. Critics may say that in this size and cost range you can only have toys, with very limited capabilities for research and even education. The work presented by researchers at the various AMiRE Symposia easily refutes this criticism. In 2001 it was true that the available microelectronic technology was not fully exploited by the mini robots of the time, and it remains true today. It is one of the AMiRE symposium goals to promote the full exploitation of the available electronics, sensor and actuator technology for making highly capable robots in a small body at a low cost.

Another argument against mini robots has been that for robots to be useful they must be commensurate with the scale of things in the human environment. It is true that a mini robot, as defined above, will not be able to open the door for a frail elderly person nor unload the groceries from your car. However, the required cognitive capabilities and physical dexterity required for performing helpful tasks are not specific to size. If the world was shrunk to match the size of the door, the groceries or the car to the mini robot the required cognitive abilities and the dexterity would remain the same for the mini robot as for the larger robot matched to a human scale environment. Useful robot behaviours that work at the mini scale can be scaled-up in most cases to the desired scale.

Mini robots in 2001 were mainly powered by 16 bit processors. Today it is possible to have several 32 bit microprocessors on a mini robot in addition to a full fledged CPU for substantial number crunching with hundreds of megabytes of memory, if not Gigabytes. Thanks to the enormous growth of the mobile phone market wireless communication is ubiquitous and cheap, and so are digital image sensors. The stage is set for a big jump in the capabilities of mini robots. The contributions in this volume provide plenty of inspiration.

Article Previews

Three, at times overlapping themes stood out at AMiRE 2011:

- Robotics for education
- Workbenches for multi-robot experimentation
- New walking, flying and balancing robot platforms that make innovative use of off-the-shelf sensing devices.

In addition to the contributed papers we were fortunate to host four outstanding keynote lecturers. Educational robotics was opened by Paolo Dario's keynote lecture on *Bioinspired Minirobotic Platforms for Educational Activities*, that highlighted the remarkable work in progress at the BioRobotics Institute of Scuola Superiore SantAnna in the Local Educational Laboratory on Robotics programe. In this program activities in robotic are not limited to teaching in science, engineering and mathematics, but also in literature, arts, popular culture, philosophy and sociology, and cultural studies. Biorobotics was also the theme of Roger Quinn's keynote lecture on *Animals as models for robot mobility and autonomy: Crawling, walking, running, climbing and flying*. Insigths into the winning strategies for RoboCup robot soccer were revealed by Manuela Veloso in her keynote lecture on *Teamwork Planning in Adversial Multi-Robot Domains*. Non-adversial cooperation of robots and humans was the focus of Naoyuki Kubota's keynote lecture on *Learning and Adaptation for Human-friendly Robot Partners in Informationally Structured Spaces*. The content of this lecture is found in the paper of the same title in this volume.

The presentation of contributed papers started with *Teaching with mini robots: The Local Educational Laboratory on Robotics* by Salvini et al., where members of Paolo Darios's BioRobotics Institute report on the remarkable robotics program for primary and secondary schools of the Valdera region of Tuscany, Italy.

In *A two years informal learning experience using the Thymio robot* Riedo et al. tell what worked and what didn't in a robotics workshop for children at the yearly robotics festival at EPFL using a purpose designed inexpensive robot kit.

Gonzalez-Gomez et al. show the educational potential of the new world of low cost *3D printing*, where everyone can print its own robot parts, in *New Open Source 3D-printable Mobile Robotic Platform for Education*.

Uribe et al. in *German LaRA: An Autonomous Robot Platform Supported by an Educational Methodology* describe and evaluate a deep learning approach for teaching robotics to tertiary students practised at the Tecnológico de Monterrey, Cuernavaca, Mexico.

In *Mutual Learning for Second Language Education and Language Acquisition of Robots*, Yorita and Kubota describe their way of assisting English language learning for Japanese speakers using robots as partners where both the pupil and the robot learn from each other.

Various kinds of small low cost toy helicopters have become commercially availabe in recent years. Pradalier et al. from the ETH (Swiss Federal Institute of Technology) in Zurich have taken this technology a step further when they report on *The CoaX micro-helicopter: a flying platform for education and research* packed with sensors that can be bought from Skybotics in Switzerland.

Also packed with processing power is the *AMiRo Autonomous Mini Robot for research and education* described Herbrechtsmeier et al. This is a small wheeled robot designed for the vision based, fully autonomous AMiRESot soccer tournament. The robot is based on a complete set of electronic modules that can also be used for robots of other physical configurations. Tetzlaff and Witkowski also report on a low budget construction of a robot for AMiRESot through carefully managed student projects in *Modular Robot Platform for Teaching Digital Hardware Engineering and for Playing Robot Soccer in the AMiREsot League*. Their two-wheeled differentially steered robot does not need castors because it can move while balancing on its two wheels. Witkowski and his students also present the first *Radar Sensor Implementation into a Small Autonomous Vehicle*.

Mini robots for research are exemplified in *The wanda robot and its development system for swarm algorithms*. In their work Kettler et al. present the 51 mm diameter low cost, yet powerful, robot aimed at investigating swarm algorithms with many tens, or even a hundred, of these robots. They also developed truthful simulation software for the robot and an experimentation arena for analysing swarm behaviours.

An upscale arena for *Multi-Robot System Validation: From simulation to Prototyping with Minirobots in the Teleworkbench* is described Tanoto et al. In a 4m by

4m area the Teleworkbench can capture and evaluate the video record of multirobot experiments from several cameras simultaneously. Di Paola et al. describe *An Experimental Testbed for Robotic Network Applications* that is a small scale and low cost arena with associated software for multirobot experimentation using Khepera II, E-Puck and Lego NTX robots.

Demanding vision tasks such as real time object recognition and visual navigation are now within reach of mini robots thanks to ingenious use of off-the-shelf hardware. Hafiz and Murase show in their paper *iRov: A Robot Platform for Active Vision Research and as Education Tool* how to combine the iPhone cameras with special optics to obtain panoramic, peripheral and foveal vision and achieve real time object recognition. Lange et al. report in *Autonomous Corridor Flight of a UAV Using a Low-Cost and Light-Weight RGB-D Camera* how they obtain the 3D structure of an indoor environment for autonomous navigation of a quad-copter with the low cost Microsoft Kinect sensor.

The following two papers address the important topic of selective attention. Maire et al. describe a method for *Segmentation of Scenes of Mobile Objects and Demonstrable Backgrounds* useful for selective attention in robot vision.

In *A Real-Time Event-Based Selective Attention System for Active Vision* Sonnleithner and Indiveri describe a vision system using spiking neurons on VLSI chips for the extremely fast localisation of regions of change in the visual field.

There are several examples of ingenuous use of small robots as a tool for research in other science and engineering fields. In *The ARUM Experimentation Platform : an "Open" Tool to evaluate Mobile Systems Applications* Severac and Roy report on using robots as carriers of wireless communication units in a large arena to investigate the performance of mobile wireless networks in controlled realistic experiments.

Very unique is the idea of Riedenklau and Petke on *Embodied Social Networking with Gesture-enabled Tangible Active Objects* of using simple small robots as mobile tangible objects for exploring new modalities in human computer interfaces.

Robots are also a research tool for biological studies. Schneider et al. designed and built *HECTOR, a New Hexapod Robot Platform with Increased Mobility - Design and Communication* as a tool for research into insect walking and also for biologically inspired locomotion for robots. It is known that insects legs cooperate with a high degree of autonomy to produce efficient walking gaits. The design of *Hector* incorporates this idea and others from the extensive knowledge of insect walking accumulated by researchers at Bielefeld University. Unlike other walking robots, *Hector* has an elastic element built into each of the eighteen joints for obtaining a compliant dynamic behaviour similar to the action of muscles.

A pure engineering approach to cooperative walking by autonomous legs is described by De Silva and Sitte in *Force Controlled Hexapod Walking* where elasticity is obtained by position and contact force feedback control.

Conclusion and Acknowledgements

The papers in this volume reflect the state-of-the-art of mini robotics research and technology and surely will make useful reading for anyone with an interested in this field. Last but not least we wish to thank the Cognitive Interaction Technology Excellence Centre (CITEC), funded by the German Research Foundation (DFG), and the Centre for Interdisciplinary Research (ZIF) of Bielefeld University for their financial and material contribution to the success of this highly productive meeting.

Ulrich Rückert
Joaquin Sitte
AMiRE 2011 Co-chairs

AMiRE 2011 Organisation

General Co-chairs

Ulrich Rückert	Bielefeld University, Germany
Joaquin Sitte	Queensland University of Technology, Australia

Advisory Board

Illah Nourbakhsh	Carnegie Mellon University, USA
Naoyuki Kubota	Tokyo Metropolitan University, Japan
Tucker Balch	Georgia Institute of Technology, USA
Francesco Mondada	Swiss Federal Institute of Technology Zürich, Switzerland

Technical Co-chairs

Felix Werner	Bielefeld University, Germany
Ulf Witkowski	South Westphalia University of Applied Sciences, Germany

Technical Committee

Michael Angermann	German Aerospace Center, Germany
Erik Berglund	Örebro University, Sweden
Thomas Braeunl	The University of Western Australia , Australia
Ansgar Bredenfeld	Fraunhofer Institute, Germany
Gilles Caprari	Swiss Federal Institute of Technology Zürich, Switzerland
Jeremi Gancet	Space Application Services, Belgium
Shlomo Geva	Queensland University of Technology, Australia
Roberto Guzman	Robotnik Automation, Spain

Giacomo Indiveri	Swiss Federal Institute of Technology Zürich, Switzerland
Narongdech Keeratipranon	Dhurakij Pundit University, Thailand
Frederic Maire	Queensland University of Technology, Australia
Lino Forte Marques	Universidade de Coimbra, Portugal
Nikolaos Mavridis	United Arab Emirates University, United Arab Emirates
Dylan Richard Muir	University of Zürich, Switzerland
Kazuyuki Murase	University of Fukui, Japan
Fiorella Operto	School of Robotics, Italy
Jacques Penders	Sheffield Hallam University, United Kingdom
Fernando Ramos	Tecnologico de Monterrey, Mexico
Axel Schneider	Bielefeld University, Germany
Serge Stinckwich	Universite' de Caen Basse Normandie, France
Hartmut Surmann	Fraunhofer Institute, Germany
Mattias Wahde	Chalmers University of Technology, Sweden

Local Organiser

Cordula Heidbrede	Bielefeld University, Germany

Website

Peter Christ	Bielefeld University, Germany

Contents

Keynote Lectures

Papers

Bioinspired Minirobotic Platforms for Educational Activities

Paolo Dario

Abstract

Since the early 1990s the BioRobotics Institute (BRI) of Scuola Superiore SantAnna has been active in the field of educational robotics by participating and organising numerous events and activities at the national and international levels. Recently, the BRI has increased its activities in this emerging field of robotics, by launching a new educational initiative called the Local Educational Laboratory on Robotics (LELR), in collaboration with the local schools and the municipalities of the Valdera area in Tuscany (Italy). The LELR main goal is to foster the development of scientific and technological knowledge in the Valdera community starting from school level. Drawing on preliminary activities and experiences with LELR, a few projects about teaching with robots in primary and secondary schools will be reported and discussed. A second very important aspect of our activities in educational robotics is the attention we gave to the use of biomimetic and bioinspired design as a method to promote scientific and interdisciplinary education in children. Many cases will be analysed and reported on this approach, that we call the Robotics Zoo of the Scuola Superiore SantAnna. Finally, activities related to the use of robots for theatre representations will be discussed.

Biography

Paolo Dario is a Professor of Biomedical Robotics at the Scuola Superiore SantAnna in Pisa, Italy. He is also a Visiting Professor at Waseda University, Japan, and at

Paolo Dario
The BioRobotics Institute
Scuola Superiore Sant'Anna
Polo Sant'Anna Valdera
Viale Rinaldo Piaggio 34
56025 Pontedera (Pisa), Italy
e-mail: p.dario@sssup.it

Zhejiang University, China. He is the Director of The BioRobotics Institute of the Scuola Superiore SantAnna, comprising a team of about 140 researchers, including 80 PhD students.

His main research interests are in the fields of medical robotics, bio-robotics, mechatronics and micro/nanoengineering. He is the coordinator of many national and European projects (including the current FET-Flagship Pilot on Robot Companions for Citizens), the editor of two books on the subject of robotics, and the author of more than 220 scientific papers (more than 180 on ISI journals). He is Editor-in-Chief, Associate Editor and member of the Editorial Board of many international journals. He has been a plenary invited speaker in many international conferences.

Prof. Dario has served as President of the IEEE Robotics and Automation Society in the years 2002-2003. He has been the General Chair of the IEEE RAS-EMBS BioRob06 Conference and of the 2007 IEEE International Conference on Robotics and Automation (ICRA07). Prof. Dario is an IEEE Fellow, a Fellow of the European Society on Medical and Biological Engineering, a Fellow of the School of Engineering of the University of Tokyo, and a recipient of many honors and awards, including the Joseph Engelberger Award. He is also a member of the Board of the International Foundation of Robotics Research (IFRR).

Animals as Models for Robot Mobility and Autonomy: Crawling, Walking, Running, Climbing and Flying

Roger Quinn

Abstract

The biorobotics program at Case Western Reserve University (CWRU) has been active for more than 20 years. This presentation highlights many of the projects undertaken during that time and describes how neuromechanical principles have benefited a number of robots. As this list of principles grows, so does the functionality and performance of the biorobots.

We use biological inspiration to incorporate neuromechanical principles of locomotion and autonomy into robot designs. The dual goals are to develop useful robots and also to develop neuromechanical models of animals to test hypotheses about their design, movement and control. These goals are complementary. Better models lead to more efficient experiments and new neuromechanical knowledge, which points the way to improved robot designs and animal models.

A robot that captures the leg designs important for cockroach locomotion will be extremely agile and therefore suitable for many missions. For example, the after action report for the robot search and rescue mission at the World Trade Center recommends that legs be used instead of tracks or wheels because they can better adapt to complex terrain. However, before a robot with the intricate leg designs of a burrowing animal such as an insect can be deployed some technical issues must be solved. Therefore, the Quinn-Ritzmann groups are using two complementary approaches to develop mobile robots. Using the direct approach we have developed a series of robots that are each more similar to cockroach. These have multi-segmented legs requiring a controller that captures neurobiological principles. Models of insect legs are being used to understand animal leg control circuits and how descending

Roger Quinn
Biologically Inspired Robotics Group
Case Western Reserve University
Glennan 418
Cleveland, OH 44106-7222, USA
e-mail: rdq@case.edu

commands from the brain interact with intermediate and local networks to profoundly change leg movements and coordinate legs. This knowledge is simplifying the control circuits for our legged robots and making them more robust.

In the more abstract biorobotics approach the fundamental principles of cockroach locomotion are applied using existing technologies. Robots called Whegs have mechanical designs that passively solve lower level motor control problems and their subsequent agility makes them suitable for many applications in the near term. Small robots called Mini- Whegs can run rapidly over relatively large obstacles and even jump up stairs. A Mini- Whegs with specially designed legs and animal inspired adhesive feet can climb vertical glass walls. It places each of its adhesive feet on the wall, propels itself through the stance phase, and peals its feet from the wall mimicking insect foot motions. Mini- Whegs has also been integrated with a micro air vehicle to form MALV (micro air and land vehicle). A new robot called DIGbot uses a biologically inspired concept called Distributed Inward Gripping (DIG) to walk inverted.

The WTC report also recommended that search and rescue robots should be capable of autonomous locomotion. A long term goal is to develop an artificial insect head with sensors and a guidance and stabilizing system. Preliminary research resulted in a Whegs robot autonomously climbing obstacles using tactile antennae and avoiding obstacles using ultrasonic sensors in a bat-inspired configuration. CWRUs Urban Challenge vehicle, Dexter, and our autonomous lawnmower, CWRU Cutter, benefit from animal inspired control architectures. Animals that have soft bodies can very effectively locomote and manipulate materials in their environment. For example, worms, leeches and slugs are all capable of moving through complex environments. The Chiel-Quinn groups have developed peristaltic robots and a soft gripper device. The peristaltic robots are hollow to allow fluid to pass through them.

Biography

Roger D. Quinn is the Arthur P. Armington Professor of Engineering at Case Western Reserve University. He joined the Mechanical and Aerospace Engineering department in 1986 after receiving a Ph.D. (1985) from Virginia Tech and a M.S. (1983) and B.S. (1980) from the University of Akron. He has directed the Biorobotics Laboratory since its inception in 1990. His research, in collaboration with Roy Ritzmann, Hillel Chiel, Mark Willis at CWRU and other biologists, is devoted to the development of robots and control strategies based upon biological principles. He has more than 200 publications and several patents. His biology-engineering collaborative work on behavior based distributed control, robot autonomy, and human-machine interfacing have each earned IEEE awards. His work on robot autonomy is resulting in the development of an inexpensive autonomous lawnmower that can edge obstacles and mow patterns.

Teamwork Planning and Learning in Adversarial Multi-Robot Domains

Manuela Veloso

Abstract

We have been participating in the RoboCup robot soccer small-size league for more than ten years. Such multi-robot domain offers a variety of challenges, including perception, control, actuation, and teamwork. In this talk, I will present an overview of the advances we have made along these years, namely in efficient vision processing, in physics-based motion planning, in adapting to the opponent, and in coordination and teamwork. At the end, I will outline the directions of our current work, in particular in building large teams of small robots, and analyzing and learning from data logged from the robots fast execution.

Biography

Manuela M. Veloso is Herbert A. Simon Professor of Computer Science at Carnegie Mellon University. She directs the CORAL research laboratory, for the study of agents that Collaborate, Observe, Reason, Act, and Learn, www.cs.cmu.edu/ coral. With her students, Professor Veloso has successfully participate in multiple RoboCup robot soccer competitions since 1997. Professor Veloso is IEEE Fellow, AAAI Fellow, AAAS Fellow, the President of the RoboCup Federation, and the President Elect of AAAI. She recently received the 2009 ACM/SIGART Autonomous Agents Research Award for her contributions to agents in uncertain and dynamic environments, including distributed robot localization and world modeling, strategy selection in multiagent systems in the presence of adversaries, and robot learning from demonstration. Professor Veloso is the author of one book on "Planning by Analogical Reasoning" and editor of several other books. She is also an author in over 250 journal articles and conference papers.

Manuela Veloso
Computer Science Department
School of Computer Science
Carnegie Mellon University
Pittsburgh, PA 15213-3890, USA
e-mail: `mmv@cs.cmu.edu`

Human-Friendly Robot Partners in Informationally Structured Space

Naoyuki Kubota

Abstract

Recently, as the number of elderly people rises, much more caregivers are required for the support to them in the aging society. In general, the mental and physical care is very important in order to avoid the progress of dementia of elderly people living alone in home, but such elderly people have little chances to talk with other people and to perform daily physical activity. It is really ideal that human caregivers should play the roles in mental and physical care, but the number of caregivers and therapists is not enough in the current situation of highly aging society. The introduction of human-friendly robots instead of people is one of possible solutions to realize the mental and physical care for elderly people. The capabilities on social communication are required for human-friendly robots such as robot pets, robot partners and robot-assisted therapy to realize natural communication with people. For example, the conversation capability of a robot can be applied for preventing dementia of elderly people. Robotic conversation can activate the brain of such elderly people and improve their concentration and memory abilities. It is difficult, however, for a robot to converse appropriately with a person even if various contents of the conversation are designed in advance. The daily new information such as daily news and whether forecast should be announced to elderly people everyday. Furthermore, in addition to verbal communication, the robot should understand non-verbal communication e.g. gestures. In general, human communication is restricted by their environmental states, and furthermore, a human assumes the other human perceives the shared environment according to the relevance theory. Accordingly, a human often utters with words as few as possible. To realize the social communication with a person, the robot should acquire the environmental information required for human interactions, while should understand the meanings of gestures. Therefore, we propose

Naoyuki Kubota
Graduate School of System Design
Tokyo Metropolitan University
Tokyo, Japan
e-mail: kubota@tmu.ac.jp

an information support system based on conversation to elderly people by integrating robot technology, network technology, information technology, and intelligence technology in this research.

The accessibility to information resources within an environment is essential for both people and robots. Therefore, the environment surrounding people and robots should have a structured platform for gathering, storing, transforming, and providing information. Such an environment is called informationally structured space. The structuralization of informationally structured space realizes the quick update and access of valuable and useful information for people. The information is transformed into the useful form and style suitable to the specific features of robot partners and people. Furthermore, if the robot can share the environmental information with people, the social communication with people might become very smooth and natural. In this talk, we explain (1) the robot partners used in this study, (2) infromationally strucutred space and sensor netwrok systems, and (3) intelligent technologies used for human-friendly interaction and communication, and (4) application examples of human-friendly robot partners. Finally, we discuss the futhre vision to realize human-friendly robot partners.

Biography

Naoyuki Kubota is currently an associate professor of the Department of System Design, Tokyo Metropolitan University, Japan. He graduated from Osaka Kyoiku University in 1992, received the M.E. degree from Hokkaido University in 1994, and received the D.E. degree from Nagoya University, Japan in 1997. He was an assistant professor and lecturer at the Department of Mechanical Engineering, Osaka Institute of Technology, Japan, from 1997 to 2000. He joined the Department of Human and Artificial Intelligence Systems, Fukui University, Japan, as an associate professor in 2000. He joined the Department of Mechanical Engineering, Tokyo Metropolitan University, Japan, as an associate professor in 2004. He was a visiting professor at University of Portsmouth, UK, in 2007 and 2009, and has been an invited visiting professor at Seoul National University since 2009. His current interests are in the fields of human-friendly robot partners, intelligent robotics, computational intelligence, and informationally structured space. He has published more than 200 refereed journal and conference papers in the above research fields. He received the Best Paper Award of IEEE IECON'96, the Best Paper Award of IEEE CIRA'97, and so on.

He has been an associate editor of the IEEE Transactions on Fuzzy Systems since 1999, and an Editorial Board Member of Journal of Advanced Computational Intelligence and Intelligent Informatics since 2004, Editorial Board Member of Advanced Robotics from 2004 to 2007, an Associate Editor of Journal of Control, Measurement and System Integration (JCMSI) since 2007, and the IEEE CIS Intelligent Systems Applications Technical Committee, Robotics Task Force Chair from 2007 to 2009, and others. He was Program Co-Chair of CIRAS 2003, Special Session Co-Chair of IEEE CIRA2003, Program Co-Chair of AMiRE 2005, Program

Chair of SCIS & ISIS 2006, Program Co-Chair of IFMIP 2006, IFMIP 2008 and IFMIP 2010, Regional Program Co-Chair of IEEE CIRA2007, Technical Co-Chair of AMiRE 2007, Tutorial and Workshop Co-Chair of IEEE Robio 2007, Technical Co-chair of IEEE CEC 2007, Program Co-chair of ICIRA2008, Program Co-Chair of IEEE RiiSS 2009 and IEEE RiiSS 2011 in IEEE SSCI, Special Sessions Area Co-chair of FUZZ-IEEE 2009, Regional Program Co-Chair of CIRA 2009, Area Co-chair of ICIRA2010, Special Session Co-chair of FUZZ-IEEE 2011, and others.

Learning and Adaptation for Human-Friendly Robot Partners in Informationally Structured Space

Naoyuki Kubota

Abstract. This chapter discusses the learning, adaptation, and cognitive development for human-friendly robot partners. First, we explain the history of studies on intelligent technologies and cognitive development for robotics from the viewpoint of Cybernetics. Next, we explain the cognitive development of robot partners based on the concepts of learning and adaptation. Furthermore, we explain informationally structured space to extend the cognitive capabilities of robot partners based on environmental systems. Finally, we explain intelligence technologies used for human-friendly interaction and communication based on the informationally structured space, and discuss the future direction on this research.

Keywords: Robot Partners, Informationally Structured Space, Learning and Adaptation, Computational Intelligence.

1 Introduction

As the number of elderly people rises, much more caregivers are required to support them in the aging society. In general, the mental and physical care is very important in order to avoid the progress of dementia of elderly people living alone at home, but such elderly people have little chances to talk with other people and to perform daily physical activity. Ideally, human caregivers should play the roles in mental and physical care, but the number of caregivers and therapists is not enough in the current situation of a rapidly aging society. The introduction of human-friendly robots instead of people is one possible solution to realize the mental and physical care for elderly people. Capabilities for social communication are required for human-friendly robots such as robot pets, robot partners and

---

Naoyuki Kubota
Graduate School of System Design, Tokyo Metropolitan University
6-6, Asahigaoka, Hino, Tokyo, Japan
e-mail: kubota@tmu.ac.jp

robot-assisted therapy. For example, the conversation capability of a robot can be applied for preventing dementia of elderly people. Robotic conversation can activate the brain of elderly people and improve their concentration and memory abilities. It is difficult, however, for a robot to converse appropriately with a person even if rich contents of the conversation are designed in advance. The new information, e.g., daily news and whether forecast should be announced to elderly people everyday. Furthermore, in addition to verbal communication, the robot should understand non-verbal communication e.g. facial expressions, emotional gestures and pointing gestures.

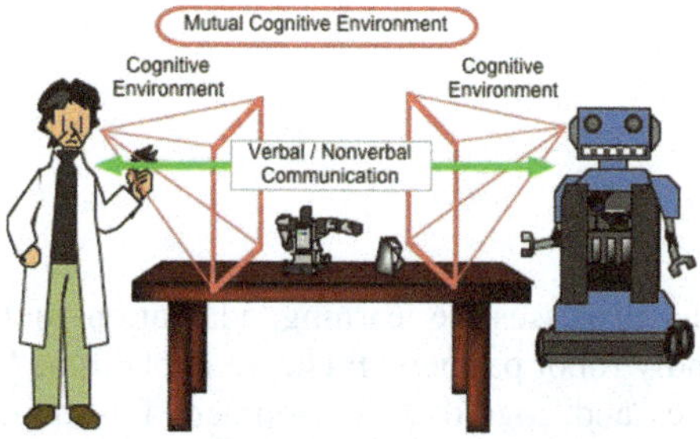

Fig. 1 Mutual cognitive environments in natural communication

In general, human communication is restricted by their environmental states, and furthermore, a human assumes the other human perceives the shared environment according to the relevance theory [1]. In the relevance theory, human thought is not transmitted but is shared between two people. Each person has his or her own cognitive environment. Although they speak different languages, one person can understand the meaning of an unknown term spoken by the other through communication because the person makes the recognized symbolic term correspond to the percept. Furthermore, an important role of utterances or gestures is to make a person pay attention to a specific target object, events, or person. As a result, the cognitive environment of the other person can be enlarged. The shared cognitive environment is called a mutual cognitive environment (Fig.1). A human-friendly robot partner also should have such a cognitive environment, and the robot should keep updating the cognitive environment according to current perception through the interaction with a person in order to realize the natural communication.

We have proposed an information support system using human-friendly conversation to elderly people by integrating robot technology, network technology, information technology, and intelligence technology based on the concept of cognitive development through mutual cognitive environments [2,3]. Since we can obtain huge amounts of data through sensor network devices and robot partners, we should extract the important information suitable and required for people and robot partners. In order to share their cognitive environments between a person and robot partner, the environment surrounding people and robot partners should have a structured platform for gathering, storing, transforming, and providing information. Such an environment is called informationally structured space (Fig.2) [4,5]. The intelligent structuring of informationally structured space realizes the quick update and access of valuable and useful information for people. The information is transformed into the useful form and style suitable to the specific features of robot partners and people. Furthermore, if the robot can easily share the environmental information with people, the social communication with people might become smooth and natural.

In this chapter, we explain (1) the history of intelligent technologies and cognitive development for robotics, (2) the robot partners developed in this study, (3) informationally structured space and environmental systems, and (4) intelligent technologies used for human-friendly interaction and communication, and (5) application examples of human-friendly robot partners. Finally, we discuss the future direction to realize human-friendly robot partners.

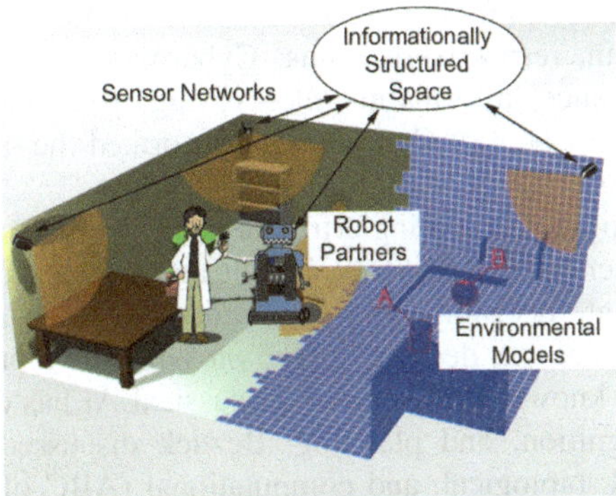

Fig. 2 Informationally structured space for people and robot partners

2 Learning, Adaptation, and Cognitive Development for Robotics

2.1 Intelligence for Robotics

Intelligence has been discussed since ancient days, but this section only focuses and discusses concepts on the intelligence for robotics. To build an intelligent system, various methodologies have been proposed and developed by simulating human behaviors and by analyzing human brains. To begin, we discuss the intelligence from the viewpoint of Cybernetics. Generally, Cybernetics is considered as the theoretical study of communication and control processes in biological, mechanical, and electronic systems [6-8]. The traditions of cybernetics can be traced back to three different approaches; Wiener's Cybernetics [7], Turing's Cybernetics [9], and McCulloch's Cybernetics [10]. Wiener's Cybernetics is the study of control systems based on the concept of feedback. The feedback analysis is used to discuss the stability of a control system. Especially, the idea of homeostasis discussed by Bernard, Cannon, and Ashby is defined as the ability to maintain internal equilibrium or to keep internal balance within suitable ranges by adjusting its physiological processes in a dynamic or open environment. Therefore, ecological, biological, and social systems are also considered as homeostatic systems. Turing's Cybernetics is the study of the intelligence on calculation and machines, based on computability. A Turing machine is a theoretical model of a computer. Turing's original aim is to provide a method for evaluating whether a machine can think or not, and Turing discussed digital computers as discrete state machines and learning machines based on an educational process. McCulloch's Cybernetics is the study of neuroscience. McCulloch and Pitts suggested a mathematical model of a single neuron as a binary device performing simple threshold logic. The brain is a network of neurons, and this is considered as the first model of connectionism. Furthermore, the McCulloch tradition in Cybernetics led to the development of second order cybernetics [6]. Recently, these research topics of Wiener's Cybernetics, Turing's Cybernetics, and McCulloch's Cybernetics have been considered as a dynamic equilibrium system, symbolic logic system, and self-organizing system, respectively. Thus, Cybernetics has influenced control theory, computer science, information theory, cognitive science, and artificial intelligence, and the research of Cybernetics has formed the basis of autonomous and/or intelligent robots.

Various methodologies concerning artificial intelligence (AI) have been developed in order to describe and build intelligent agents that perceive an environment, make appropriate decisions, and take actions [11]. In a classical point of view, an intelligent agent was designed based on symbolic representation and manipulation of explicit knowledge. Especially, classical AI has dealt with symbolic search, pattern recognition, and planning. Bezdek discussed intelligence from three levels: artificial, biological, and computational (ABC of Intelligence) [12]. In the strictest sense, CI depends on numerical data and does not rely on explicit knowledge. Furthermore, Eberhart defined CI as a methodology involving computing [13]. We also summarized CI as follows [14]: CI tries to construct intelligence by the bottom-up approach using internal description, while classical AI tries to construct intelligence by the top-down approach using external (explicit) description. Furthermore, we can find the individual streams of Wiener's Cybernetics, Turing's Cybernetics, and McCulloch's Cybernetics in BI, AI, and CI, respectively.

Basically, it is hard to define intelligence, because the intelligence itself is very abstract and conceptual. Pfeifer referred to many definitions on intelligence, and defined intelligence as the ability to survive [15]. This definition is very simple, but the ability to survive in natural environments needs the other abilities such as abstract thinking, adaptation to environments, social communication, and the acquisition of knowledge and skill. On the other hand, for example, in the Cambridge International Dictionary of English a robot is defined as a machine used to perform jobs automatically which is programmed and controlled by a computer. However, Brady defined robotics as the intelligent connection of perception to action [16]. This definition is also interesting, because this definition does not include mechanical and electrical terms such as sensors, actuators, and computers. Furthermore, the term of 'intelligent' is used to define robotics. In

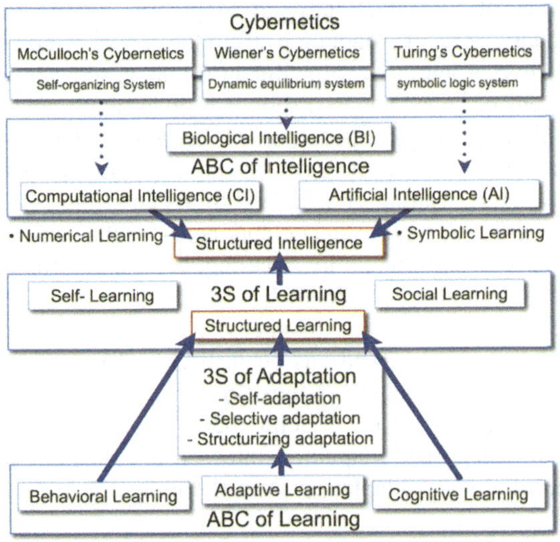

Fig. 3 Learning and adaptation toward intelligence for robotics

general, a robot, which can learn and apply knowledge or skill, is called intelligent. In the next section, we discuss the concept of learning, evolution, and adaptation used for intelligent robots.

2.2 Learning, Evolution, and Adaptation

The concepts of learning, evolution, and adaptation have been discussed from various points of view until now. We have also discussed these concepts simply from the viewpoint of CI. As a common feature, these concepts include changes of knowledge, rules, behaviors, shapes, and others. Changes are done after evaluations in the learning, while evaluations are done after changes in the evolution. This means the learning is a target of changes, while the evolution is a result of changes. Furthermore, we discussed the difference between reinforcement learning and evolutionary learning based on the concept of states, action, and situations [17]. Because the aim of learning is to acquire knowledge and skills, we can categorize the learning into behavioral learning and cognitive learning form the physical and embodimental point of view. The cognitive learning mainly acquires language, knowledge, perceptual information, while behavioral learning mainly acquires skills and behaviors based on sensory-motor coordination.

On the other hand, in the adaptation, changes are done after evaluations based on the matching with an environment. This means the concept of adaptation requires the explicit interaction with environments. Both cognitive and behavioral learning requires the explicit interaction with environments, but if environmental conditions change, the robot must deal with the change. Therefore, this is called adaptive learning in this chapter. In this way, we can categorize learning into adaptive, behavioral, and cognitive (ABC of Learning). Furthermore, we can categorize the types of adaptation into self-adaptation, selective adaptation, and structurizing adaptation (3S of Adaptation) from the spatial or environmental point of view (Fig.3). Most of learning in intelligent robots is self-learning. In the self-adaptation, an agent or robot adapts itself to a given environment by changing behaviors, recognition, or physical components without changing the given environment. On the other hand, in the structurizing adaptation, a robot adapts to a given environment by rebuilding or structurizing its environments without changing behaviors, recognition, and physical components. In the selective adaptation, a robot adapts to a given environment by selecting its local environments suitable to cognitive, physical and behavioral conditions without changing behaviors, recognition, physical components, and the given environments. The selective adaptation is similar with bird imigration, and has been discussed in the research fields of collective robotics and distributed robotics. In general, an intelligent robot combines these adaptations to adapt itself facing the environment. In addition to 3S of adaptation, we can categorize adaptation into sensory adaptation, physiological adaptation, and evolutionary adaptation from the temporal or biological point of view.

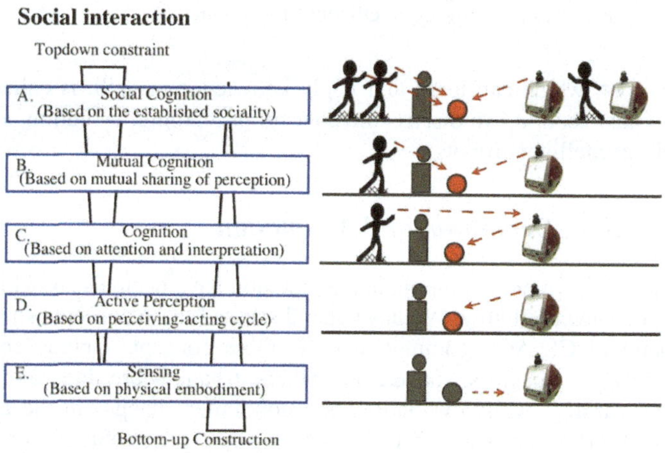

Fig. 4 Cognition level in social interaction and interaction with environments

We have proposed the concept of structured learning [17-20]. The structured learning emphasizes the importance of interdependent linkage among structurally coupled learning modules and functions through adaptive learning, behavioral learning, and cognitive learning. Structured learning is a learning methodology, and we must discuss how to apply structured learning in real environments. We have categorized the learning for intelligent robots in self-learning and social learning according to the availability of teaching signals [21]. In general, we can categorize the learning with teaching signals as a supervised learning, but we use the term of social learning because we focus on intelligent robots based on cognitive development, and the intelligent robot can use human models as teaching signals in the context of supervised learning. On the other hand, if teaching signals are unavailable, the robot must perform self-learning by trial and errors or by generating tentative teaching signals by the robot itself. In this way, intelligent robots need structured learning as a learning strategy, and need self-learning and social learning as learning tactics (3S of learning).

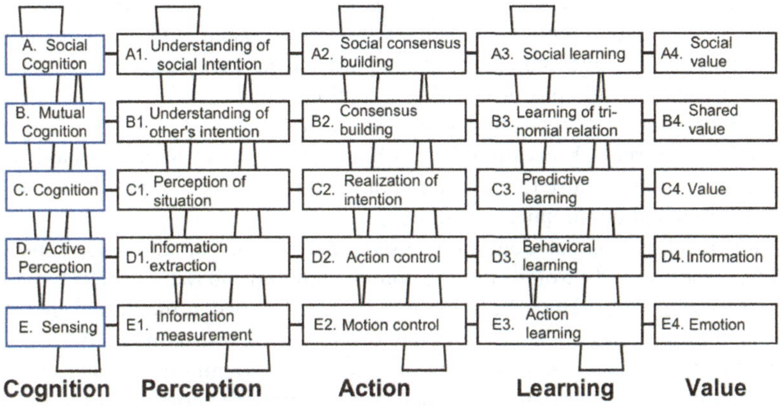

Fig. 5 Conceptual levels from the interaction with the environment to the social interaction and their corresponding methods for perception, action, learning, and values

2.2 *Perception to Cognition*

The previous section discussed ABC of learning and 3S of learning, but we should discuss the cognitive development of intelligent robots based on the concepts of learning, evolution, and adaptation to realize human-friendly robot partners. In this paper, we define the cognitive development as the refinement of relationships, e.g., the relationship between perception and action, the relationship between perceptual information and symbolic information, and the relationship among symbolic labels or variables in the associative memory [2, 22]. Figures 4 and 5 show the conceptual levels from the interaction with the environment to the social interaction and their corresponding methods on perception, action, learning, and values. First, we discuss the interaction of robots with their environment based on the concept of perceiving-acting cycle in ecological psychology [23, 24]. Basically, reactive motions based on sensing (E) and sensory-motor coordination are performed as the bottom level or in the most direct way without high level decision making. The robot measures the necessary environmental data as sensory inputs, and performs its corresponding motion control. The subsumption architecture might be categorized into this level because each layer is selective according to sensory inputs [25]. However, the original subsumption architecture includes high level decision making in the selection mechanism.

The next level is based on active perception (D). In ecological psychology, the smallest unit of analysis must be the perceiving–acting cycle situated in an intentional context [23, 24]. The perceptual system extracts perceptual information to be used for making action outputs. Here the importance is to extract perceptual information over a series of motions to take an intentional action based on selective attention. In addition, the output of the action system constructs the spatiotemporal context for the specific perception with the dynamics of the environment. Therefore, the perceptual system must search for and select perceptual information required by a specific action.

The next level is based on cognition (C). Situated perception enables prediction suitable to the spatiotemporal context of the environment. Furthermore, predictions of human goal-directed behaviors may arise from knowledge and experience on human cognitive and physical abilities. Action is defined as a motion sequence observed by internal observation, while behavior is defined as motion sequence observed by external observation. The robot should extract human behavior patterns in finite time because the prediction of the human behavior patterns is important to interact and communicate with people.

In the level of mutual cognition based on relevance theory (B), consensus building is performed through the communication and interaction between a person and robot partner. Here the intention is shared between them, and therefore, they can perform the cooperative behaviors based on a trinomial relation. The highest level is based on social cognition. The social cognition is related with ontology and the social learning in this sense is to obtain the commonsense or criteria of social values based on cognitive learning using natural languages. The aim of the social communication of robot partners is to reflect the social common senses or social values to robots or to construct the social identity of robots.

In this figure, it is most important to consider the bottom-up construction and top-down constraints. For example, the information extracted in D1 uses the measurement data sensed at E1, while the meaning of the extracted information is restricted by the situation perceived at C1. As another example, the consensus building in B2 is done according to the inference of other's intention in C2 under the constraint of the social commonsense in A2. In this way, top-down constraint and bottom-up construction clarify the mechanism of the cognitive development in the structured learning.

2.4 Prediction-Based Perceptual System

This section shows an example of cognitive development of a robot partner based on a prediction-based perceptual system [20]. The research on a prediction-based perceptual system corresponds to the category of C1: Prediction of situation in Fig.5. The prediction-based perceptual system is composed of four layers: 1) the input layer (I-layer); 2) clustering layer (C-layer); 3) prediction layer (P-layer); and 4) perceptual module selection layer (S-layer) shown in Fig.6.

The I-layer is composed of spiking neurons [28] used to recognize a specific state. The S-layer is also composed of spiking neurons used to select the perceptual modules and to control the sampling interval of each perceptual module. Here spiking neurons for the I-layer and for the S-layer are called SN-I and SN-S, respectively. Each perceptual module generates the inputs to SN-I from sensory inputs according to the spike output of the SN-S corresponding to the perceptual module. Therefore, the time series of spike outputs from the SN-I and SN-S construct the spatiotemporal pattern of the perception. Since the change of the firing patterns indicates the dynamics of perception, the robot can select perceptual modules to be used in the next perception by learning the changing patterns as a prediction result.

The C-layer performs unsupervised learning [29] based on the spike outputs of the SN-I by using reference vectors. As a result of unsupervised learning, each neuron at the C-layer acquires the relationship among the sets of perceptual information. Here, a clustered perceptual state is called a perceptual mode. The

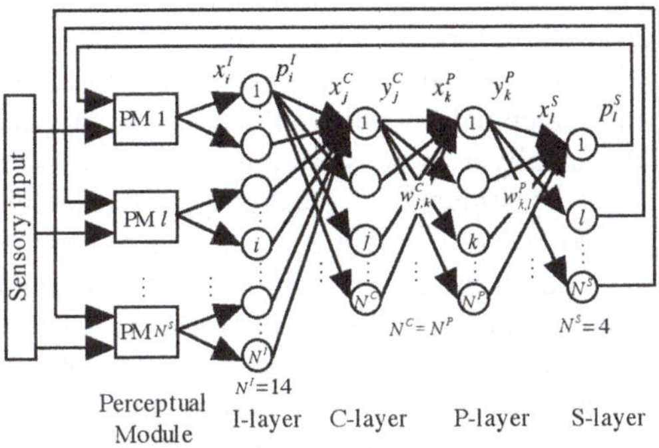

Fig. 6 A prediction-based perceptual system composed of the I-layer, C-layer, P-layer, and S-layer

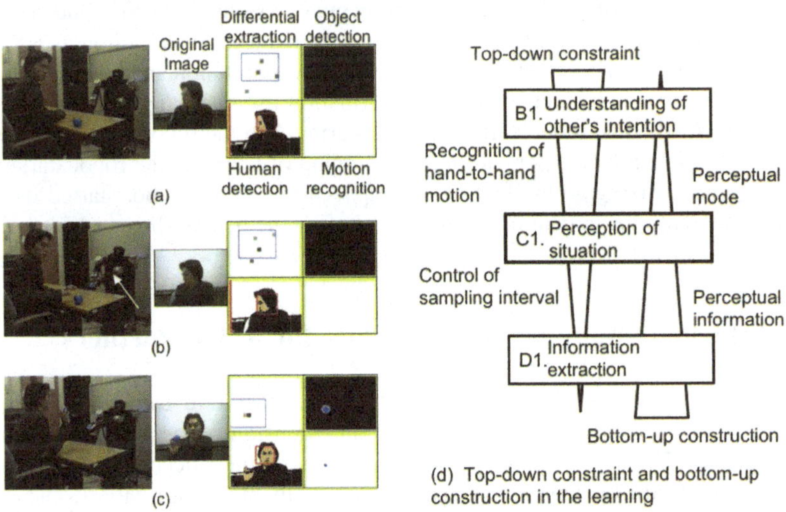

Fig. 7 An example of the prediction based on visual information of a prediction-based perceptual system.

dimension of a reference vector is the same as the number of the SN-I. Each perceptual mode relates to a specific combination of perceptual modules to extract its perceptual information.

The transition of perceptual modes can represent the dynamics of human interactions and environments. Therefore, the robot should select perceptual modules suitable to the dynamics of human interactions in the next perception. Next, the robot learns the transition among perceptual modes to select perceptual modules for the next perception. The P-layer calculates the mode transition probability among perceptual modes. According to the prediction, the S-layer selects the perceptual modules for extracting perceptual information required in the next perception.

The proposed method simultaneously learns several intelligent modules, i.e., 1) the clustering of perceptual information (the extraction of spatial patterns); 2) the prediction of transition among the clusters (the extraction of temporal patterns); and 3) selection of perceptual modules (the control of sampling intervals). Because this is the mutually nesting structure, the proposed method enhances learning by updating the learning rates according to the learning state of other functions. This idea is based on the structured learning. When the learning state in the clustering of perceptual information is not good, the learning rate of prediction should be small. Afterward, the learning rate of prediction can be increased gradually according to the progress of clustering. On the other hand, when the accuracy of prediction is not improved, the clustering should be improved. As a result, the learning rate in the clustering is temporally increased. In this way, the learning rate of each module is updated according to the learning state of other modules.

bot partner named Hubot II. In this example, the robot partner predicts the next human behavior from the visual information of executing a hand-to-haFigure 7 shows an example of prediction-based perceptual system using the rond motion. The top-down constraint of C1 is to carry out the hand-to-hand motion as B1: understanding of other's intention in Fig.5, and the prediction in C1 controls the sampling interval to extract perceptual information used as a top-down constraint to D1 (Fig.7 (d)). The robot starts moving its arm when the robot predicts the next object to be shown by the person. In the beginning of learning, the robot started its arm after recognizing a blue ball or red cup. Afterward, the robot started its arm after recognizing the human facial direction before the person shows a blue ball or red cup (Fig.7 (b)).

3 Informationally Structured Space for Robot Partners

Informationally structured space can extend the cognitive capability of robot partners based on sensor networks. This is an extension of the embodiment on a robot partner. The learning and adaptation of robot partners depends deeply on the availability of information in the cognitive environments shared with people. This section explains the informationally structured space used for natural communication of robot partners with people.

3.1 Robot Partners

We have used various types of robot partners such as MOBiMac, Hubot, Apri Poco, palro, miuro, and others (Fig.8) for the support to elderly people, rehabilitation support, and robot edutainment [2-4,17-22,26,27]. In order to popularize robot partners for home use, the price of a robot partner should be as low as possible. We are developing on-table small size of robot partners called iPhonoid and iPadrone (Figs.8 (d) and (e)). Since iPhones are equipped with various sensors such as gyro, accelerometer, illumination sensor, touch interface, compass, two cameras, and microphone, the robot itself is enough to be equipped with only cheap range sensors, e.g., Microsoft Kinect sensors.

(a) Apri Poco (Toshiba) (b) parlo (Fujisoft) (c) miuro (ZMP)

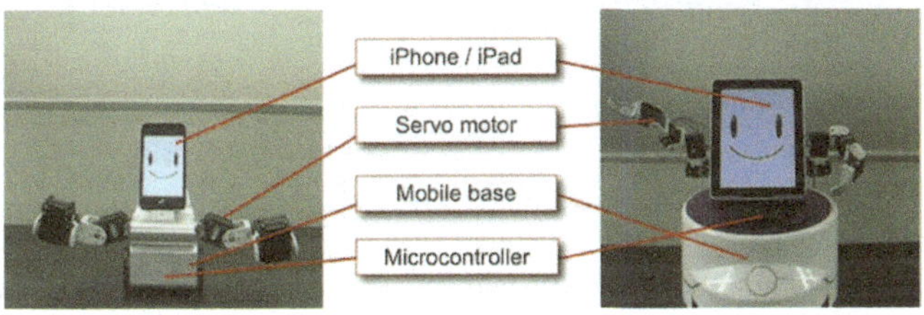

(d) iPhonoid (e) iPadrone

Fig. 8 Robot partners

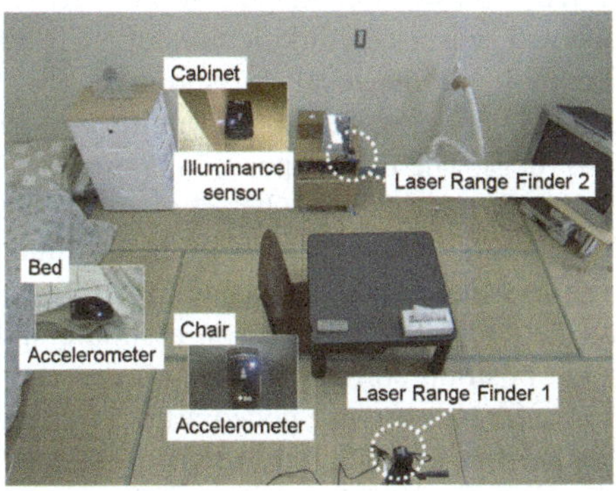

Fig. 9 An environmental system composed of sensor network devices in a living room

3.2 Environmental Systems

The environmental system using informationally structured space is based on a ubiquitous wireless sensor network composed of sensors equipped with wall, floor, ceiling, furniture, and home appliances [5]. These sensors measure the environmental data and human motions. The measured data are transmitted to the database server, and then feature extraction is performed. Each robot partner can receive the environmental information suitable to the specification of each robot. Furthermore, since the robot partner can share the same environmental information, they can have the same conversation system, and can make the coordinated utterances by multiple robot partners based on a scenario like a drama.

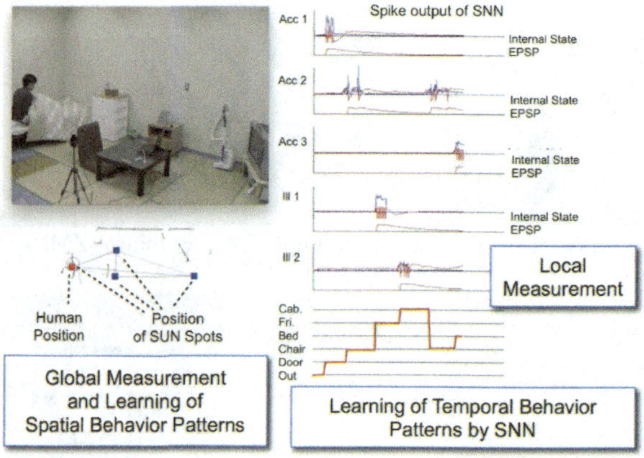

Fig. 10 An example of human localization and human behavior estimation using SNN

The most important task in the measurement is the estimation of human position to realize human-friendly interaction and natural communication. We develop intelligent sensor networks composed of global measurement system and local measurement system. We use two laser range finders (LRF) for global measurement of human position and sensor network devices of SUN Spot (Sun Small Programmable Object Technology) for local measurement (Fig.9). The LRFs can measure the distance up to approximately 4,095 [mm] in 682 different directions covering a range of 240 [deg]. The human position is estimated by the differential extraction operator with long-term memory based on the change of measured distance. The Sun SPOT has three sensors; accelerometer, illuminance sensor, and temperature sensor. The measured data are transmitted to the database server. We apply a spiking neural network (SNN) based on a simple spike response model [28] to localize a human position corresponding to a specific place such as bed, chair, and cabinet (Fig.9). SNN is a pulse-coded artificial neural network that can learn spatiotemporal patterns based on a time series of measurement data. A spiking neuron is fired when its corresponding human behavior is recognized according to the change of measurement data. For example, we assume that a person is sitting on the chair if its corresponding spiking neuron is fired by using the change of measurement data of the chair equipped with the accelerometer (see Fig.9). Furthermore, the spike outputs are also used for inputs to other spiking neurons, and the connection weight between two spiking neurons are updated by temporal Hebbian learning when temporally sequential firing occurs between two spiking neurons. In this way, SNN can learn the temporal transition of human behaviors. Figure 10 shows an example of human localization and human behavior estimation. The estimated human behavior information is stored as a life log in the informationally structured space.

The proposed method also can perform the sensor localization based on the human position measured by the LFR and the firing of a sensor network device. The environmental system sequentially updates the position of each sensor network device.

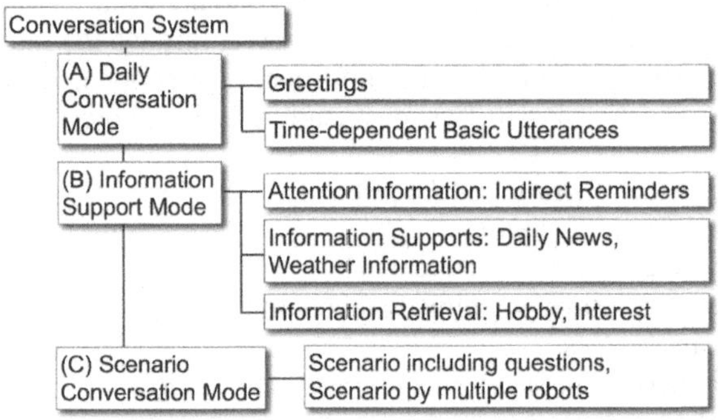

Fig. 11 Utterance modes for robot partners

3.3 Conversation System for Elderly People

The conversation system is composed of (A) daily conversation mode, (B) information support mode, and (C) scenario conversation mode shown in Fig.11 [3,30]. The environmental system refers to the states on people and environments through the database in the informationally structured space.

The daily conversation is done according to daily life pattern extracted by life logs. If we can prepare the daily life pattern or schedule for an elderly person, the robot can make time-dependent utterances.

The information support is done by three different kinds of utterances; 1) Reminder support in abnormal situations of people and rooms; 2) Daily information service on weather forecast and news for daily life extracted from web pages; 3) Information retrieval support from web pages. In the reminder support, the robot partner notifies a person of careless behaviors, e.g., when the person forgets to close doors, to turn off lights, and others. The environmental system can detect such situations in the informationally structured space by using the set of states of spiking neurons. For example, if both of neurons corresponding to the fridge and chair are fired simultaneously and continuously, this situation means that the fridge is still open and the person sits on a chair. In this situation, the robot can conduct the reminder support to the person. Daily information service is done by using the database on calendar and the data downloaded through the web pages. The web pages on daily news such as weather, traffic jams, climes, and terrorism, and disasters of earthquakes and fires, are registered. Every morning, the robot conducts the daily information service after several greetings. The information retrieval service is done by the interactive conversation. The words of the person obtained by voice recognition are used as keywords of query. Next, we use the word for retrieving with a web engine (e.g., yahoo search engine).

The scenario conversation mode is performed by three interrelated modules; topic selection modules, conversation control module, and utterance selection module. The topic selection module decides the global flow of conversation based on the selection probabilities of topics. The conversation control module controls the flow of utterances based on transition probabilities of utterances. Furthermore, scenarios are downloaded periodically from the database server through the Internet.

Figure 12 shows an example of the information support mode by robot partners; MOBiMac and Apri Poco. The conversation flow is shown in the following;

1. MOBiMac: It's time to take pills.
2. Human: OK, I will.
   (1 minute)
3. Apri poco: Have you already taken pills?
4. Human: Yes.
5. Apri poco: Where are pills in this room, MOBiMac ?
6. MOBiMac: I don't know.
7. Apri poco: Please look at the box for pills.
8. Human: Okay, I forgot to close the cabinet.

First, the robot makes a time-dependent basic utterance to make a person take pills. After 1 minute, the robot recognizes that the person forgot to close the cabinet including a pill case according to the simultaneous firing of neurons corresponding to the chair and cabinet. The most direct reminder is "Please close the cabinet." However, this kind of direct reminder sometimes makes a person angry or unpleasant. Therefore, the robot tries to remind the person that the cabinet is not closed by using the indirect reminder such as "Please look at the box for pills."

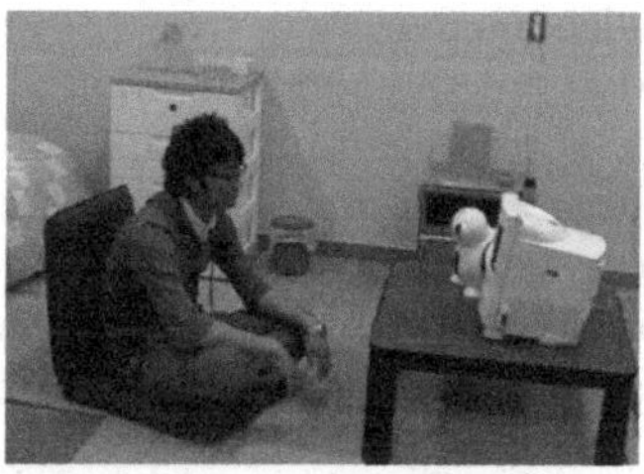

Fig. 12 An example of the information support mode by robot partners

4 Summary

In this paper, we discussed the concepts of learning, adaptation and cognitive development in intelligence for robotics, and explained a conversation system for information support to elderly people based on informationally structured space. We showed experimental results of cognitive development of a robot partner and conversation system using sensor network devices and robot partners. In the experimental results, robots could speak utterances according to the time-dependent basic utterance and attention information based on human behaviors and environmental information measured by sensor network devices. We are developing small and cheap robot partner called iPhonoid and iPadrone.

As a future work, we will integrate these robot partners with the informationally structured space to realize natural communication and human-friendly interaction. The emerging synthesis of robot technology, information technology, and network technology by intelligence technology is the most important and promising approach to realize a safe, secure, and comfortable society for the next generation.

References

1. Sperber, D., Wilson, D.: Relevance - Communication and Cognition. Blackwell Publishing Ltd. (1995)
2. Kubota, N., Yorita, A.: Structured Learning for Partner Robots based on Natural Communication. In: Proc (CD-ROM) of 2008 IEEE Conference on Soft Computing in Industrial Applications, pp. 303–308 (2008)
3. Kimura, H., Kubota, N., Cao, J.: Natural Communication for Robot Partners Based on Computational Intelligence for Edutainment. Mecatronics 2010, 610–615 (2010)
4. Kubota, N., Yorita, A.: Topological Environment Reconstruction in Informationally Structured Space for Pocket Robot Partners. In: 2009 IEEE International Symposium on Computational Intelligence in Robotics and Automation, pp. 165–170 (2009)
5. Obo, T., Kubota, N., Lee, B.H.: Localization of Human in Informationally Structured Space Based on Sensor Networks. In: Proc. of 2010 IEEE World Congress on Computational Intelligence, pp. 2215–2221 (2010)
6. Umpleby, S.A., Dent, E.B.: The Origins and Purposes of Several Traditions in Systems Theory and Cybernetics. Cybernetics and Systems 30, 79–103 (1999)
7. Wiener, N.: Cybernetics. John Wiley and Sons, New York (1948)
8. Ashb, W.R.: An Introduction to Cybernetics. Chapman & Hall, London (1999), `http://pcp.vub.ac.be/books/IntroCyb.pdf`
9. Turing, A.M.: Computing Machinery and Intelligence. Mind 59, 433–466 (1950)
10. Anderson, J.A., Rosenfeld, E.: Neurocomputing. MIT Press (1988)
11. Russell, S.J., Norvig, P.: Artificial Intelligence. Prentice-Hall, Inc. (1995)
12. Zurada, J.M., Marks II, R.J., Robinson, C.J. (eds.): Computational Intelligence - Imitating Life. IEEE Press (1994)
13. Palaniswami, M., Attikiouzel, Y., Marks II, R.J., Fogel, D., Fukuda, T. (eds.): Computational Intelligence - A Dynamic System Perspective. IEEE Press (1995)
14. Fukuda, T., Kubota, N.: Intelligent Learning Robotic Systems Using Computational Intelligence. In: Fogel, D.B., Robinson, C.J. (eds.) Computational Intelligence: The Experts Speak, pp. 121–138. IEEE Press (2003)

15. Pfeifer, R., Scheier, C.: Understanding Intelligence. The MIT Press, Hillsdale (1999)
16. Brady, M., Paul, R.: Robotics Research. In: The First International Symposium. The MIT Press (1984)
17. Kubota, N., Nishida, K.: Situated Perception of A Partner Robot Based on Neuro-Fuzzy Computing. In: Proc (CD ROM) of 2005 IEEE Workshop on Advanced Robotics and its Social Impacts (2005)
18. Kubota, N.: Structured Learning for Partner Robots. In: Proc. of 2004 IEEE International Conference on Fuzzy Systems, vol. 1, pp. 9–14 (2004)
19. Kubota, N.: Computational Intelligence for Structured Learning of A Partner Robot Based on Imitation. Information Science (171), 403–429 (2005)
20. Kubota, N., Nishida, K.: Perceptual Control Based on Prediction for Natural Communication of A Partner Robot. IEEE Transactions on Industrial Electronics 54(2), 866–877 (2007)
21. Kubota, N.: Computational Intelligence for Social Learning of A Partner Robot. In: Proc (CD-ROM) of 2nd International Symposium on Computational Intelligence and Intelligent Informatics, pp. 99–104 (2005)
22. Kubota, N.: Cognitive Development of Partner Robots Based on Interaction with People. In: Proc (CD-ROM) of Joint 4th International Conference on Soft Computing and Intelligent Systems and International Symposium on Advanced Intelligent Systems (2008)
23. Gibson, J.J.: The Ecological Approach to Visual Perception. Houghton Miffilin Company (1979)
24. Turvey, M.T., Shaw, R.E.: Ecological Foundations of Cognition I. Journal of Consciousness Studies 6(11-12), 95–110 (1999)
25. Brooks, R.A.: Cambrian Intelligence. The MIT Press (1999)
26. Kubota, N., Wakisaka, S.: Location-dependent Emotional Memory for Natural Communication of Partner robots. In: IEEE Symposium Series on Computational Intelligence, pp. 107–114 (2009)
27. Kubota, N., Wagatsuma, Y., Ozawa, S.: Intelligent Technologies for Edutainment using Multiple Robots. In: Proc (CD-ROM) of The 5th International Symposium on Autonomous Minirobots for Research and Educatinment, pp. 195–203 (2009)
28. Gerstner, W.: In: Maass, W., Bishop, C.M. (eds.) Pulsed Neural Networks, pp. 3–53. MIT Press, Cambridge (1999)
29. Bickel, P., Diggle, P., Fienberg, S., Krickeberg, K., Olkin, I., Wermuth, N., Zeger, S.: The Elements of Statistical Learning. Springer (2002)
30. Tang, D., Kubota, N.: Information Support System Based on Sensor Networks. In: Proc. (CD-ROM) of World Automation Congress (2010)

Teaching with Minirobots: The Local Educational Laboratory on Robotics

P. Salvini, F. Cecchi, G. Macrì, S. Orofino, S. Coppedè, S. Sacchini,
P. Guiggi, E. Spadoni, and P. Dario

Abstract. In this paper, we are going to present and discuss a few activities related to the application of minirobots in school education. The activities have been carried out in the framework of the Local Educational Laboratory on Robotics (LELR), which has been developed by Scuola Superiore Sant'Anna (SSSA) in collaboration with local Municipalities (i.e. Valdera Union) and a network of primary and secondary schools (i.e. Costellazione Network) in the Valdera area of Tuscany, Italy. The LELR is part of SSSA efforts to actively participate in the scientific and technological education of young generations, starting from school age. The laboratory is based on the deployment of robotics, in its several manifestations. in teaching activities. Drawing on preliminary activities and experiences, the paper will report on and discuss a few projects about teaching with minirobots in primary and secondary schools education, pointing out the relevance of promoting an interdisciplinary approach to minirobots educational activities – namely not limited to scientific and technological subjects – as well as developing a critical attitude towards scientific and technological progress in students.

1 Introduction

Autonomous minirobots have brought robotics to a wider audience. In the last decades, schools started to use them to teach fundamental subjects such as maths, physics, logic, programming language, mechanics, electronics, etc, exploiting the

P. Salvini · F. Cecchi · G. Macrì · S. Orofino · E. Spadoni · P. Dario
Scuola Superiore Sant'Anna – The BioRobotic Institute, Pontedera (PI), Italy

S. Coppedè
Istituto Comprensivo "G.Mariti" – Fauglia (PI), Italy

S. Sacchini · P. Guiggi
Scuola Secondaria Dante Alighieri, Capannoli (PI), Italy

ludic and fascinating features of robotics and so their ability to motivate students to learn. The philosophy behind educational robotics refers mainly to Seymour Papert theories [1], which described the advantages of using simple construction kits and programming tools for educational purposes. According to Papert's perspective, children, by using robotic kits could became active participants in their learning and creators of their own technological artefacts, not just users of devices that others had made for them [2]. This theory inspired the development of the Logo programming language, an easy to use programming language, which students could use to animate their technological inventions. An interesting application of Logo involved a "floor Teacher Education on Robotics-enhanced Costructivist Pedagogical Methods turtle," a simple mechanical robot connected to a computer by a log cord. Floor turtles made drawings on paper commanded by Logo programs, by using pens mounted in their bodies. In the late 1970's, with the introduction of personal computers faster and more accurate turtles were proposed for didactic laboratorial activities; these novel instruments offered more opportunities for children to investigate and solve complex mathematical problems. Successively, in the 1980's, the first microcomputers entered schools. They allowed children to explore their own ideas by building specific problems to evaluate them. Moreover, in the mid-1980's, it was introduced the LEGO/Logo technology, the first true robotic construction kit ever made available, which consisted of the combination of the popular LEGO construction kit with the Logo programming language [3] [4]. By using the LEGO kit, children could build machines by using the traditional LEGO building bricks and newer pieces like gears, motors, and sensors as well. By using the Logo programming language, children were then allowed to construct behaviours for their artefacts [5].

Although the LEGO/Logo technology was highly efficient it had some drawbacks related, for example, to the nuisance caused by the wires connecting the robot to the computer, which made it difficult for children to create autonomous and mobile robots. Some of those drawbacks were overcame by a new product: the Programmable LEGO Bricks, which appeared in late 1980's. This novel solution could run without wires providing in this way autonomous function to children's mechanical constructions [6]. The last release of LEGO kit consisted in the LEGO Mindstorms kits (http://www.legoeducation.com). They were based on research and ideas from the Lifelong Kindergarten group at the MIT Media Lab [1], [5], and were soon diffused world-wide in both elementary and secondary schools as well as in higher education programs. Lego Mindstorms kits, with respect to the previous releases, included servo-motors, new sensors and the NXT-G iconic programming software but can also be supported by a variety of other programming languages (such as NXC, NBC, leJOS NXJ, and RobotC). Moreover, combined with Crickets, which was another robotic technology, developed in parallel with Lego Mindstorms, they gave children novel and funny instruments to learn important math, science, and engineering ideas; as an example, they allowed the creation of musical sculptures, interactive jewelry, dancing creatures (http://www.picocricket.com/). The Cricket functionality was successively reinforced with the introduction of novel elements ("Display Cricket", "MIDI Cricket", "Science Cricket", "Cricket Bus system") ,which provided true analog-to-digital converters on the sensor inputs, so allowing the use of a greater variety of sensor devices [6]. The main goal of Cricket was to allow children to design their own scientific instruments for investigations which they personally found meaningful; in this way they could gain a deeper appreciation and understanding of many scientific concepts [2].

Other interesting explorations were also allowed by Cricket [5], by adding computation and other functionalities to traditional children's toys (Bitballs Project); some of those functionalities were provided by built-in microprocessor and LED or built-in electronics and infrared communication [5].

In the following we briefly describe some of the most important results of the most significant experiences of educational robotics, made around the world.

- Kärnä-Lin et al. (2006), through qualitative action research, identified various advantages, introduced by the use of educational robotics, into learning in the field of special education. They demonstrated as the robotic technologies make it possible for students to practice and learn many necessary skills, such as collaboration, cognitive skills, self-confidence, perception, and spatial understanding [6].
- Dias et al. (2005), described the positive outcomes of three higher education initiatives in Sri Lanka, Ghana, and the USA that focused on implementing robotic technologies for developing communities; they examined the intersections of robotic technologies with education and sustainable development [7].
- Pekarova et al. (2008), commented the results of the integration of Robotics in Early Childhood Education; according to their observations, developing attractive activities resulted an effective practice for learning with digital technologies at preschool age [8].
- Rossi et al. (2007), observed that robotic programmable bricks enabled students to make possible new types of science experiments for children. All these activities meet well the goals set such as an increase of the quality and impact of education in the primary schools [9].

Summarizing almost each of the activities performed till now, world-wide, on educational robotics differed very much from each other, in their target audience (e.g., primary schools, secondary schools, universities), their pedagogical goals, their organizational background; the diversity of the approaches among different studies prevented, to some extent, a coordinated approach. Moreover many of the described activities lacked of a previous identification and incorporation in the school curricula of an appropriate teaching method.

In this paper we present a further way to employ minirobots in educational research activities and applications, taking inspiration from the following key sentence in the call for papers of the AMiRE 2011 Symposia: 'autonomous minirobots are a microcosm of advanced embedded systems technology that permeates our technological culture'. Based on the fact that technological culture is starting to permeate also educational activities, we argue that educational activities with minirobots could benefit from promoting interdisciplinary activities and a critical attitude on science and technology in students. As a matter of fact, microrobots are an accessible example of what, in bigger scales and in much more complex ways, exists and will exist in our future societies. However, too often, educational activities with minirobots are centered around teaching strictly scientific and technological subjcts. In other words, they are devoid of any connection with other disciplines, such as literature, philosophy, art, or ethics, which, in our opinion, should be complementary and essential for a complete technological and scientific education. In addition, proposing interdisciplinary activities on robotics can have positive effects on creativity and innovation, can be fundamental for the development of problem solving abilities, besides eliciting a critical, as opposed to a passive, attitude towards technology.

The aims of the LELR is to participate in the education of young generations by providing schools with human and technological resources for carrying out several kinds of activities involving minirobots, based on the conviction that robotics can be a useful tool for teaching and learning in a funny and constructive way. The LELR approach to educational robotics is strongly characterised by:

1) the promotion of interdisciplinary projects: it seeks to exploit not only the technological and scientific potential of robotics, but also its connections with other school subjects;

2) the generation of a critical attitude towards technology: the assumption is that students should not be passive receivers or users of technology, but they should be taught what is inside the technology and how it works in order to generate in them a more responsible use as well as insights on the possible risks that technology may raise.

The paper, therefore, will report on a few experiences in educational activities with minirobots in school education which were carried out or planned in the framework of the LELR.

The paper is organised as follow: in the next sub-section we will introduce a few examples of connections between robotics and non strictly scientific nor technological subjects; in section 2 we will briefly describe the LELR's functions and aims; in section 3 will report on two preliminary experiences carried out in the framework of LELR in order to attempt to make a systemic integration of robotics as cross-disciplinary learning instrument in the schools from primary to secondary.

1.1 Robotics and Its Connections with Other School Subjects

Robotics is a subject with multiple educational potentialities and can be used also by involving school subjects other than science and technology, such as biology, mechanics, electronics, computer science, etc.). The following are just examples of possible connections between on the one hand Robotics and on the other Literature, Linguistics, Arts, Philosophy, Sociology and Cultural Studies, respectively.

Literature

Didactic activities involving the teacher of literature could focus on reading and analysing selected literary texts about robots, such as the play R.U.R. (Rossum's Universal Robots), the sci-fi novels by Asimov's or more classical texts such as A.

Huxley's Brave the New World (1932), Mary Shelley's Frankenstein (1818) or Samuel Butler's Erewhom (1872) and then discuss the author's view of scientific and technological progress.

Arts
(History of Art) Didactic activities on arts and robotics may start from the study of various automata built in Europe throughout several centuries, such as the mechanical clocks of the Middle Age, the toys and tricks of the Renaissance, for instance, Leonardo da Vinci's (Leonardo is supposed to have designed a humanoid robot, called "The Knight" in 1495) or the fascinating production of automata of the XVIII century. Since the history of automata is not rooted only in Western countries, but there exist remarkable traditions also in Eastern countries, such as the automata made in the 12th century by Arabian engineer Al Jaziri or the Karakuri ningyo dolls in Japan, it could be possible to design activities aimed at studying the cultural differences in the representation or acceptance of automata in different cultures. The relationship between, on the one hand, the arts and, on the other, robotics or scientific and technological progress can also be studied with reference to paintings, sculptures, theatrical performances and other artforms. Consider, for instance, the faith in technological progress that characterizes the Futurist artists or, on the contrary, the less optimistic view of scientific and technological progress in much of Postmodern artworks.

Popular Culture
There are many movies, comics, TV series and other products of popular culture about robotics, such as music videos, that can serve for didactic activities based on robotics. Many of these products can be used for studying or introducing ethical, legal, social, political and economic implications of robots, such as stereotypes, cultural differences, business interests, legal gaps, social risks, ethical dilemmas, etc. Activities could also be focused on the analysis of the different messages about robotics technologies contained in popular culture. Moreover, many of the stories told in these artform could be used to introduce the topics related to the acceptance of the different (i.e. the monster) and unknown.

Philosophy and Sociology
Didactical activities may have students reflecting on some of the current ethical and societal implications of robotic technologies and systems. They may also be requested to study the relations between philosophical theories and robotics, from the mind-body dichotomy to the current debates on artificial consciousness and intelligence. For instance, to describe a robot by referring to the parts of the human body could be debatable, as it assumes a mechanistic approach to the human being, a way of thinking very popular in the philosophy of XVIII (e.g., Descartes).

Cultural Studies
Robotics, as we have already pointed out, can be used to have students reflecting on their own cultural situatedness and background and to foster a positive relationship among different cultures, for instance, by considering the different approaches to robots in Western and Eastern countries.

2 The Local Educational Laboratory on Robotics

The Valdera area, is one of greatest economic areas of Tuscany, in Italy. The analysis of the main sectors of the local economy shows an area with great potentialities in the field of innovative technologies. This area is characterized by the strong influence of the mechanical division of PIAGGIO, the large company known for the Vespa and for other popular brands of two-wheeled vehicles. In Valdera all the Municipalities are members of the Valdera Union which has the aim to jointly exercise a variety of features and services, in order to exploit the potentially competences of the 15 municipalities associated. In particular, in the branch of Education, the Union supports and encourages the creation of a common training system in collaboration with all the institutions, agencies and associations that are present in the area. For this reason, on November 2010 a pact called "Agreement for the Education of the Community" has been signed in order to define a common educational plan to follow the trajectories of the scientific territorial development. This pact, signed by Unione Valdera; Scuola Superiore Sant'Anna, "Rete Costellazioni"- a local network of schools -, Pont-Tech, and the Municipality of Pisa, will try to encourage the creation of an integrated training system based on Local Educational Laboratories with a shared planning in order to improve education in public schools. The first laboratory that will start will be the one on Robotics that aims to promote and to share the scientific knowledge among the students and among teachers. The choice of Robotics is not accidental in fact the economy of the Valdera area relies heavily on mechatronic skills and technologies.

The Local Educational Laboratory on Robotics (LELR) has started its activities since December 2011. The laboratory involves six pilot schools: 2 high schools, 2 secondary schools, and 2 primary schools. About 10 tutors, among which PhDs students in biorobotics, robotics researchers and technical staff of the BioRobotics Institute of Scuola Superiore Sant'Anna, have made themselves available for collaborating with teachers in designing and developing robotics related activities. Usually a number of 5/6 meetings between SSSA tutors and school teachers are planned in order to design and carry out the activities. Tutors may be invited to collaborate during school time in teaching activities together with teachers. A final public event held at the end of the school year (June 2011) will conclude all the laboratories activities. During the final event, students will have the possibility to present their works to a wide audience outside the school. What is remarkable is that all the activities carried out, which span from 20 to 40 hours, as considered as extra activities both for SSSA people as well as for teachers. No funding or other financial support is expected in the initial phases of the Laboratory. Besides human resources, SSSA is making available to schools its educational robotic platforms, which consists of three robotic dogs AIBOs (Sony), one robotic Dinosaur Pleo (by e-Motion), one humanoid robot I-Droid (by DeAgostini), one humanoid robot Nao (by Aldebaran Robotics) and five robotic kits RoboDesigner (distributed by RoboTech srl). However, many of the activities planned with schools will not be based on commercially available robotic platforms but will consist in the creation of new robotic mechanisms (such as a the realization of a mechanical clock and the application of actuators to a school skeleton) or in the exploitation of the results and materials produced in some research activities carried out in the BioRobotics Institute, such as the European Union funded project Lampetra (http://www.lampetra.org/index.php).

3 Preliminary Experiences with LELR

In the following, we present two projects carried out in the framework of LELR in a primary and secondary schools. Unfortunately the projects has started only recently and it is not possible to provide many details on their implementation and results. However, what characterizes both projects is that robotics is used in connection with other schools subjects.

3.1 Bio-inspired Minirobots for Learning about Nature in Primary Schools

This project, which is called 'Atelier of the curious minds' started in January 2011 and was devised by prof. Silvia Coppedè in a primary school of the G. Mariti Institute located in Fauglia (Pisa, Italy), in collaboration with SSSA tutors. It is based on the belief that robotics can be useful for teaching and learning about nature in school activities. The project started with the observation of a living being, i.e. a lamprey. Students were requested to study the animal living environment, its morphological features, the way it moves and behaves, etc. In the second phase, students were asked to observe and study the same features they observed in the real animal, in a robotic version of lamprey, the one realized by SSSA in the framework of the European funded project Lampetra (http://www.lampetra.org/index.php). A small scale version of a lamprey robot was realized based on the previous model developed by SSSA. In this way, students were given the possibility to learn basic concepts, by building or manipulating their model, which can reproduce the main functions of the real animal. A parallelism was established between the observed living being and its robotic double in order to facilitate learning about robotics and nature. The activities were crossed disciplinary in that they involved different subjects, such as linguistics, anthropology, logics, mathematics, creativity and expression, and technological and scientific subjects.

About 15-20 students of different ages, from seven to eleven years old, were involved in this school project. This activities were carried out in a mixed laboratory group where cooperative learning were implemented: children worked on mini robotics platforms divided into small groups of different ages in which personal competence, skills, knowledge were enhanced and amplified.

3.2 Secondary School: From Thinking to Practice with Minirobots

As far as secondary schools are concerned, we report on the project carried out by the school Dante Alighieri, located in Capannoli (Pisa, Italy). The project leaders were Prof. Patrizia Guiggi and Prof. Simona Sacchini, both at their first experience with robotics. The laboratory activities were carried out in the framework of the European project Comenius (http://ec.europa.eu/education/lifelong-learning-programme/doc84_en.htm) which started in 2009. The laboratory involved 15 students aged 12-13. The robotic platform used was the RoboDesigner. The project was characterized by an interdisciplinary approach to robotics, whit a good balance between humanistic and technological/scientific subjects. As a matter of fact, the project was carried out as laboratory activity in collaboration with teachers of other subjects (i.e. mathematics, art, foreign language, technology, literature and even motor activities). The teachers of Italian literature, for instance, proposed to have students read fables/legends (e.g. The Golem), sci-fi short stories and/or novels (i.e. Asimov, Philip Dick, Frederic Brown) or watch excerpts from some popular science-fiction movies, such as Blade Runner, Frankenstein, or Edward scissorshands). In addition, creative writing activities (i.e. inventing and telling tales with robots as characters) were carried out. All these humanistic activities were aimed at eliciting discussions on ethical implications of robotics applications, such as social consequences of human-robot interaction, changes in interpersonal relationships among human beings, acceptance of the different, use of robots instead of modern slaves, etc. Artistic and creative activities were carried out too, in which students were asked to imagine and depict bad or good robots. As to technology and mathematics, Leonardo's machines were studied and taken as models to design and develop simple microrobots and implementing simple programs in C language using a commercially available robotic kit: i.e. the hardware and software of RoboDesigner.

4 Discussion and Conclusions

It is widely acknowledged that teaching with minirobots can be an effective way to have students learn scientific and technical subjects. Futhermore, offering students interdisciplinary activities about minirobots can foster creativity and elicit a critical attitude, especially in relation to the pervasive presence of technology in our societies. We have reported on two activities carried out in a primary and secondary schools in which the activities about minirobots were connected with neithertechnological nor scientific subjects, but on the contrary, they were based on literature, art, and philosophy. Such an interdisciplinary approach required a considerable efforts both from the parts of the teachers as well as that of students: it required a strong flexibility by the teachers and a strong motivation to collaborate by the students. In fact, both students and teachers had to "learn" together to acquire new competences and skills, which are not strictly connected to the traditional school subjects or to their background knowledge.

Moreover, the study of robotics elicited educational methodologies based on laboratory activities and constructivism, in which "doing is thinking". It changed the ways of learning, but also the ways of thinking. In fact, students had to observe an event first, and then to make some hypothesis, to validate his/her own ideas, to design and create. The experimental component was fundamental in almost all activities, students built the robot and thus avoided extreme abstractions and because the robot gave an immediate feedback, that feedback represented an incomparable educational reinforce. As regards to experimental activities, robotics offers teachers a multidisciplinary and highly flexible and effective tool.

In addition, the LELR activities were often planned by dividing students in groups, and this had countenanced the cooperative learning in which personal competence, skills, knowledge were enhanced and amplified.

In the primary school laboratory, the presence of the tutor was aimed at promoting the discussions and the curiosity about the main characteristics of the lamprey. In addition, the tutor designed the school activities taking into account the age of the children and planned the activities in the form of a game. The children showed their fantasy and creativity in the drawings in which they drew the lamprey robot, taking into account not only its aesthetics features, but also the basic components of the robot.

Finally, we would like to make an example of why it is necessary to develop in students a critical attitude. One of the possible risks in using minirobot kits with very young students is related to what can be defined as "the robot as perfect model problem". In other words, if the robot behaviour is not understood or its real nature is not clearly explained by the teacher, there might be the risk that pupils can see the robot as perfect and themselves as non perfect. This is even more so, if we consider that the idea of perfection is usually associated with the cold qualities of machines, i.e. rationality, perfection, precision, reliability and not with the warm qualities which are usually associated with human beings: humanity, faculty of feeling, faculty for sensation. According to a survey carried out by Arras and Cerqui, 'humans are better assessed in case they have cold qualities, normally linked to machines' [10]. The authors points out that 'from an anthropological point of view this means that the "warm" qualities are no longer those which are considered best in our society' [10].

In conclusion we believe that educational activities with minirobots should promote and develop in students:

- an interdisciplinary approach and vision of robotics. As a matter of fact, robotics is an multidisciplinary subject. As we have seen previously, it can be easily linked not only to scientific and technical subjects, but also to humanistic subjects, such as literature, history, philosophy, art, etc. Fostering an interdisciplinary approach in educational activities based on robotics is important in order to overcome rigid divisions between subjects, which on its turn may elicit in students a "systemic vision" of reality, critical thinking, curiosity, creativity and improve the management of complexity [11].
- appropriate technological and scientific knowledge as well as "critical instruments" fitted to an increasing complex, ever changing and scientifically and technologically permeated world. This can be achieved by fostering critical reflections on techno-scientific progress and about the not always positive implications on the natural environment and all living beings.

References

[1] Papert, S.: Mindstorms. Children, Computers and Powerful Ideas. Basic books, New York (1980)

[2] Martin, F., Mikhak, B., Resnick, M., Silverman, B., Berg, R.: To Mindstorms and Beyond: Evolution of a Construction Kit for Magical Machines. In: Robots for Kids: Exploring New Technologies for Learning. Morgan Kaufmann Series in Interactive Technologies, pp. 9–33 (2000)

[3] Resnick, M., Ocko, S.: LEGO/Logo: Learning Through and About Design. In: Harel, I., Papert, S. (eds.) Constructionism. Ablex Publishing, Norwood (1991)

[4] Resnick, M.: Behavior Construction Kits. Communications of the ACM 36(7), 64–71 (1993)

[5] Resnick, M.: Technologies for Lifelong Kindergarten. Educational Technology Research & Development 46(4) (1998)

[6] Kärnlä-Lin, E., Pihlainen-Bednarik, K., Sutinen, E., Virnes, M.: Technology in finnish special education: toward inclusion and harmonized school days. Informatics in Education 6(1), 103–114 (2007)

[7] Bernardine Dias, M., Ayorkor Mills-Tettey, G., Mertz, J.: The TechBridge World Initiative: Broadening Perspectives in Computing Technology Education and Research. Robotics Institute School of Computer Science (January 1, 2005)

[8] Pekárová, J.: Using a Programmable Toy at Preschool Age: Why and How? In: Intl. Conf. on Simulation, Modeling and Programming for Autonomous Robots Workshop Proceedings of SIMPAR 2008, Venice, Italy, November 3-4, pp. 112–121 (2008)

[9] Rossi, S.: La Robotica educative nella scuola primaria. Master Degree Thesis, Scienze delle Formazione Primaria, Università degli Studi di Firenze (2007)

[10] Arras, K.O., Cerqui, D.: Do we want to share our lives and bodies with robots? A 2000-people survey Technical Report Nr. 0605-001 Autonomous Systems Lab Swiss Federal Institute of Technology, EPFL (June 2005)

[11] Manifesto on Creativity and Innovation (2009), http://ec.europa.eu/education/lifelong-learning-policy/doc/year09/manifesto_en.pdf

A Two Years Informal Learning Experience Using the Thymio Robot

Fanny Riedo, Philippe Rétornaz, Luc Bergeron, Nathalie Nyffeler, and Francesco Mondada

Abstract. Technology is playing an increasing role in our society. Therefore it becomes important to educate the general public, and young generations in particular, about the most common technologies. In this context, robots are excellent education tools, for many reasons: (i) robots are fascinating and attract the attention of all population classes, (ii) because they move and react to their environment, robots are perceived as close to living beings, which make people attracted and attached to them, (iii) robots are multidisciplinary systems and can illustrate technological principles in electronics, mechanics, computer and communication sciences, and (iv) robots have many applications fields: medical, industrial, agricultural, safety ... While several robots exist on the market and are used for education, entertainment or both, none fits with the dream educational tool: promoting creativity and learning, entertaining, cheap and powerful. We addressed this goal by developing the Thymio robot and distributing it during workshops over two years. This paper describes the design principles of the robot, the educational context, and the analysis made with 65 parents after two years of use. We conclude the paper by outlining the specifications of a new form of educational robot.

1 Introduction

Robots have already proved to be successful educational tools. At university level, the e-puck robot [12] is a standard open-source educational tool that is used as

---

Fanny Riedo · Philippe Rétornaz · Francesco Mondada
Ecole Polytechnique Fédérale de Lausanne
e-mail: `firstname.lastname@epfl.ch`

Luc Bergeron
Ecole Cantonale d'Art de Lausanne
e-mail: `luc.bergeron@ecal.ch`

Nathalie Nyffeler
Haute Ecole d'Ingénierie et de Gestion du Canton de Vaud
e-mail: `Nathalie.Nyffeler@heig-vd.ch`

a platform in dozen of universities and in several courses. Other robots like the Roomba Create from iRobot[9], the Robotino from Festo [2], the Khepera robot [13] and its successors are robots used for engineering education. At a younger age, as robotics is quite a new topic and is generally not part of the school's curriculum, interested students learn by themselves with kits such as Lego[1] Mindstorms [7], the BoeBot [4], or more generic kits like the Arduino [5]. Nevertheless, there is increasing interest for this field and more and more courses are available to teenagers, either out of school or as facultative lessons. The Roberta Initiative, for instance, has provided both materials and methods for teaching robotics in a gender-balanced way [10, 1]. Those workshops allow especially but not only girls to realize that technology is not as complicated and inaccessible as they might think.

To give opportunity to a wide audience to discover more about technology and robots, we have organized a robotics festival at EPFL since 2008 [6]. It has attracted a wider audience every year: 3'000 people in 2008, 8'000 in 2009, 15'000 in 2010. The festival is a free event including exhibitions, shows, talks, industry and lab presentations, and workshops for children. Those workshops include a wide range of activities such as robot building, soldering, or programming. For this occasion, we wanted to create affordable workshops introducing robotics to young children. In that aspect, engineering tools like the e-puck robot are too complicated and expensive for children. The Mindstorms NXT is better suited for a young audience but is still very expensive, and while schools can afford them, most parents cannot afford to buy such tools for their children. Therefore we decided to develop a robotic kit at very low price, adapted for young children, promoting creativity and learning. Combining this requirement with some ideas generated by the School of Art of Lausanne (écal), we developed the Thymio robot, a modulable robot for children.

2 The Thymio Robot

The main goal of this development was to create a mobile robot encouraging creativity and promoting the understanding of technology. At the same time, the robot had to have a sufficiently low cost to be distributed among the participants of the workshops for a very reasonable price (less than 50 Euros). Finally the robot had to be entertaining with some non-trivial behaviors.

During a joint workshop between EPFL and écal, Julien Ayer and Nicolas Le Moigne had the idea of a kit allowing children to build their own robot out of any type of object, without having to program or solder anything. Figure 1 illustrates this first conceptual idea. The inspiration behind this idea was the concept of Mr. Potato[8]. With basic elements (legs, arms, nose and eyes) any object, even a potato, can be turned into a character. Similarly, having a kit with some sensors, wheels and buttons should allow to create a robot around anything.

A first functional kit (Figure 2 left) was designed at EPFL, consisting of five electronic boards connected by cables. The five modules were two wheels, one infrared proximity sensor module, one battery module and one button module for switching

[1] Lego is a registered trademark of Lego Corporation.

Fig. 1 The first abstract idea was to have a connected set of sensors and actuators that could be adapted around any type of shape to create a new form of robot.

on and off and for choosing some parameters. The robot performed line following and a track was available to children to test their creations. This first prototype was used in 2008 at the robotics festival workshops, but as there were only 15 kits, children had to destroy their robots to give back the electronics at the end of the session. Because of several technical choices such as removable connectors and sensors, it was difficult for children, even with their parents, to assemble the robot correctly without the help of a trained teacher. The workshops were however hugely popular.

Based on this first success, we decided to build a kit simpler to operate and better respecting the regulations for toys. In particular we designed plastic cases around the components to fulfill the security requirements (see Figure 2 right). Despite the better design and simplified interface, this second prototype had the same problem as the first: assembling a robot is not a trivial task. Putting the wheels and the sensors in the right position, balancing the weight etc. is a non-trivial task for beginners. Therefore we decided to keep the modular concept but change the assembling approach. Instead of having a kit becoming a robot, we decided to have a robot that could become a kit. This resulted in the final version we called "Thymio", illustrated in Figure 3, top. Thymio is a small mobile robot which is approximately 15 cm wide, 12 cm long and 4 cm high and can be separated into four parts. Two parts drive the wheels, one includes the batteries and the speaker, and a last one has the main control button and the infrared proximity sensors. When disassembled, the parts have a transparent side to reveal the internal components, supporting technological explanations. The disassembled parts can then be reassembled around something else, such as a cardboard structure.

The infrared proximity sensors -five instead of three in the first prototype- are now oriented towards the user for a more interactive behavior. While the first prototype would just follow a line, Thymio has three different behaviors, all based on obstacle detection by the infrared sensors and presented as "moods": it can be

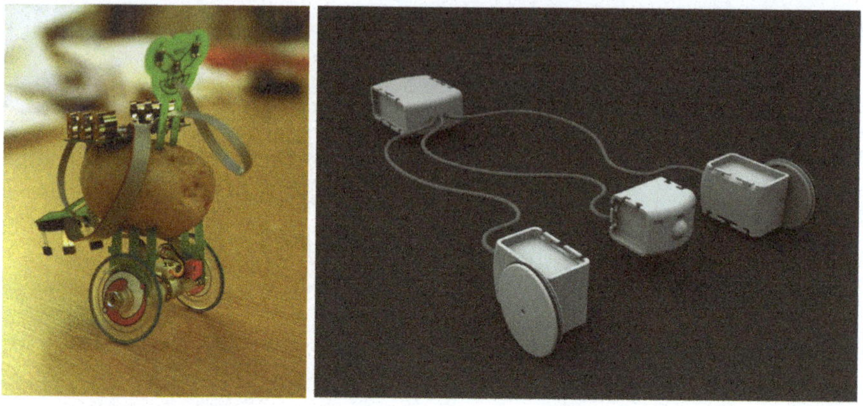

Fig. 2 The first functional prototype and the first design prototype were based on the original idea illustrated in Figure 1, but were difficult to assemble.

either (i) friendly, following the object in front of it still keeping a given distance, (ii) curious, making obstacle avoidance, or (iii) shy, going backs if one tries to touch it on the front. In addition, a series of LEDs and simple sound effects enhance those moods. Switching between the moods is made with the button. Those behaviors give a pet-like aspect to the robot and help attracting the children's attention. They are also used as example to explain to children how we can implement different controls on a robot based on the same set of sensors and actuators.

Finally, as we wanted to promote creativity, the robot case is white and can be used as drawing support. It has also the advantage of being gender-neutral. The box itself is also white and has clever dimension that makes it convenient to use as a construction base for the robot.

The final production price, including packaging, was close to 20 USD, therefore perfectly in our target price.

3 Workshops

Workshops at the EPFL robotics festival provide an excellent opportunity to introduce robotics to children in an informal and fun environment. They allow us to create a framework where we give the basic information about Thymio and we encourage the children to learn more about robotics by building their own robot. The target children age is between 6 and 12 years.

Workshops take place in a classroom, organized specifically for the workshop. In the middle of the classroom there are work places for the kids. Each table is shared by two children and supervised by a trained staff member. On tables against the walls children can choose and take DIY materials such as cardboard, color papers, feathers, etc. The front part of the room serves as playground for the children so that they can test their robots.

Fig. 3 The Thymio robot is introduced as a full mobile robot (top left) that can be disassembled into four main part representing sensors, motors and batteries (top right). This kit can be assembled around several objects to create very different robots like those illustrated in the three bottom images.

Children can chose the workshop based on a short description and a picture. Workshop places can be booked in advance via the festival's website or on the day of the festival at the welcome desk. The subscription fee for all Thymio-based workshops, including materials and the robot kit, was 49 swiss francs (around 50$), the workshop itself being free of charge. When the children arrive in the workshop, they receive a new robot in its box each. The parents are welcome to stay but do not need to, as we provide strong coaching. The workshop starts with some information an basic explanation about the robot. After this, the most important part of the workshop is dedicated to building and decorating Thymio with the provided materials. Once the participants are finished, they test their robots and play with the other kids on the playground.

In 2009, we had only one workshop with 50 children, and a duration of one hour. This was quite short and the room, though quite big, was obviously overcrowded. No specific directions were given on what to build with the robot. We noticed that most children would build the same basic robot using the packaging box and decorate it (see Figure 4), probably influenced by the assistants and the lack of time. So in 2010, we decided to make three different workshops with different topics:

1. Animals: This workshop was about making a robot with the shape of an animal, mostly by decorating the Thymio robot. The workshop goal was to promote creativity in this robotic context.
2. Obstacle passing: This was closer to an engineering challenge. Kids had to adapt Thymio to help him passing bigger obstacles, learning about techniques to improve this particular capability.
3. Cardboard workshop: This third workshop was centered around learning the specific technique of cardboard-based construction (see as example the bottom right robot in Figure 3).

The rooms were smaller and offered only 15 places each. The duration was extended to one hour and a half. This proved to be successful, as the kids would learn a specific aspect and could finish their robot before the end of the workshop, allowing to test it on the playground. In 2009 this has not been possible, because we had to rush them out of the room to make space for the next group. The three different themes naturally attracted different age and interest groups.

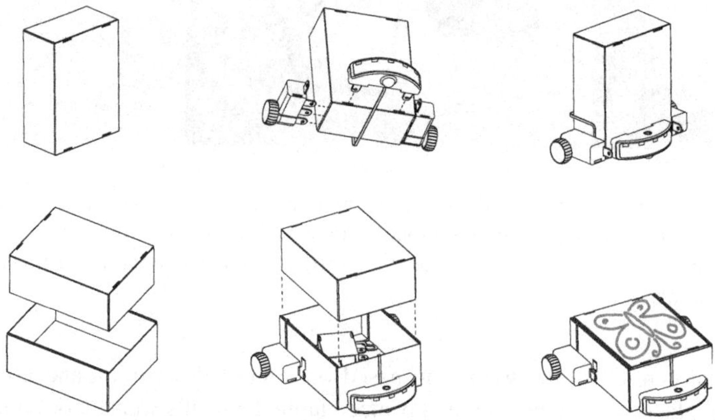

Fig. 4 Some basic constructions using the robot's packaging box as alternative body.

4 Evaluation and Analysis

Within the festival, the workshops were in general a huge success every year. For each edition we had a general survey to understand which activity (among exhibitions, shows, workshops, talks) was the most appreciated. In all three editions the workshops have been ranked first in the preferences.

The Thymio-based workshop were particularly appreciated. Several people came back from year to year and almost all the workshop places were taken (356 out of 400 places in 2009, 341 out of 360 in 2010). After two years of use of Thymio, a study was conducted in collaboration with the School of Engineering and Business Vaud (HEIG-VD) by sending an online questionnaire to 346 parents who had brought their children to the Thymio workshop. We got 65 answers. The data shows

	Strongly disagree to somewhat disagree	Somewhat agree to strongly agree	TOTAL
The worskshop is of good quality	3.1%	96.9%	100%
Your child is satisfied with the workshop	4.6%	95.4%	100%
The workshop is cheap (CHF 50 including the robot)	4.7%	95.3%	100%
From your point of view, the workshop is educational	10.8%	89.2%	100%
From you point of view, the workshop is fun	15.4%	84.6%	100%
After the workshop, your child has shown greater interest for robotics	21.5%	78.5%	100%
All together	**10.0%**	**90.0%**	**100%**

Fig. 5 The questionnaire shows that the visitors are very satisfied with the workshop.

that the parents are very satisfied with both the quality and the price of the workshop (see Figure 5). To the question "Do you intend to return to the festival in 2011?" 63.9% answered yes, 29.5% answered maybe and only 6.6% said no.

The workshops themselves have evolved through the years: now every child can go home with his own robot and the different topics allow to work on a specific competence, addressing more specifically the needs of different age groups. The duration of one hour and a half has proven to be a good choice as most children are naturally finished with their construction and start playing with the other participants and their robots.

However we identified several problems. First, we suspected that once they got back home, the children would not really play anymore with their robot. Contrary to our basic idea that they would create other robots with their kit, apparently once they got back home they would display their robot somewhere but they did not want to destroy it. This was confirmed by the study: most children play occasionally or rarely with their Thymio at home (see Figure 6). Even for the children who play often the playing sessions are quite short, generally under 30 minutes.

We investigated the reason for this loss of interest. What came out clearly from the parent's feedback is that the 3 behaviors are not sufficient. In the workshops already there were many demands for changing or reprogramming the behaviors. Though during the activity the kids would enjoy the behaviors, this was motivated by the fact that there are other robots to interact with, making the behavior richer. For example, they always found that a train can be created by putting one Thymio in obstacle avoidance mode and several others behind it in follower mode. Once at

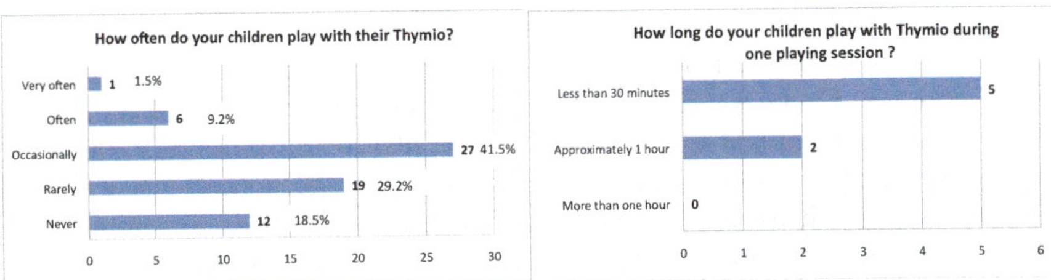

Fig. 6 The children rarely play with their Thymio at home.

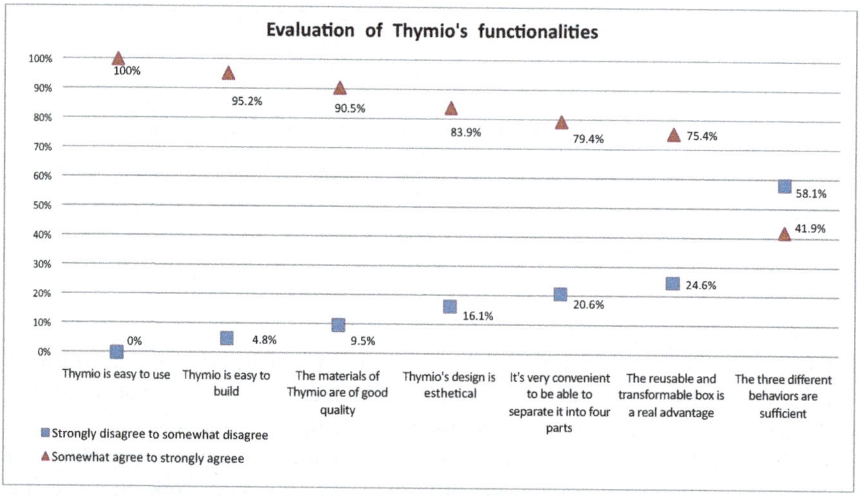

Fig. 7 The modularity and and usable carbord box are not as sucessful as the other characteristics, but what stands out clearly is that the 3 behaviors are not satisfying.

home there is nothing more to discover with the behavior, so the kids use it more as decoration. However, it was stated by several parents that children get really attached to their Thymio and are for example reluctant to let someone else use it.

Related to this problem, a technical issue made the robot consume energy even while off, so when it was left for several days the batteries needed to be changed before the child could play again. This waste of disposable batteries was a serious drawback for the parents and for us. And because the robot was decorated, the exchange of batteries often required the destruction of part of the decoration, discouraging the kids to change the batteries.

Another surprising effect was that the children would not be very impressed by the modularity of the robot. Most of them built a very basic shape with the packaging box (see Figure 4), and many would simply decorate the robot without separating the parts. For the parents also, the modularity and usage of the box is among the less popular features (see Figure 7).

The parents were also asked what new features they would add to the robot. The most popular ideas were compatibility to lego parts, possibility to program it, and use of rechargeable batteries instead of disposable ones. (see Figure 8). The Lego compatibility was tested in a focus group realized in a class of 14 seven-year-old children (see Figure 9). They were given modified robots that had Lego bricks added on top of them and lots of Lego brick to play with. As a result, the kids were immediately comfortable with building the robot. We noticed that with the Lego, they can easily assemble something, then destroy it to build something else. They do not have the same scruple as with destroying something they built out of cardboard and paper.

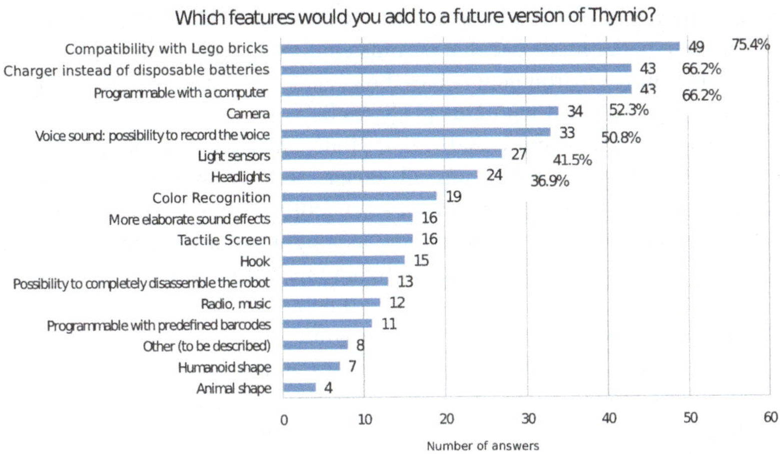

Fig. 8 People would like compatibility with the Lego system and reprogrammability of the robot. They also care about the waste of batteries. Other attractive features would be a camera or the possibility to record one's voice. It is interesting to note that a humanoid or animal shape -very frequent in robot toys- is not appealing to them.

Fig. 9 Left: a creation with a modified Thymio and Lego made by children during the focus group. Right: a creation with the new Thymio II and Lego.

5 Toward a New Thymio Version

After this first experience with Thymio we designed a new version, Thymio II, taking advantage of the lessons learned, which generated a new set of specifications. Some aspects of Thymio were convincing and were not changed: the price range, the neutral shape and color, the geometry of the wheels, the size, the activities using the robot in a specific context with clear learning goals.

As the modular shape did not bring sufficient added value, it was abandoned. Simple Lego connectors are cheaper and offer the support for construction, allowing moreover an easier destruction and reuse. The budget for the electronics was increased (total production cost around 40 USD instead of 20) to improve the programming possibilities: a larger set of sensors (accelerometer, more proximity sensors in different directions, temperature, IR receiver module, microphone... see Figure 10) and a better processor that permits confortable programming.

The programming environment is based on Aseba, which was successfully validated at the robotic festival with kids [11]. Aseba Studio allows easy modification of the behaviors and real-time visualization of all sensors and actuators. To further support understanding and debugging, all sensors are highlighted by an LED to illustrate their activity. This way, the sensor's activation is shown both as a numerical value in the programming environment and as a light instensity next to the sensor itself. Finally the batteries were replaced by an internal accumulator rechargeable by a USB connector, supporting at the same time a link with the computer where the program is developed. By adding the programming possibilities while keeping the price range, we plan to keep the activities with young kids making simple decorations, but enabling at the same time activities with older kids who can start programming. This new version has been distributed at the EPFL Robotics festival 2011 to 300 children and has encountered a real success.

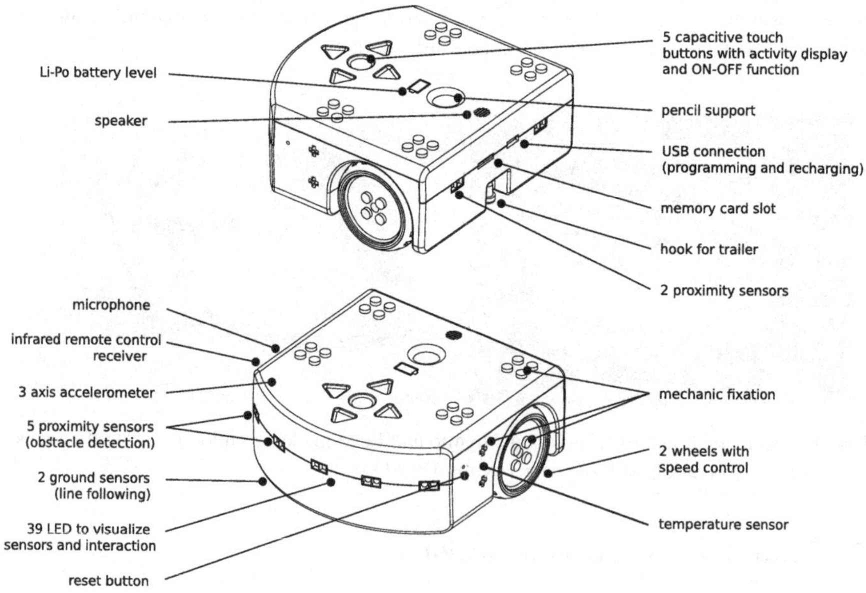

Fig. 10 The new Thymio II has a much larger set of sensors and actuators.

In parallel, we will develop teaching materials such as instructions on how to reproduce activities like the festival workshops in class, providing ready-to-use exercises for the teacher. We set up a wiki containing several examples of use along with all basic information on usage and programming of Thymio II [3]. Also the basic software is downloadable for free for teachers and children who wish to try new applications.

6 Conclusion

This two year adventure allowed us to have a first experience with a widely distributed robot (approximately 900 Thymio I robots were distributed during two festivals and other smaller events) and collect the feedbacks to generate the specification for a new generation of robots. The workshops were successful and proved to be a good way to introduce technology to children. The vast majority of the interviewed parents are very satisfied with the quality of the workshop and its price. However we identified two main problems: the children do not reuse their robot at home and the modularity is not the incredible feature we had hoped it would be. Our goal to promote creativity and understanding of technology was partly fulfilled by the success of the workshop. But we really want now to add value to the robot and give it a long-term edutainment potential.

To achieve this we released a new version of Thymio that is reprogrammable, rich in sensors and actuators, rechargeable and supporting Lego modularity. Finally, we hope to create a community around the robot so that users could find new programs to run on their robot and share their ideas. This new robot should serve as a basic tool, accessible to all, to teach robotics or other fields to young generations.

Acknowledgements. We would like to thank Stéphane Magnenat, Michele Leidi, Laurent Soldini, Nicolas Le Moigne, Julien Ayer, Andrea Suriano, Michael Bonani, Omar Benjelloun, Gilles Caprari and Daniel Burnier for their support in the several developement phases of this robot. We thank Mariza Freire and Rémy Pilliard for the organization of the festival and the workshops. This research was supported by the Swiss National Science Foundation through the National Centre of Competence in Research Robotics.

References

1. Roberta: Basics and experiments. Roberta Series, vol. 1. Frauenhofer IRB Verlag, Stuttgart (2009)
2. Festo didactic web site (2011), `www.festo-didactic.com`
3. Thymio-aseba wiki (2011), `http://www.thymio.org`
4. Balogh, R.: Basic activities with the Boe-Bot mobile robot. In: Proceedings of Conference DidInfo (2008)
5. Cuartielles, D.: Arduino board, project web site (2011), `http://www.arduino.cc`
6. EPFL: Robotic festival web site (2011), `http://festivalrobotique.epfl.ch`
7. Ferrari, M., Ferrari, G., Astolfo, D.: Building Robots with LEGO Mindstorms NXT. Syngress Media Inc. (2007)
8. Hasbro: Mr potato web site (2011), `www.hasbro.com/playskool/en_US/mrpotatohead`
9. iRobot: Company web site (2011), `http://www.irobot.com`
10. Jeschke, S., Knipping, L., Liebhardt, M., Muller, F., Vollmer, U., Wilke, M., Yan, X.: What's it like to be an engineer? robotics in academic engineering education. In: Canadian Conference on Electrical and Computer Engineering, CCECE 2008, pp. 941–946 (2008), doi:10.1109/CCECE.2008.4564675

11. Magnenat, S., Noris, B., Mondada, F.: ASEBA-Challenge: An Open-Source Multiplayer Introduction to Mobile Robots Programming. In: Markopoulos, P., de Ruyter, B., IJsselsteijn, W.A., Rowland, D. (eds.) Fun and Games 2008. LNCS, vol. 5294, pp. 65–74. Springer, Heidelberg (2008)
12. Mondada, F., Bonani, M., Raemy, X., Pugh, J., Cianci, C., Klaptocz, A., Magnenat, S., Zufferey, J.C., Floreano, D., Martinoli, A.: The e-puck, a Robot Designed for Education in Engineering. In: Gonalves, P., Torres, P., Alves, C. (eds.) Proceedings of the 9th Conference on Autonomous Robot Systems and Competitions, pp. 59–65. IPCB: Instituto Politcnico de Castelo Branco, Portugal (2009), http://www.est.ipcb.pt/robotica2009/
13. Mondada, F., Franzi, E., Ienne, P.: Mobile robot miniaturization: A tool for investigation in control algorithms. In: Yoshikawa, T., Miyazaki, F. (eds.) Proceedings of the Third International Symposium on Simulation on Experimental Robotics, ISER 1993. LNCIS, vol. 200, pp. 501–513. Springer (1993)

A New Open Source 3D-Printable Mobile Robotic Platform for Education

J. Gonzalez-Gomez, A. Valero-Gomez, A. Prieto-Moreno, and M. Abderrahim

Abstract. In this paper we present the Miniskybot, our new mobile robot aimed for educational purposes, and the underlying philosophy. It has three new important features: 3D-printable on low cost reprap-like machines, fully open source (including mechanics and electronics), and designed exclusively with open source tools. The presented robotic platform allows the students not only to learn robot programming, but also to modify easily the chassis and create new custom parts. Being open source the robot can be freely modified, copied, and shared across the Internet. In addition, it is extremely cheap, being the cost almost exclusively determined by the cost of the servos, electronics and sensors.

1 Introduction

Mobile robotics is increasingly entering the curricula of many technical studies. Robotics is gaining terrain in industry and consequently more firms are recruiting candidates with experience in robot programming. For this reason, many universities are teaching robotics in their master and degrees programmes[15, 13].

A common approach when teaching robot programming is the use of simulations, in which the user can create different robot configurations with low effort. These ad-hoc robots can be also shared with other people, multiplying the number of *out-of-shell* platforms [3, 5]. Furthermore, the cost is zero, you

J. Gonzalez-Gomez · A. Valero-Gomez · M. Abderrahim
Robotics Lab, Carlos III University of Madrid, Spain
e-mail: {jggomez,avgomez,mohamed}@ing.uc3m.es

A. Prieto-Moreno
Autonomous University of Madrid, Spain
e-mail: aprieto@uam.es

Fig. 1 Left: The new Miniskybot v1.0 robot. Right: The educational skybot robot

may have as many robots as you want, and they will never break. But this solution has one drawback: simulated robots are not like real robots. Things working on simulation may not work the same on a real platform. In addition, students will not enjoy the same testing their ideas on a real robot than on a simulated one.

Should it not be great that robots could be shared in the same way that code is shared (like in simulation)? If this could be possible researchers, professors and students could share their open source robots through the Internet, exchange ideas with other research groups, compare prototypes, test their algorithms on different configurations, evolve proposals from others..., such an idea is now possible and affordable thanks to the open source Reprap-like 3D printers[4].

This opens a new way of teaching robotics with the following advantages:

- Fast prototyping of robotic platforms.
- Low cost printing of robot parts.
- Easy reconfiguration and adaptation of the platform (evolution).
- Easy sharing of robot models among people.
- Motivation for students not only to implement algorithms on an existing platform but also to design and build new platforms.

In this paper, our new 3D printable Miniskybot robot platform is presented (shown in figure 1). It is fully open source (both the mechanical and electronics parts) and exclusively designed with open source tools (Openscad, Freecad and Kicad). The parts were furtherer printed in a Makerbot Cupcake 3D printer.

2 Motive and Problem Statement

Among the commercial educational platforms we can find a great variety of opportunities, starting with the well known Lego Robot, and going through the Meccano Robot, the RoboRobot robotic kit[1], or the OWI Robot Arm Edge[2]. These products are quite extended in the educational environment, they are affordable, and easy to use. They usually come with associated software, which allows users to interface with the robot, having access to sensors and actuators, program them, and so forth. These platforms have been present for some years now in the educational environment. In [10] the authors demonstrated the idea of a children's league for RoboCup, using robots constructed and programmed with the LEGO MindStorms kit to play soccer. Since then, RoboCupJunior has evolved into an international event where teams of young students build robots to compete in one of three challenges: soccer, rescue and dance [11, 14]. Goldmand et al.[6] presented an educational robotics curriculum to enhance teaching of standard physics and math topics to middle and early high school students. This project was also centered around the Lego MindStorm.

The major disadvantage of these platforms is that they are close. The users can hardly adapt them to their necessities, and instead, they must adapt to them. The reconfiguration of the platform may be a great advantage in order to be able to deploy all the initiative of the researchers, professors or students. The Lego MindStorm inherits the "build-it-yourself" of the Lego traditional toys, but users are constrained to use the sensors provided by the manufacturer, as well as the development software. An effort could be done to work around this limitation, but this goes beyond the original design of the platform. A work trying to meet the "open source" and the "non-free" directions is done by O'Hara et al[12].

Ad-hoc mini-robots have been built by research groups or university spin-offs mainly for educational purposes. These solutions overcome the limitations of the commercial robots, providing cheaper and more adapted solutions. Efforts have been done with the intention of developing effective and low-cost robots for education and home use, designed and built to fit the particular requirements of a teaching programme. Examples are those of IntelliBrain-Bot [3], Martin F. Schlogl's robots[4], the TankBot[5], the Trikebot [8] among many others.

This had been also our way of teaching robotics during many years, with our Skybot[6] platform (shown in figure 1). In our courses, the students build the Skybot from scratch and then program it. Sometimes they are so motivated that they propose wonderful modifications to the robot design. Even though some modifications are known beforehand that will not work well, we would like the student to discover it by himself. But in any case, it is not

[1] http://roborobo.koreasme.com/educational-robot-kit.html
[2] http://www.owiroboticarmedge.com/
[3] http://www.ridgesoft.com/intellibrainbot/intellibrainbot.htm
[4] http://www.mfs-online.at/robotics.htm
[5] http://profmason.com/?p=320
[6] http://goo.gl/MdRJs

possible to implement these modifications during the course due to the time it takes to the manufacturer to build the parts. At the end we had to keep the platform, or in the best case, change it for the next course with new students.

To summarize, the classical way of teaching robotics must focus, by necessity, on the programming of the robotic agent given a particular platform. Even if only this can be quite challenging and inspiring, with our current proposal of open source printable robots, the teaching programme must not be focused any more *only* on the robotic agent, but it may also include its mechanical design. Beginning with a basic platform, like the Miniskybot, students can be guided through the design and programming process. In this way, they may discover the tight relation between hardware and software, and how each of them can and must, adapt to the other requirements in order to achieve a precise task. They may learn that a particular mechanical design suits better a precise task, test different alternatives, and so forth. And something that is hardly considered in robotic programmes, students may learn that a change in the mechanical design could solve a problem better, faster, and more robustly than a software solution.

3 On Low-Cost 3D Printers

Bradshaw et. al [2] have recently made a study on low-cost 3D printing. They briefly run through the history of 3D printing, beginning in the late 1970s. These more than thirty years have driven to affordable 3D printers for individuals[1], and allow them to print complex engineering parts entirely automatically from design files that it is straightforward to share over the Internet. While open source software development has been studied extensively, relatively little is known about the viability of the same development model for a physical object design. 3D printers are offering new possibilities of sharing physical objects. As they can be defined using code, researchers can share their own parts, evolve them and "build" them straight forward using 3D printers. This allows for a decentralized community to independently produce physical parts based on digital designs that are shared via the Internet. Apart from improving the device, dedicated infrastructures were developed by user innovators. As Bruijn shows in his master thesis [4], a considerable improvement of hardware are proposed by people sharing parts and having access to 3D printers. This hardware modifications are relatively easy for others to replicate. As it has been the case with software for many years, currently, there are also on-line repositories of parts, where people can download and upload their designs[7].

In figure 2 four of the most important open source 3D-printers are shown. The origin of these kind of printers was the reprap project[8] [9] started by Adrian Bowyer in 2004. The aim of this project was to develop an open-source

[7] http://www.thingiverse.com

[8] http://reprap.org/wiki/Main_Page

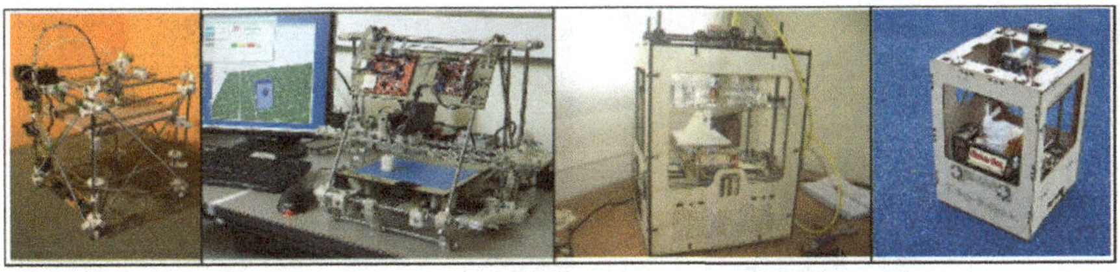

Fig. 2 Pictures of some open source 3D printers. From left to the right: RepRap Darwin, the first generation (May, 2007); Reprap Mendel (Sep, 2009) the second generation; Makerbot Cupcake (April, 2009), the first commercial open-source 3D printer; Makerbot Thing-o-Matic (Sep, 2010), second version

self-replicating machine. In May 2007 the first prototype, called Darwin was finished and some days later, in May 29th the first replication was achieved. Since then, the reprap community (original reprap machines and derived designs) has been growing exponentially[4]. The current estimated population is around 4500 machines. The second reprap generation, called Mendel, was finished in September 2009. Some of the main advantages of the Mendel printers over Darwin are bigger print area, better axis efficiency, simpler assembly, cheaper, lighter and more portable.

Initially, both Darwin and Mendel were not designed for the general public but for people with some technical background. As the reprap project was open-source, small companies were created to start shelling these 3D printers, as well as derived designs. The first company was Makerbot Industries[9], who shipped a first batch of their Cupcake CNC in April 2009. By the end of 2009 they had shipped nearly 500 complete kits. After operating for a year they had sold about 1000 kits in April 2010. Their latest design is the thing-o-matic printer, announced in September 2010. It is really easy to build and use, and their cost is around 950€.

Currently, at the System Engineering and Automatic Department of Carlos III University of Madrid we have one thing-o-matic available for the students, shown in figure 3. It was fully assembled by the students. Anyone has free access to it so that they can print whatever designs they want. Our main goal is to stimulate their imagination and enhance their creativity.

In addition we have started a project, called "Clone wars"[10], in which a group of students are building their own reprap printers from the scratch. All the parts are being printed in our thing-o-matic, which has been named MADRE (that means mother in spanish). We have chosen the Prusa Mendel model as the design to build, because it is very well documented and it is rather easy to assemble. In figure 3(on the right) the first prototype is shown. In total the students are building 20 of them.

[9] http://www.makerbot.com/

[10] http://asrob.uc3m.es/index.php/Proyecto:_Clone_wars

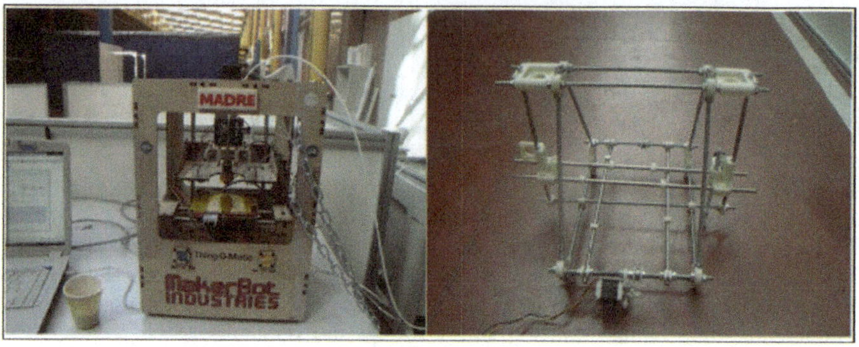

Fig. 3 Our open source 3D printers at Carlos III University of Madrid. On the left: the Makerbot Thing-o-matic, called MADRE. On the right: a Prusa Mendel prototype, being built by the students

4 The Miniskybot Mobile Robot Platform

4.1 Introduction

The new Miniskybot robotic platform[11] is open source: all the mechanical and electronic design has been released with a copy-left license. Furthermore, only open source software tools have been employed. This is important because in doing so it is guaranteed that anyone will be able to read, understand and modify the design files without license issues and using their preferred computer platform (Linux, Mac, BSD, Windows...).

The Miniskybot is a differential drive robot composed of printable parts and two modified (hacked) hobby servos. It has been designed so that it can be printed on open source reprap-like 3D-printer. Two mechanical designs have been developed: the minimal version and the 1.0.

4.2 Minimal Version

The first prototype developed was a minimal robot chassis. The idea was to design a printable robot with the minimal parts, a kind of "hello world" robot. It is shown in figure 4. It consist of only four printable parts: the front, the rear and two wheels. They are all attached to the servos by means of M3 bolts and nuts. Standard O-rings are used as wheel tires. For making the robot stable, the rear part has two support legs that slide across the floor. Therefore this prototype is only valid for moving on smooth flat surfaces. The goal of this first design was to show the students a minimal fully working mobile robot for stimulating their minds. They were encouraged to improve this initial design.

---

[11] http://www.thingiverse.com/thing:7989

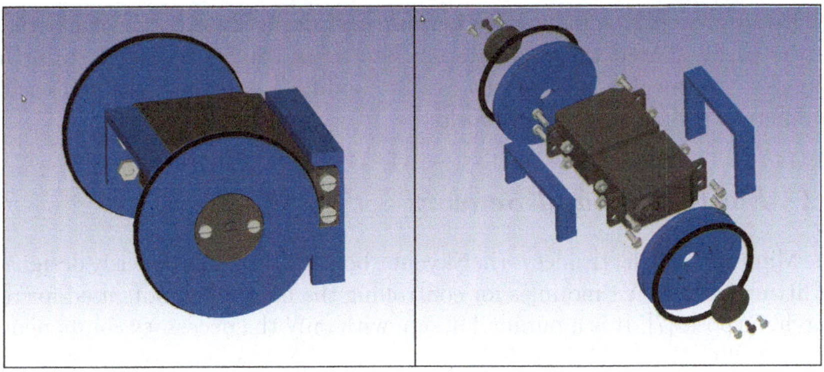

Fig. 4 Miniskybot. Minimal version

4.3 Miniskybot 1.0

The version 1.0 chassis is an evolution of the previous design (figure 5). It consist of nine printable parts: the front, the rear, two wheels, the battery compartment, the battery holder and the castor wheel. An important feature is that the parts have been parameterized, just changing some parameters new parts are obtained. For example the battery compartment is automatically changed if the parameter battery type is set from AAA to AA. In this case a new compartment capable of holding AA batteries (instead AAA) is generated.

The parametric feature is possible thanks to the open source Openscad[12] software used for designing the pieces. The parts themselves are not graphical meshes but scripts that determine how they are built by primitive geometric forms. When these scripts are "compiled" the graphical part is generated and rendered on the screen, and later exported as an STL file for 3D printing.

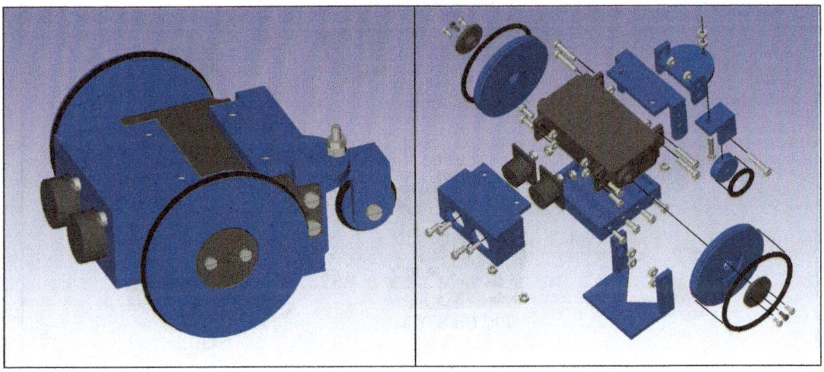

Fig. 5 Miniskybot Robot. Version 1.0

[12] http://openscad.org

This approach is very flexible because the parts are ASCII scripts that can be easily shared through Internet, stored in repositories and so forth. Therefore the mechanical designs can be modified, used, and printed easily by different people around the world.

4.4 Electronics and Sensors

The Miniskybot's electronics is the Skycube board[13]. It was previously designed for fitting into the Y1 modules for controlling the modular robots used for research purposes[7]. It is a minimal design with only the necessary components for controlling the robot. It includes an 8-bit pic16f876a micro-controller, headers for connecting the servos, an I2C bus for the sensors, serial connection to the PC, a test led and a switch for powering the circuit (figure 6).

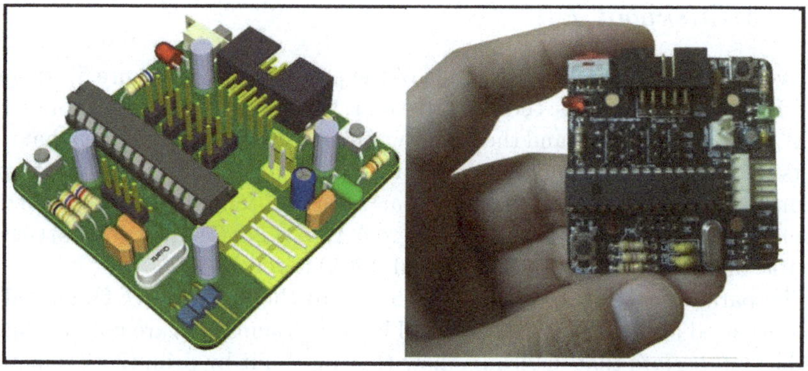

Fig. 6 Electronics. Skycube board

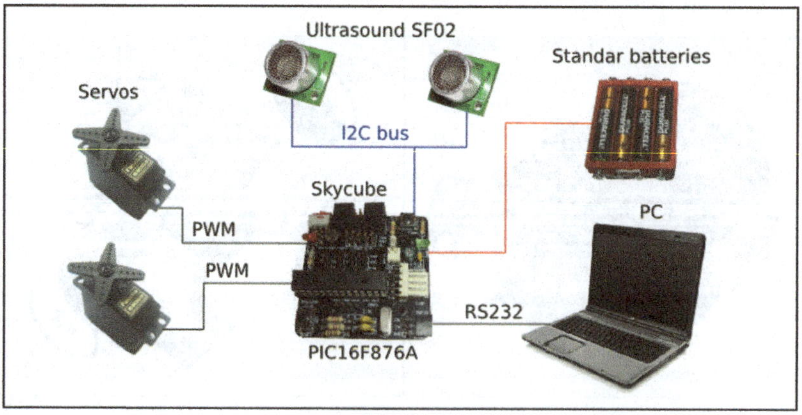

Fig. 7 Electronic diagram

[13] http://www.iearobotics.com/wiki/index.php?title=Skycube

An electric connection diagram is shown in figure 7, where the servos are connected directly to the board. The speed is set by means of two PWM signals. The two ultrasound sensors in the robot's front are connected thought the I2C bus. Robot version 1.0 have two ultrasound sensors, but as they are connected to the I2C bus, more sensors can be easily added. For the power supply four AAA type standard batteries are used. The board can be connected to the PC by a serial RS232 connection for downloading the firmware. The PCB has been designed with the open source Kicad tool.

The robot is programed in C language using the open source SDCC cross compiler and the binary files are downloaded into the board by means of a serial cable. Previously a bootloader firmware needs to be burned in the flash memory by means of the ICSP connector. Loading the firmware this way the students do not need to use any programming hardware but just a simple cable. Also, the download is done very fast, where it takes only a few seconds to complete the whole process.

5 Derived Designs from Miniskybot

In contrast to our previous Skybot robot which remained unmodified for many years, the MiniSkybot has inspired the imagination of the students which have developed new designs in record time. There were two main reason for this motivation, according to the students: 1) Full access to the Miniskybot "source code", 2) Being able to turn their thoughts on real physical objects very fast, thanks to the 3D printer. The former let the students to fully understand a real robot and realize that it is not so difficult to design the mechanical parts. Instead of starting from the scratch, they just simply start modifying the Miniskybot parts. The latter is related to the strong feeling of happiness and power that the students have when they see their designs become a reality.

In the following sections two new derived design are presented, fully created by second year undergraduate engineering students with no special knowledge on mechanics.

5.1 Caterpillator

The Miniskybot robot uses two drive wheels for moving. Two students wonder if it was possible to design a robot with tracks instead of wheels. Inspired by this chain an pinions design in thingiverse[14], Olalla Bravo and Daniel Gomez decided to create the first printable track for mobile robots. After some initial failed tries, they succeed in building a parametric track[15]. The beauty of this

[14] http://www.thingiverse.com/thing:5656
[15] http://www.thingiverse.com/thing:7209

Fig. 8 The caterpillator robot

design was not only its functionality but its property of being parametric. Just changing some parameters, different tracks can be obtained, as well as the necessary pinions. In addition, 3mm plastic spool was used as pins for the links. Therefore no special screws and nuts were necessary.

The latest version is shown in figure 8. It is also available in thingiverse[16], along with some videos showing how it moves.

5.2 UniTrack and F-Track

A different approach was taken by Jon Goitia. He focused on designing robots with articulated tracks. The first design was Unitrack[17], shown in figure 9 (on the left and in the middle). It is an autonomous track driven by a hacked Futaba 3003 servo (the same servo used for the Miniskybot). It consist of two wheels attached to the servo and five standard o-rings used as tracks. Another o-ring is used as the transmission system between the servo and one wheel. Unitrack is also parametric, therefore the wheel's diameter and number of o-rings can be easily changed. This innovative design was for one month the first most popular thing on thingiverse, which is not easy to achieve (currently there are more than ten thousand things!).

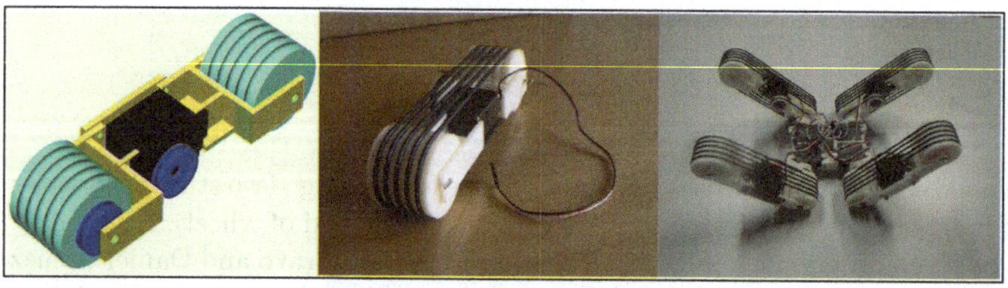

Fig. 9 Unitrack (left) and F-track (right) robots

[16] http://www.thingiverse.com/thing:8559

[17] http://www.thingiverse.com/thing:7640

Once Unitrack was fully functional, the F-track robot was created, shown in figure 9 (on the right). It consist of four articulated independent Unitracks joined to a body. This design is an example on how the creativity emerges from some students when they are stimulated.

6 A New Design Paradigm: Evolutionary Robots

Our new robotic platform combines two important features. On one hand it is open hardware, so that anyone can study, modify and distribute the robot. On the other hand the robot is printable making it very easy for the people to materialize it. The result is that anyone in the wold with access to Internet and to an open-source 3D printer can copy the robot, improve it or create derived design.

These features allow the emergence of a new design paradigm in robotics: Evolutionary robots. The robots can now be evolved by the community in the same way the open source movement creates and maintain in a distributed way new software applications, such as the Linux kernel, gnu tools, wikipedia, firefox and so forth. Now it is possible to bring these ideas into the robotics world.

In the previous section we have shown the derived robots created by a group of local students from the Miniskybot robot. It is difficult to imagine and foreseeing the wonderful robots that can be developed by thousand of people around the world collaborating together.

With the Miniskybot robot we have planted the seed. We have already gotten some indications of the potential of this idea: some weeks after the Miniskybot were published on thingiverse, at least three derived design were built. The first was printed by people from Makerbot at the RoboFest 2011 in Baltimore, the second at the FUBAR hacklab space in New Jersey. They are using Roboduino as electronics. The third one was built by CW kreimer[18] for teaching robotics at the Pittsburgh boy scout high tech camp.

7 Results

The Miniskybot robot has been successfully printed on a Makerbot Cupcake 3D printer in ABS plastic (Acrylonitrile Butadiene Styrene). The machine is equipped with the MK5 extruder and a heated build platform. It is very affordable with a total cost of 680€.

All the parts have been printed without raft. The software used was Replicator-G 0023 with Skeinforge 35. In figure 10 a red prototype is shown, along with all the parts needed for assembling the robot.

The total printing time is 2 hours and 50 minutes and the total robot cost is around 57€, as shown in table 1. It can be seen that the cost of the chassis

[18] http://www.youtube.com/watch?v=2EqvuPXYKf0

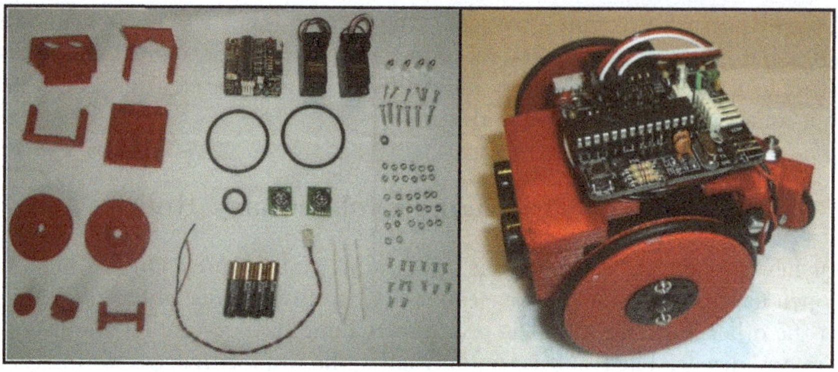

Fig. 10 The MiniSkybot robot printed in red ABS

(the printable parts) is marginal: less than 1€. Therefore, the robot cost is only determined by the cost of the electronics, motors and sensors.

Although this kind of 3D printers are not meant for production but for prototyping, they can be used for building small series of robots for giving courses on robotics to small groups. Given that every 3h the parts for a new robot are built and if the machine is working without interruption, 8 robot chassis per day can be printed. In figure 11 a group of six Miniskybots is shown, printed in different colors.

Table 1 Printing time and cost of the MiniSkybot v1.0 robot

Parts	Printable	Printing time (min)	Cost (€)
Wheels	yes	2x24	2x0.05
Battery compartment	yes	30	0.07
Front	yes	30	0.07
Rear	yes	16	0.04
Battery holder	yes	14	0.03
Castor wheel part 3	yes	12	0.03
Castor wheel part 2	yes	6	0.01
Castor wheel part 1	yes	4	0.01
Wheel O-rings	no	——	2x0.5
Castor Wheel O-ring	no	——	0.4
SRF02 ultrasound sensor	no	——	11.8
Skycube board	no	——	20
Servo Futaba 3003	no	——	2x9
4 AAA batteries	no	——	2.5
Nuts and bolts	no	——	2.5
Total:		**170 min (2h, 50min)**	**56.6€**

Fig. 11 A group of six MiniSkybot robots (v1.0) in different colors

8 Conclusion and Future Work

Using the latest open source 3D-printers a new printable robotic platform has been designed, built and tested. Our results confirm the viability of these new printable robots. They offer new important features for educational purposes. First, they are very flexible where the students can design new custom pieces easily which are printed and tested very fast. Therefore the robot can be mechanically evolved during the courses. Second, the robot can be thoroughly studied, modified, copied and distributed by anyone. This way the robot can evolute not only in our university but also around the world. This feature is enhanced by the fact that the mechanical parts are Openscad ASCII scripts, like any other software. Consequently, they behave like open source software and can be distributed and shared in a similar way. Finally, the total cost is very low, depending almost exclusively on the servos, electronics and sensors. The Miniskybot v1.0 costs 57€ and the printing time is around 3h, which means that eight robot chassis can be printed per day.

As a future work we are planning to continue evolving the robot in collaboration with our students, designing new parts for adding more sensors as well as creating new derived robots. Currently we are working on a new electronic board, with the same size than the Skycube but compatible with the Arduino software that is becoming more and more popular. In addition, we are developing a new idea on how to design mechanical parts using object oriented programing. We have called it as *object oriented mechanics language* (OOML).

Acknowledgements. We would like to thank Prof. Luis Moreno, head of the department, for supporting this educational initiative. Also we would like to thank all the students who have been involved in building our 3D printer MADRE. Special thanks to: Olalla Bravo, Jon Goitia and Daniel Gomez for their great contributions to the Miniskybot Robot.

References

1. Bowyer, A.: The Self-replicating Rapid Prototyper, Manufacturing for the Masses. In: 8th National Conference on Rapid Design, Prototyping & Manufacturing (June 2007)
2. Bradshaw, S., Bowyer, A., Haufe, P.: The Intellectual Property Implications of Low-Cost 3D Printing. SCRIPTed 5 (2010)
3. Carpin, S., Lewis, M., Wang, J., Balakirsky, S., Scrapper, C.: Usarsim: a robot simulator for research and education. In: 2007 IEEE International Conference on Robotics and Automation, ICRA 2007, pp. 1400–1405 (2007)
4. de Bruijn, E.: On the viability of the open source development model for the design of physical objects. Lessons Learned from the Reprap Project (November 2010)
5. Gerkey, B.P., Vaughan, R.T., Howard, A.: The player/stage project: Tools for multi-robot and distributed sensor systems. In: 11th International Conference on Advanced Robotics, ICAR 2003, Portugal, pp. 317–323 (June 2003)
6. Goldman, R., Eguchi, A., Sklar, E.: Using educational robotics to engage inner-city students with technology. In: Proceedings of the 6th International Conference on Learning Sciences, ICLS 2004, pp. 214–221. International Society of the Learning Sciences (2004)
7. Gonzalez-Gomez, J.: Modular robotics and Locomotion: application to limbless robots. PhD thesis (December 2008)
8. Hsiu, T., Richards, S., Bhave, A., Perez-Bergquist, A., Nourbakhsh, I.: Designing a low-cost, expressive educational robot. In: Proceedings of 2003 IEEE/RSJ International Conference on Intelligent Robots and Systems, IROS 2003, vol. 3, pp. 2404–2409 (2003)
9. Jones, R., Haufe, P., Sells, E., Iravani, P., Olliver, V., Palmer, C., Bowyer, A.: RepRap-the replicating rapid prototyper. Robotica 29(01), 177–191 (2011)
10. Hautop Lund, H., Pagliarini, L.: Robot Soccer with LEGO Mindstorms. In: Asada, M., Kitano, H. (eds.) RoboCup 1998. LNCS (LNAI), vol. 1604, pp. 141–151. Springer, Heidelberg (1999)
11. Lund, H.H., Pagliarini, L.: Robocup jr. with lego mindstorms. In: Proceedings of the 2000 IEEE International Conference on Robotics and Automation, ICRA 2000, April 24-28, pp. 813–819. IEEE, San Francisco (2000)
12. Keith, J., O'Hara, K.J., Kay, J.S.: Investigating open source software and educational robotics. J. Comput. Small Coll. 18, 8–16 (2003)
13. Rawat, K.S., Massiha, G.H.: A hands-on laboratory based approach to undergraduate robotics education. In: Proceedings of 2004 IEEE International Conference on Robotics and Automation, ICRA 2004, vol. 2, pp. 1370–1374 (2004)
14. Sklar, E., Eguchi, A., Johnson, J.: RoboCupJunior: Learning with Educational Robotics. In: Kaminka, G.A., Lima, P.U., Rojas, R. (eds.) RoboCup 2002. LNCS (LNAI), vol. 2752, pp. 238–253. Springer, Heidelberg (2003)
15. Verner, I.M., Waks, S., Kolberg, E.: Educational robotics: An insight into systems engineering. European Journal of Engineering Education 24(2), 201 (1999)

Germán LaRA: An Autonomous Robot Platform Supported by an Educational Methodology

R. Francisco, C. Uribe, S. Ignacio, and R. Vázquez

Abstract. This article presents the evolution of a learning methodology, from its conception ten years ago until its maturity, grounded on Robotics. This methodology is based on building robots within long term projects, which involves key features to encourage students to engage in research and development. This work identifies two phases in the Tec de Monterrey Campus Cuernavaca Robotics' history. In the first one, students assimilated commercial platforms and worked on programming tasks and some adaptations of control systems. In the second phase, students design, develop and integrate mechanical parts, hardware architectures and programming modules. The current work within this educational framework is presented as Germán Lara, an autonomous robot platform developed by undergraduate students, as part of a Greenhouse Automation project. Finally, an assessment based on a qualitative analysis of the student's skills shows the difference between both phases.

1 Introduction

A practical approach, as a didactic technique, is a simple way to introduce the challenge of describing insights about real-world problems to students. With some basic principles, a teacher can provide convenient heuristics so that students actually build artificial systems (e. g. robots). Building systems is crucial because we, as researchers and scientists in computer science, want to design and construct

R. Francisco · C. Uribe
Tec de Monterrey Campus Cuernavaca, Autopista del Sol km 104,
Colonia Real del Puente, Xochitepec, Morelos, México, 62790
e-mail: a00375673@itesm.mx

S. Ignacio · R. Vázquez
Tec de Monterrey Campus Cuernavaca, Autopista del Sol km 104,
Colonia Real del Puente, Xochitepec, Morelos, México, 62790
e-mail: a01126782@itesm.mx

intelligent artificial systems so that we can understand intelligent systems in general: this is the synthetic methodology [9].

As a basic methodology of the new artificial intelligence (also mentioned in literature as nouvelle or embodied artificial intelligence [6]), it can be summarized as learn by building. For example, if we are trying to understand human spatial navigation, the synthetic methodology requires that we build an actual navigating robot with sensors and actuators that enable it to move and interact within an environment.

Building a real physical system always yields the newest insights. A real-world navigating agent, like a human or a physical robot, has to somehow deal with the inaccuracy of its actions and the unpredictability of the real-world environments, while these problems can be partially or totally ignored in a simulation.

By actually building agents (i.e., real robots) students can learn about the nature of the phenomenon, develop an analytical sense about the problem and understand the applicability or making meaning [4] of the possible solution(s) with a synthetic or integrative criteria in fields so diverse as computer or cognitive science.

In the Laboratory of Robotics and Automation (LaRA), we have followed the methodology of learning by building. This has been possible thanks to professors like Dr. Fernando Ramos Quintana that in the last 15 years has greatly encouraged students to engage in research. Project after project the grade and majors of the students have come from Ph. D. students to first-year undergraduate students.

The success of this approach could be said to reside in the robots themselves. They allow concrete testing of ideas in a very objective way: a robot either works or it does not. Moreover, robots serve as excellent platforms for transdisciplinary research [8] and cooperation between people with different backgrounds and ages.

The most relevant Robotics achievements of the Tec de Monterrey Campus Cuernavaca are presented in Section 2. Then, the educational and academic framework of LaRA as well as the Program for Undergraduate Research Encouragement with the current Greenhouse Automation project, are described in Section 3. After that, the research on computer science that is using Germán Lara, an autonomous robot, as platform is introduced in Section 4. Finally, conclusions and a qualitative assessment are presented in Section 5, and future work is discussed in Section 6.

2 Related Work

The Tec de Monterrey, as an institution of higher education and research, was funded in Mexico more than 60 years ago. Tec de Monterrey Campus Cuernavaca is relatively younger, almost half the age of the first campus in Monterrey. However, it took half of this time, 15 years, to consolidate a strong group of researchers in electronics, control and computer science to start a fresh 15-year history in Robotics.

This group of researchers and academic leaders started in 2000 to establish two-year projects, increasingly involving younger undergraduate students. The most trascending difference between the projects, besides the degree of the students, is the nature of the platform used. There is a general division between the projects that

Table 1 Smart Hexapods project's output: Ph. D. and Master in Computer Science theses

Author	*Year*	*Type*	*Title*
Edgar E. Vallejo C.	2000	Ph.D.	Evolución del comportamiento en agentes autónomos: una contribución a la locomoción de insectos. [15]
Alejandro Vargas H.	2001	Master	Aprendizaje por refuerzo en línea para la locomoción de un hexápodo.
Arnulfo Martínez P.	2002	Master	Locomoción de un hexápodo combinando andaduras en superficies planas e irregulares.

used commercial robotic platforms (students as users) and those that used prototype platforms (students as designers and developers).

Navigation algorithms need to handle multiple and varied information to develop proper behaviors. Since the interaction in the real-world is time-constrained, an agent cannot generate a completely new behavior for every situation. Rather, it shall exploit its experience in order to adapt its behavior based on prior knowledge.

The **Smart Hexapods** project used a Lynxmotion[1] Quadropod and Hexapod robots to compare and develop insect-inspired locomotion models. Ph. D. and Master in Computer Science students approached locomotion as an emergent property of the interactions between the control mechanisms of individual legs [5]. Table 1 summarizes the scientific production of this project that lasted two years.

Navigation algorithms need to cope with the intrinsic uncertainty of the real-world in order to increase the behavioral diversity of the agent overtime. Master in Computer Science students developed architectures that enabled artificial agents to explore novel characteristics of a situation. Thus, the agents could react appropriately to certain situations in real-time **Robot Soccer** matches.

After the Smart Hexapods project, a new platform came onto scene: Yujin[2] Mirosot Robot. Table 2 summarizes the theses obtained from the research effort of this dynamic and competitive challenge. Additionaly, to their effort, students were encouraged to participate in international events. Which resulted in a 5th place in simulated and 6th place in real, both Mirosot League at the 2002 FIRA[3] Cup Korea, and a 4th place in simulated Mirosot League at the 2003 FIRA Cup Austria.

A combination of the machine learning algorithms from the Smart Hexapods project, the multi-agent approach from the Robot Soccer project, and the empowerment of the students, as well as the entire Campus, with the FIRA Robot Soccer World Championship, brought another platform: ZMP[4] E-Nuvo Walking robot.

Nine Mechatronics Engineering fourth-year students were integrated in two teams each with a humanoid robot. By taking advantage of the mobile robotics

[1] http://www.lynxmotion.com
[2] http://www.yujinrobot.com
[3] http://www.fira.net
[4] http://www.zmp.co.jp

Table 2 Robot Soccer project's output: Master and Ph. D. in Computer Science theses

Author	*Year*	*Type*	*Title*
A. Dizan Vázquez G.	2002	Master	Soccer robótico: análsis crítico del área y diseño e implementación de una arquitectura para robots.
Huberto Ayanegui S.	2002	Master	Una arquitectura multicapas para el control de agentes de soccer robótico.
David A. Gómez J.	2004	Master	Desarrollo de algoritmos robustos para el sistema de visión de Mirosot FIRA.
Moisés Memije R.	2005	Master	Desarrollo de un módulo de estrategias para un equipo de fútbol robótico.
Huberto Ayanegui S.	2008	Ph. D.	Reconocimiento y descubrimento de patrones de comportamiento en sistemas multi-agente cooperantes. [1]

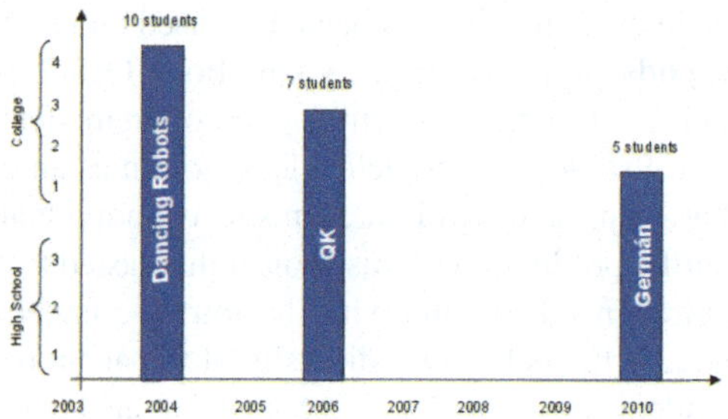

Fig. 1 Undergraduate students' projects. X axis shows the year in which each project was started. Y axis represents the average grade year of the students from each project.

specialization[5] of the students, an attractive and challenging project was proposed: **Dancing Humanoids** for the 6th International Robot Olympiad[6] 2004, in Korea.

Even though there were no graduate theses involved in this project, the fourth-year students could collaborate and cooperate with the research faculty. A simulator [10] was developed in order to analyze the motion behavior of the humanoid robots while dancing. From this interaction, the students successfully programmed different dancing routines for each robot. Both teams participated at the Robo League in the RoboDancing category of the IRO 2004, achieving the 1st and 3rd places.

With the inertia from the International Robot Olympiad and the fact that undergraduate students could perform outstandingly, fulfilling the requirements and exigencies of a researcher, a final dare was pending: to design and develop a robot platform. Another difference from the previous project was that instead of graduate or fourth-year students, the project would be undertaken by third-year students.

5 E.g. mathematical methods for rigid bodies kinematics, and dynamics and control of robots.

6 http://www.iroc.org

The 8th International Robot Olympiad 2006, was held in Gold Coast, Australia. A team of seven Mechatronics Engineering third-year students designed, manufactured, developed and tested the 18-Degrees-Of-Freedom lab prototype robot, named **QK**, for forest fire surveillance. Encompassing from the design of the mechanics and electronics, to the manufacturing and printed circuit boards construction, from the wireless and vision sensors selection, to the algorithms and protocol testing. The team won the 1st place at the Robo League in the Creativity category of the IRO 2006. This, matured a Robotics laboratory as well as an educational framework.

3 LaRA

The Laboratory of Robotics and Automation (LaRA) is part of the facilities of the Tec de Monterrey Campus Cuernavaca. In contrast with other laboratories that support the academic life, there are no scheduled courses during the semester in LaRA.

In other laboratories, like the Manufacturing or Electronics, there are classes related to, or that need resources from them. Each teacher assigns a final project as part of the evaluation of the course. The students have to develop a project related to a subject of the course and deploy it by the end of the semester. In the worst case, during the semester the students could have up to six different projects.

This approach has the disadvantage that, the next semester, the students will work in different short-time projects. Also, the students only have the opportunity to work with students from the same class, that mainly are of the same major and age.

LaRA, aside from supplying resources to teachers and students like the other laboratories, has long-term projects, giving advice and support to the students about technology and projects, and offers a full-time available workplace where students can collaborate and interact with people of different ages, majors and professional interests. LaRA serves students from high school, college and graduate school that study, or are interested in, careers in engineering, management or design. Most of them are curious about mechanics, electronics, programming or entrepreneurship.

Most of the students working in LaRA are enrolled in a long-term project, where they match the requirements of their course projects to the needs of the LaRA's projects. Also, there are students that work in LaRA as a scholarship-service. Generally, they match their projects (personal or from a course) to the LaRA's projects needs as well, but they are required to work in the lab at least five hours per week.

3.1 PROFIL

The Program for Undergraduate Research Encouragement (PROFIL [11], according to its initials in Spanish[7]) emerged as part of the academic framework of Tec de Monterrey Campus Cuernavaca. The main goal is to boost a research culture in the scholastic community based on innovation, technological development and applied research, within an educational framework in LaRA.

[7] Programa para el Fomento a la Investigación en Licenciatura.

PROFIL has been the key to change from commercial platforms to functional prototypes. Thus, younger students have engaged in long-term and ambitious projects, becoming designers and developers, instead of users. Robotics is particularly effective because of the complexity of mechanics, electronics and software development. The activities involved with the design and development, interact with the curiosity of the students giving an unique experience to any student and team. The learning processes happen more effectively when the students build a prototype using their own ideas, ingenuity and creativity.

PROFIL involves the students within an enviroment of self-learning, advanced topics and long-term projects with a transdisciplinary approach. This constrasts with most universities that offer research opportunities to students [13], which mainly comprise working in a research lab during the summer, internships or occasionally during the academic year, with intense and focussed research work.

These are fundamental aspects that have changed the perspective of the undergraduate students' capacities and roles. In addition to this enrichment of the students' experience, this program is a scaffolding for students that are interested in or could become future researchers. The educational framework is based on:

- **Long Duration Projects:** A two-year duration gives the students the opportunity to get closer to a graduate-level aim and aspiration.
- **Participation in International events:** Motivation enables the students to be autonomous, so an international context pushes their creativity and enthusiasm.
- **Attractive and Challenging Projects:** Robotics is a dynamic field proper for technological development, which animates the students and makes them feel part of something important, allowing a high level of commitment.
- **Adaptation of the Curriculum:** Students need to match their course requirements, as well as their professional interests, with the project needs.
- **Adoption of a Formal Learning Model:** Understanding by building is a learning strategy that frames the academic and educational frameworks.

3.2 Germán

As a natural evolution of PROFIL and with an environment, health and safety focus of the LaRA, the current project recruited second-year students and seeks for a field robot prototype for greenhouse automation. A greenhouse enviroment, from a Robotics perspective, is a dynamic and challenging place. The irregular terrain and the hallways make it complex to navigate. The weather and the chemical products have to be considered when designing a device for greenhouse automation.

Germán is a cheap field prototype that intends to cover all of the requirements for working in a greenhouse. Additionally, Germán is intended as a research platform, which requires modularity, flexibility, to be easy to modify and adapt to different tasks. The project is organized in four areas:

- **Mechanics**

  Design, manufacturing and testing of the robot structure and mechanisms. This involves chosing the materials in order to cover needs such as strength, chemical or water-resistance, heat transfer, availability and cost.

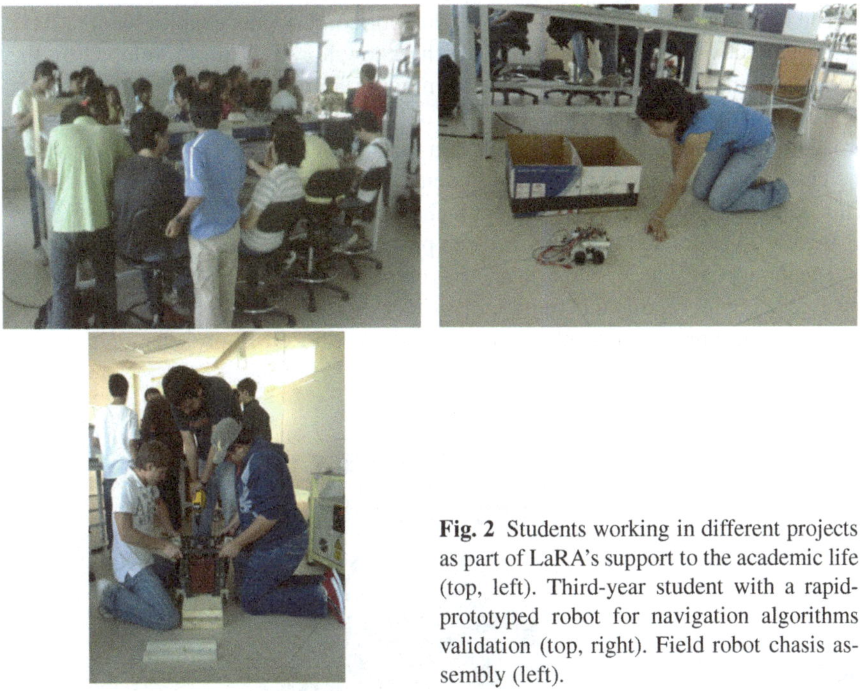

Fig. 2 Students working in different projects as part of LaRA's support to the academic life (top, left). Third-year student with a rapid-prototyped robot for navigation algorithms validation (top, right). Field robot chasis assembly (left).

- **Electronics**
  Design, building and verification of circuits and PCBs for sensors (e.g. laser, infrared and sonar range finders), actuators (e.g. motors), controllers (e.g. Microchip[8] dsPIC), and power supply.
- **Software Development**
  Design, development and validation of the architecture, programs and routines for motion control, sensor data adquisition and analysis, motion planning, communication between software layers, and information about the state and goals.
- **Business Plan**
  A field robot must be competitive, not only technically, but also commercially. The price of the components, the working time of the team, as well as advantages such as energy saving, multipurpose design and a robust platform, are considered into a sales strategy to position Germán in a commercial market.

The main idea is to design and build as much as possible, considering the resources and limitations of the LaRA. In several previous projects, one major problem was that at certain stage of the development it was necessary a change in the software, electronical or mechanical module. Most of the time these kinds of adjustments were not possible due to a lack of technical experience, issues with warranty policies, or because of the functional design of the platform itself.

[8] http://www.microchip.com

The learning process of the students who work on Germán enables them to have a more effective understanding because the know-how is being developed by the students themselves. Younger students commonly approach the engineering aspects of project work with enthusiasm, but often only pay lip service to the methodologies, procedures, materials and use of tools and machinery. The most interesting part of the process is when the experienced students show and teach younger students [2]. This is more natural than classic approaches that immerse undergraduate students to already consolidated graduate research programs [3].

The main contribution of Germán, as a robotic platform, is that the same students are able to repair, modify and replicate it. This work in progress is the base for future field and entrepreneurial developments in LaRA and PROFIL projects.

Fig. 3 Germán LaRA with a thermonebulizer for Greenhouse fumigation.

3.2.1 Current Work

Germán is in the development and building phase. The mechanical structure has been manufactured and assembled (see Figure 3). Germán was designed to weigh 40 kg maximum in order to carry 40 kg. The team working in mechanics is formed by three second-year, five first-year and two high-school students.

The electronic design for power supply and motor control has been protoyped and tested. The controller for velocity is being tested. This work has been done by four second-year students. The business plan, done by two second-year students, has

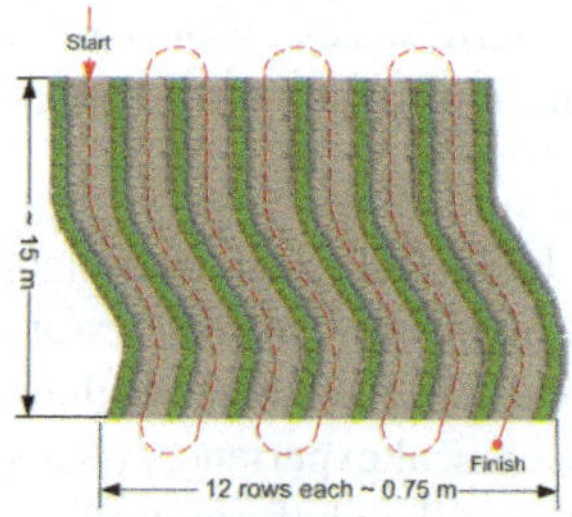

http://ilf.rz.tu-bs.de/index.php/tasks.html

Fig. 4 The robot navigation between crop rows is a classic task for greenhouse automation evaluated the strenghts and weaknesses of the project, and it is being benchmarked against other automation products and services for greenhouses.

4 Incremental Learning for Autonomous Navigation

One of the most challenging goals in Robotics lies in the area of autonomous navigation. This research field aims for an artificial agent (e.g. robot) moving in the real world, which includes reasoning about uncertainty to build reliable sensory-based motion strategies [7]. In order to plan its motions, the agent must compute collision-free paths among possibly moving objects, which requires to:

To cope with the complexity and to integrate the constraints concerned with real-world situations, there are machine learning techniques like Reinforcement Learning [12] (RL) that stress learning from interaction. Interaction endows an agent with a wealth of information about cause and effect, about the consequences of actions, and about what to do in order to achieve goals. Recent research efforts [14] focus on RL, together with optimal stochastic tools, for motion control learning.

Interacting with the environment also involves uncertainty in sensing because of noise, signal delay and range limitations. As well as inaccuracy of the agent's actions and unpredictability of the real-world environments. Altogether make the information regarding the current situation to be partially accesible or incomplete. Probabilistic models, in this case, are appropriate to learn and predict, allowing to refine prior knowledge on the basis of the available observations.

A proposal for designing and building agents that autonomously plan and control their own motions, while reasoning about the uncertainty in control, sensing and prediction, is an hybrid system capable of learning from its experience about the environment, and to increasingly develop and perform more sophisticated behaviors.

The current research will develop a "learn and predict" structure integrated to an onboard sensor-based motion planning. The aim is to enable an agent to improve RL policies based on the same observations that are used to predict motion as it happens, thus, contributing to the autonomous navigation improvement.

This work is part of a Ph. D. thesis, and will be using Germán as a platform for testing navigation algorithms. This collaboration between graduate and undergraduate students will benefit the participation of Germán in international contests for agricultural robots, such as the Field Robot Event[9]. This agricultural robot with autonomous navigation capacities shall be able to perform more complex tasks like fumigating, pollinating, harvesting and diagnosing crops.

The collaboration between graduate an undergraduate students, besides the technical basis and the educational framework, has helped to manage the continuity of Germán with a student population that changes from semester to semester.

The Ph. D. thesis and PROFIL have been the bridge in LaRA to have permanent personal support for the students to ensure the maintenance of the focus on Germán.

[9] http://www.fre2011.dk

5 Conclusions

In this paper we have reviewed the evolution of an educational framework until its consolidation. As a result, Germán is a robot platform designed and developed in LaRA, and grounded on the PROFIL methodology. PROFIL has evolved naturally to engage younger students in applied research.

During the first phase, undergraduate students teamworked with graduates in Computer Science. The projects were developed using commercial platforms and focusing on programming tasks (e.g. simulators). Some technical, but not challenging, adaptations were made in mechanical and electronic parts.

The second phase was characterized by an important motivational factor: students became designers and developers in order to meet the challenge of building robots for solving real-world problems. Also, funds were provided to the undergraduate teams, which needed to manage the available resources. The transition from the first phase to the second one required the development of skills, which are the essential difference between both phases.

Such skills are qualitatively described as follows: 1) students understood the relevance of the *integration* process, immersed into a transdisciplinary environment, for a project; 2) they worked *collaboratively*; 3) improved *problem-solving* skills e.g. analysis, synthesis, abstraction and structuring; 4) undergraduate students were responsible of *administrative* tasks for managing funds; 5) they developed *self-directed* learning attitudes.

Although major teaching institutions, as Tec de Monterrey, have developed different didactic techniques as academical frameworks, such as Problem-based learning, Collaborative learning, Project-oriented learning, Cases, Service-learning, and Research-based learning. The real technological developments need more time and effort than what an undergraduate student can dedicate to a single course project.

Laboratories as LaRA that follow a synthetic methodology as understanding by building, and educational frameworks as PROFIL, allow the students to enhance their experience with applied research. Consolidating a laboratory or organizing the research faculty is not an easy task, but is a long-term investment for an Academia committed to high quality education.

Every year new generations arrive to universities, and it is more difficult to encourage them to develop an analytical and critical sense. This paper describes the case of Tec de Monterrey Campus Cuernavaca that after ten years consolidated an educational framework. Since QK, five years ago, there have been three generations involved in PROFIL, which means an average of five students per generation.

From the first generation, four are graduate students participating actively in research and technological developments. It is hard to evaluate quantitatively the difference between PROFIL students and standard ones since PROFIL is a Campus program, open to any student, and not specific for a group of classes.

Those students that have been enrolled in PROFIL show stronger abilities in problem solving, communication, cooperation and general technical knowledge.

This can be more clearly seen in the interest and success of the second generation on more ambitious programs such as international stays at the Massachusetts Institute of Technology and at Texas A & M, both in U.S.A, and at the Université du Maine and at Grenoble INRIA research centre, both in France. Also, they have been motivated to contribute to scientific articles as in the present work.

The third and second generation are working together in Germán, which has been vinculated with at least 12 courses in three semesters, raising the standards of the courses and the expectations of the professor. Finally, what is most important, students work in these kinds of projects because of curiosity and ambition, not for a grade, which results in a more natural experience of learning and understanding.

6 Future Work

The cycle that follows long-term projects together with the inclusion of younger students has been a complete success. Novice students get enrolled in a project, and they learn from the experienced ones and understand useful concepts for their courses. Then they become experienced and teach new students. A system to track this cycle for continuity is needed.

Also, younger people, as high school students, is required to be involved for the next project. The first step has been taken giving specialization courses like Robotics, Biomechatronics and Automotive Mechatronics. Just as it happened five years ago with fourth-year Mechatronics students.

Even that PROFIL is a campus program, it has lacked of a formal adoption in the curricular structure of the university. There should be classes and courses marked as candidates for PROFIL projects, so professors from different subjects know in advance what is going to be the semester project. This way, the research faculty will focus on specific characteristics of the long-term project, while they adapt the course material for the students in the present semester.

Greenhouse and agricultural robots must be able to learn new motion strategies while interacting with the environment in order to relate and cooperate more naturally with other robots and people. Germán shall be capable of incrementally learning new behaviors, toward an online adaptation within the environment dynamics.

References

1. Ayanegui, H., Ramos, F.: Recognizing patterns of dynamic behaviors based on multiple relations in soccer robotics domain. In: Ghosh, A., De, R.K., Pal, S.K. (eds.) PReMI 2007. LNCS, vol. 4815, pp. 33–40. Springer, Heidelberg (2007)
2. Chin, C., Spowage, A.: Project Management Methodology Requirements for use in Undergraduate Engineering Research Projects. In: IEEE ICMIT (2008)

3. Dahlberg, T., Barnes, T., Rorrer, A., Powell, E., Cairco, L.: Improving retention and graduate recruitment through immersive research experiences for undergraduates. In: SIGCSE (2008)
4. Ditcher, A.: Effective Teaching and learning in higher education, with particular reference to the undergraduate education of professional engineers. Int. J. of Engng. Ed. (2001)
5. Ferrell, C.: A Comparision of Three Insect-Inspired Locomotion Controllers. Robotics and Autonomous Systems 16, 135–159 (1995)
6. Iida, F., Pfeifer, R., Steels, L., Kuniyoshi, Y. (eds.): Embodied AI. Springer, Berlin (2004)
7. Latombe, J.-C.: Robot motion planning. Kluwer Academic Publishers (1991)
8. Moulton, B., Johnson, D.: Robotics education: a review of graduate profiles and research pathways. World Transactions on Egineering and Technology Education (2010)
9. Pfeifer, R., Bongard, J.: How the body shapes the way we think. MIT Press (2007)
10. Ramos, F., Salinas, S., Rebón, J.C., Arredondo, A.: A Simulator for Learning Symbolically about the Behavior of Motions in Bipedal Robots. Int. J. of Engng. Ed. (2009)
11. Ramos, R., Zárate, V., Álvarez, J.: PROFIL: the development of robotics projects to introduce the applied research culture in the undergraduate level of the engineering area. International Journal of Human Welfare Social Robots. KAIST, South Korea (2005)
12. Sutton, R.S., Barto, A.G.: Reinforcement learning: an introduction. MIT Press (1998)
13. Taraban, R., Blanton, R.: Creating effective undergraduate research programmes in science: The transformation from students to scientist. Teachers College Press (2009)
14. Theodorou, E., Buchli, J., Schaal, S.: Reinforcement learning of motor skills in high dimensions: A path integral approach. In: ICRA, pp. 2397–2403 (2010)
15. Vallejo, C., Edgar, E., Ramos, F.: Evolving Insect Locomotion using Non-uniform Cellular Automata. In: Genetic and Evolutionary Computation Conference, vol. 869 (2000)

Mutual Learning for Second Language Education and Language Acquisition of Robots

Akihiro Yorita and Naoyuki Kubota

Abstract. Recently, language education has great demand from elementary school to adults. The robot is used as a teaching assistant in Robot-Assisted Language Learning (RALL). It is very effective to use robots for language education. But robots have some problems. One of the problems is to get bored with interacting with robots. This paper discusses the role of robots based on mutual learning in language education. Next we explain the concept of self-efficacy using evaluation for learning condition of robots. We propose a conversation system for language education. The essence of the proposed method is mutual learning of humans and robots. The experimental results show the applicability of the system used for education.

Keywords: Human-Robot Interaction, Robot Assisted Instruction, Second Language Learning, Language Acquisition, Self-efficacy.

1 Introduction

English education is being done more enthusiastically than ever. Japanese government decided to introduce English education to elementary school and some Japanese companies use English as an official language. We can talk with native speakers on the video call recently, but when we talk with them we cannot talk very well because we may be nervous. Therefore we need to practice conversation with robots.

In robot-assisted language learning (RALL) shown in Table 1 [1-5], a humanoid robot named Robovie has taught English at an elementary school in two weeks [1]. It is an effective way to motivate students learning English, although it is less effective than educational software.

Akihiro Yorita · Naoyuki Kubota
Tokyo Metropolitan University
e-mail: yorita-akihiro@sd.tmu.ac.jp, kubota@tmu.ac.jp

In Korea, RALL has been the major way to learn English. It is also called r-learning. The robot helps human teachers and does role-playing with students. Robot IROBI is used as a home robot and teaching assistant in a classroom [2].

The robot was used to examine the learning effect on children. Using robots is good compared with using books and tapes, or computers. Robots are also used as native teachers in rural areas [3]. As the teachers prefer not to leave big cities, the students have few opportunities to take classes by them.

Table 1 Robot-Assisted Language Learning

Type	Teaching Assistant				Learning Companion
Robots	IROBI [2]	Telepresense robot [3]	Mero and Engkey [4]	Humanoid robot [5]	Robovie [1,6]
Aim	Interest	Interest	Listening	Cheering	Motivation
	Concentration	Confidence	Speaking	Conversation practice	Long-term Interaction
	Achievement	Motivation			
Case	In class				Recreation hour
Country	Korea			Taiwan	Japan

People regard virtual agents and robots as intelligent life [19]. They appear intelligent at first, but humans discover patterns gradually. Then people may stop communication. Therefore it is difficult to realize long-term interaction.

In [6], pseudo-development and confidential personal matters enable the robot to do long-term interaction. Here the robot changed interaction patterns along with each child's experience, the robot seems as if it learns something from the interaction. In fact, the robot can learn words [7]. In order to adapt to an open environment, a robot will have to learn the language dynamically. The system for Noun Concepts Acquisition (SINCA) forms utterances about an image, but SINCA is a language acquisition system not robot [8].

We propose a method of learning words between a robot and a human. Our target is to develop the method using it in everyday life. Especially learning and growing with robot is important.

Human symbiotic robots are utilized in various fields. In welfare, Paro and ifbot are representative robots [20,21]. In entertainment, AIBO is the most famous pet robot and miuro plays music with dance adapting to human preference [22, 23].

In education, students usually build a robot. Sometimes a communication robot is used as a teacher or a teaching assistant as noted above, but not a friend who learns together. In Japanese animation "Doraemon" [27], Doraemon is sent back from future to look after Nobita. It is preferable to call them human remediation robots instead of human symbiotic robots. Humans are not good at repetitive tasks, but robots can do. Then it is useful that we give a desire to learn English to the robot and the robot encourages us to learn. Therefore the robot controls his desire by self-efficacy and changes its utterances by the value of self-efficacy so that humans continue to learn. Here, we define self-efficacy for language learning and examine the change during conversation.

First, we explain the background of robots used in education. Next, we explain the education system, and explain computational intelligence technologies. Furthermore, we discuss how to support students using robot partners. Finally, we discuss the future vision toward the realization of educational partner robotics.

2 Robot Edutainment

2.1 Various Roles of Robots in Education

Various types of robots have been applied to the fields of education with entertainment (Edutainment). Basically, there are three different aims in robot edutainment. One is to develop knowledge and skill of students through the project-based learning by the development of robots (Learning on Robots). Students can learn basic knowledge on robotics itself by the development of a robot [9,26]. Lately, low cost 3D printers have been developed and students can easily make robots that they hope [24]. The next one is to learn the interdisciplinary knowledge on mechanics, electronics, dynamics, biology, and informatics by using robots (Learning through Robots). The Local Educational Laboratory on Robotics proposes that it is good to learn about nature in primary school and to think and understand humans in secondary school with using minirobots [25]. In the Robockey Cup, the students studied how to use motors, circuits, microcontrollers, and so on [10]. Moreover, to make a humanoid robot gives opportunities to understand voice recognition and image processing for communication between humans and robots [11]. The robot is also useful for children with autism. The last is to apply human-friendly robots instead of personal computers for computer assisted instruction (Learning with Robots).

A student seldom shows physical reactions to a personal computer in the computer-assisted instruction (CAI), because the student is immersed into 2-dimensional world inside of the monitor. However, a student aggressively tries physical interactions to a robot, because the robot can express its intention through physical reactions.

We showed the effectiveness of the learning with robots in the previous works [12,13]. A robot partner in educational fields cannot be the replacement of a human teacher, but the replacement of a personal computer. Of course, the robot also should play the role of a personal computer. Therefore, we propose the concept of robot-assisted instruction (RAI) to realize the style of education based on the learning with robots. A robot can be not only an assistant, but also a partner or collaborator in RAI. A personal computer is useful to collect, access, edit, and store data, but agent-like communication capability is low in a personal computer. Therefore, a robot partner can be replaced with a personal computer.

2.2 Robot Partners for RAI

We have developed PC-type of physical robot partners called MOBiMac (Fig.1) in order to realize human-friendly communication and interaction. This robot has two CPUs and many sensors such as CCD camera, microphone, and ultrasonic sensors. Furthermore, the information perceived by a robot is shared with other robot by the wireless communication. Therefore, the robots can easily perform formation behaviors. We have applied steady-state genetic algorithm (SSGA), spiking neural networks (SNN), self-organizing map (SOM), and others for human detection, motion extraction, gesture recognition, and shape recognition based on image processing [14]. Furthermore, the robot can learn the relationship between the numerical information as a result of image processing and the symbolic information as a result of voice recognition [15]. MOBiMac can be also used as a standard personal computer.

We used Voice Elements DTalker 3.0, which was developed by EIG Co., Ltd., Japan, for voice recognition and synthesis in the robot [18]. It was able to perform voice recognition using a sound segment network that made speaker-independent recognition possible. In addition, with the number of words that are recognized dependent on the memory, it achieved a recognition rate of 96.5% (for 200 words).

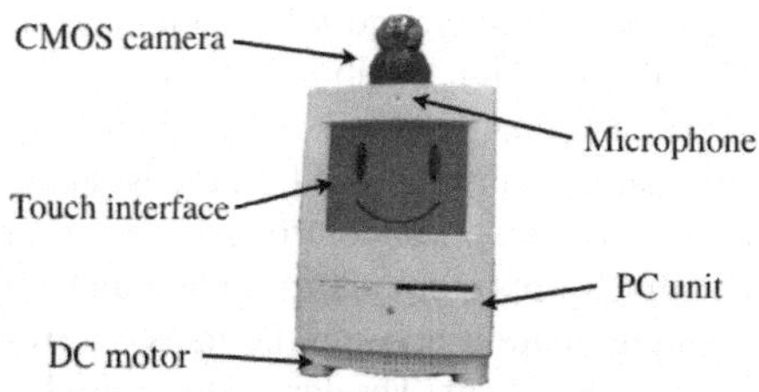

Fig. 1 Human-friendly partner robots; MOBiMac

We have used apple iPad as pocket robot partners, because we can easily use the touch interface and accelerometer in the program development. In this paper, we use iPad as a face of the robot and interaction with students. Figure 2 shows the overview of interfaces used in iPad.

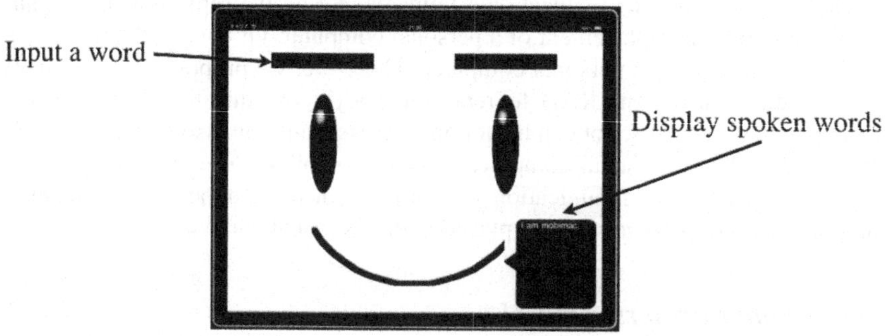

Fig. 2 iPad as a face of MOBiMac

3 Conversation System for Language Education

3.1 Learning Words and Conversation System

Figure 3 shows the total architecture of the perception, decision making, learning, and action. First, the voice recognition and image processing are performed to extract visual and verbal information through the interaction with a person. In addition, the function of word input is used for learning of the robot (Fig.4). In this paper, the robots use perceptual modules for various modes of image processing, such as differential extraction, human detection, object detection, and human hand-motion recognition (Fig.5).

After that, the robot selects the conversation mode from (1) scenario-based conversation, (2) daily conversation, and (3) learning conversation. In the scenario-based conversation mode, the robot makes utterances sequentially according to the order of utterances in a scenario. In the daily conversation, the robot uses a long-term memory based on spiking neural network. The robot selects an utterance according to the long-term memory corresponding to the internal states of spiking neurons. In the learning conversation, the robot updates the relationship between spiking neurons used in long-term memory by associative learning. Finally, the robot makes utterance. In the following sections, we explain the image processing based on steady-state genetic algorithm, and associative learning between perceptual information and verbal words (Fig.6).

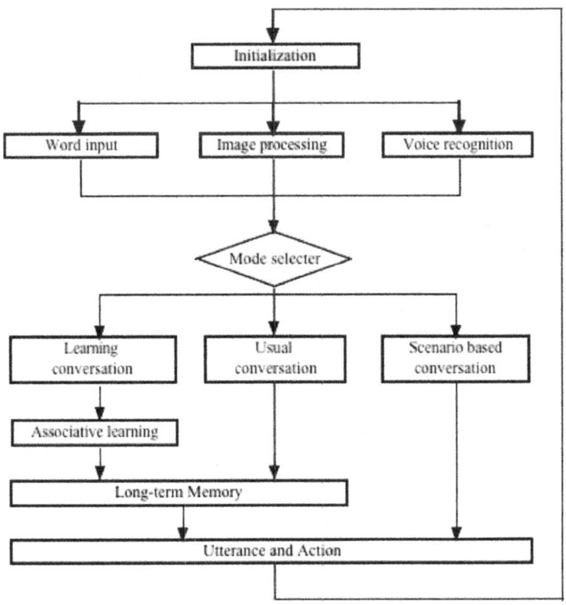

Fig. 3 Flow chart of conversation

Fig. 4 The screenshot of inputting a word to the robot

We use a simple spike response model to reduce the computational cost for associative learning. First of all, the internal state $h_i(t)$ is calculated as follows:

$$h_i(t) = \tanh(h_i^{syn}(t) + h_i^{ext}(t) + h_i^{ref}(t)) \quad (1)$$

Here, a hyperbolic tangent is used to avoid the bursting of neuronal fires, $h_i^{ext}(t)$ is the input to the ith neuron from the external environment, and $h_i^{syn}(t)$, which includes the output pulses from other neurons, is calculated by

$$h_i^{syn}(t) = \gamma^{syn} \cdot h_i(t-1) + \sum_{j=1, j \neq i}^{N} w_{j,i} \cdot h_j^{EPSP}(t) \quad (2)$$

Furthermore, $h_i^{ref}(t)$ indicates the refractoriness factor of the neuron, $w_{j,i}$ is a weight coefficient from the jth to ith neuron, $h_j^{EPSP}(t)$ is the excitatory postsynaptic potential (EPSP) that is approximately transmitted from the jth neuron at the discrete time t, N is the number of neurons, and γ^{syn} is the temporal discount rate. The presynaptic spike output is transmitted to the connected neuron according to the EPSP, which is calculated as follows:

$$h_i^{EPSP}(t) = \sum_{n=0}^{T} \kappa^n p_i(t-n), \quad (3)$$

where κ is the discount rate ($0<\kappa<1.0$), $p_i(t)$ is the output of the ith neuron at the discrete time t, and T is the time sequence to be considered. If the neuron is fired, R is subtracted from the refractoriness value in the following:

$$h_i^{ref}(t) = \begin{cases} \gamma^{ref} \cdot h_i^{ref}(t-1) - R & \text{if } p_i(t-1) = 1 \\ \gamma^{ref} \cdot h_i^{ref}(t-1) & \text{otherwise} \end{cases} \quad (4)$$

where γ^{ref} is the discount rate. When the internal potential of the ith neuron is larger than the predefined threshold, a pulse is outputted as follows:

$$p_i(t) = \begin{cases} 1 & \text{if } h_i(t) \geq q_i \\ 0 & \text{otherwise} \end{cases} , \tag{5}$$

where q_i is the threshold for firing. The weight parameters are trained based on the temporal Hebbian learning rule as follows:

$$w_{j,i} \leftarrow \tanh(\gamma^{wgt} \cdot w_{j,i} + \xi^{wgt} \cdot h_j^{EPSP}(t-1) \cdot h_i^{EPSP}(t)), \tag{6}$$

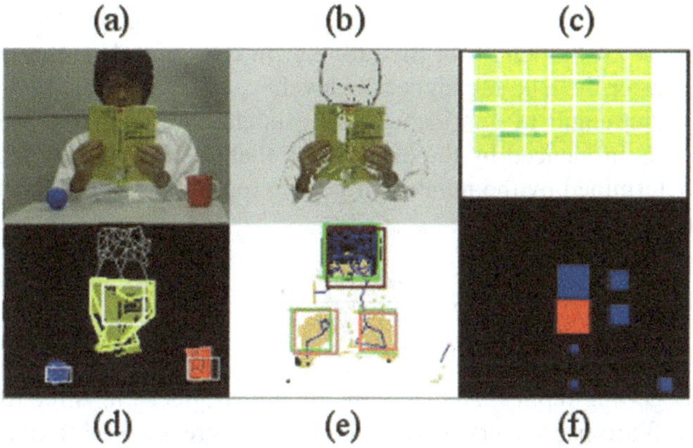

Fig. 5 The robot performs associative learning interacting with the person. (a) the original image, a photograph, (b) differential extraction, (c) the reference vectors of SOM corresponding to gestures, (d) object recognition results by SSGA-O, (e) human detection results by SSGA-H, the green box indicates the candidates for human face position produced by SSGA-H, the red box indicates the face position produced by human tracking, and the pink box indicates the hand position and (f) EPSP of the spiking neurons. which indicates the spatiotemporal pattern captured from the subject's hand motion. The red rectangle is EPSP, and it gradually diminishes, turns blue, and becomes smaller.

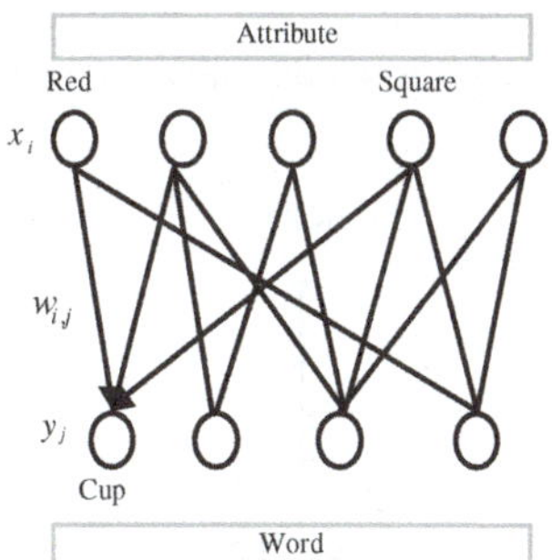

Fig. 6 Learning relationship with SNN

3.2 Self-efficacy of the Robot

When the robot learns with humans, it is desirable for them to learn similarly. Then to evaluate his own learning state, the robot uses self-efficacy proposed by A.Bandura [16].

Self-efficacy is represented by level, strength, and generality (Fig.7).

$$S = S_L + S_S + S_G , \tag{7}$$

where S_L is a level of the action to put the difficulty etc. together on the achievement of the action. It shows the difficulty of speaking English. The difficulty is different because of the length of the talk e.g. in case of only one word, or sentences. The easier the talk, the higher the level.

S_S refers the strength of confidence that executes how much possibility is in each action. It is influenced by the expectation that gets replies and praises. Concretely it is determined by the number of getting replies and neglects when the robot speaks to humans.

$$S_S = \frac{\alpha n_R - \beta n_N}{n_I} \tag{8}$$

n_R is the time getting replies, and n_N is the time of neglects, n_I is the time of interaction. α and β are parameters between 0 and 1. There are 3 rules. If a person was talked in Japanese and answered, n_R increases. If a person was talked in English and answered in English, n_R increases. If answered in Japanese, n_N increases. S_G means the generality of contents adapting to similar circumstances. In comparison with Japanese conversation, the robot thinks how well it can speak English. Concretely, the number of English words is compared to the number of Japanese words.

$$S_G = \frac{n_E}{n_J} \tag{9}$$

Here n_E means the number of English words, and n_J is the number of Japanese words.

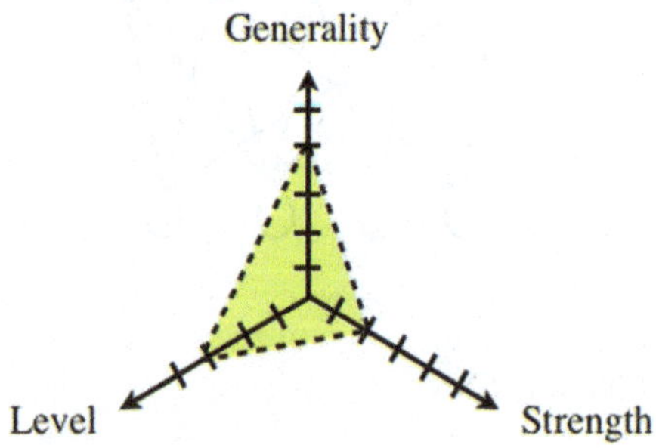

Fig. 7 The dimension of self-efficacy

Basically self-efficacy is enhanced by teaching words. If humans do not reply, it becomes weak. Therefore the conversation becomes fluid.

By enhancing self-efficacy, the robot tries to speak English actively if the robot thinks it can get replies. We think self-efficacy is high, we are willing to communicate [17]. The robot estimates the English skills of humans which will be improved in this way. Self-efficacy is used as a criterion for judgment to speak English or Japanese. Figure 8 shows the concept of self-efficacy in conversation. Outcome expectation is that the robot can get a reply and efficacy expectation decides that the robot speaks English or Japanese.

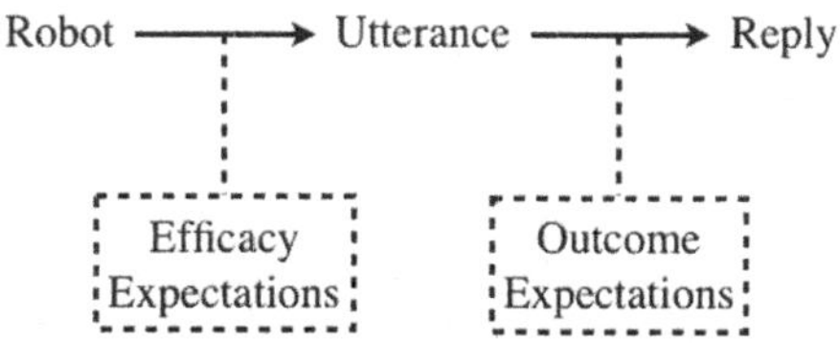

Fig. 8 Representation of self-efficacy in conversation

4 Experimental Results

This section shows experimental results of the proposed method for the language education. We did experiments in children's house. The subjects are some students in elementary school. There are some objects using playing house around the robot. The interaction starts from the conversation in daily life. The subjects merely talk to the robot and show objects then the robot responds to that. When the subjects speak in Japanese, the robot speaks Japanese as well. In learning conversation mode, the subjects teach objects to the robot, the robot learns language and speaks English about the objects. In developmental psychology, there are two kinds of child (the one who tries to remember the name of the thing and the one who tries to memorize the word concerning person's appearance) when the word is memorized. Here we developed the system making conversation by trying to learn the object name.

Next we compared the difference of interaction with the robot (Fig.9). There are 3 patterns, (a) use no touch display and objects, (b) use touch display, (c) use both touch display and objects. The conversation contents are a lot of varieties in pattern (c). Therefore we concluded the pattern (c) is the most appropriate style of robot-assisted language learning. We did not do questionnaire because the children could not listen English words the robot spoke. This was attributable to children's capacity.

Next we show the value of self-efficacy of the robot (Fig.10). We set n_J=30, α=1.0, β=0, threshold for self-efficacy θ=0.5. It was changed by the contents of conversation of a person.

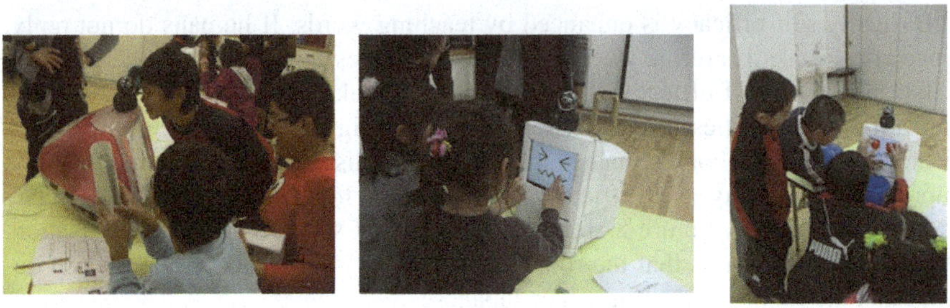

(a) No touch display and objects (b) Using touch display (c) Using touch display and objects

Fig. 9 Learning with the robot

The condition of case 1 are n_I=12, n_R=7 in the final state. It is difficult to talk in English if we did not teach words at first. In the talk of the robot, it is also difficult to talk in Japanese. Therefore it is important that the interaction to teach words to the robot.

In the case 2, we talked to the robot with teaching words up to n_E=10. The value of self-efficacy was monotonically increasing and became high. In this case, the robot always spoke English because the value was high. If we want to make a conversation fluid, we would need to set the value of ß high. And we also need to think the way of communication after self-efficacy is high.

In the case 3, we had talked a little bit longer term than other cases. When the value was low, the robot had spoken only Japanese. Then a person taught English words to the robot and did conversation with the robot, the value was rising. After that the robot had spoken English.

Therefore the robot can learn English as the same pace with the person. If the value were too high, the value was down when the person did not answer the robot's question. As a result of conversation, self-efficacy is effective for robot-assisted language learning.

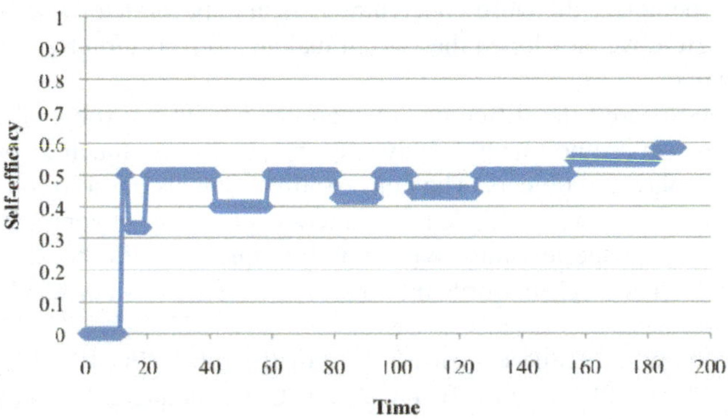

(a) Case 1

Fig. 10 The value of self-efficacy

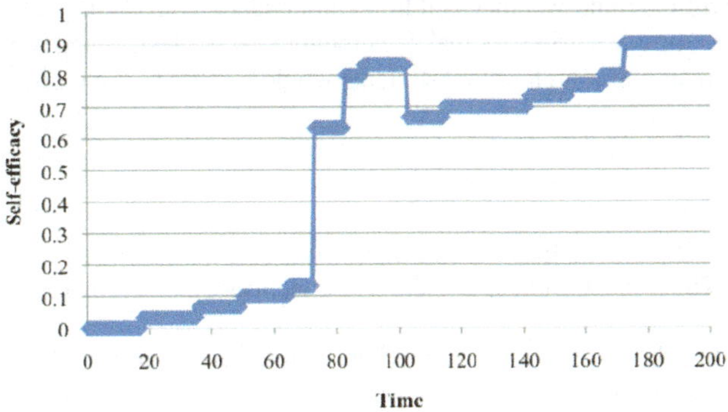

(b) Case 2

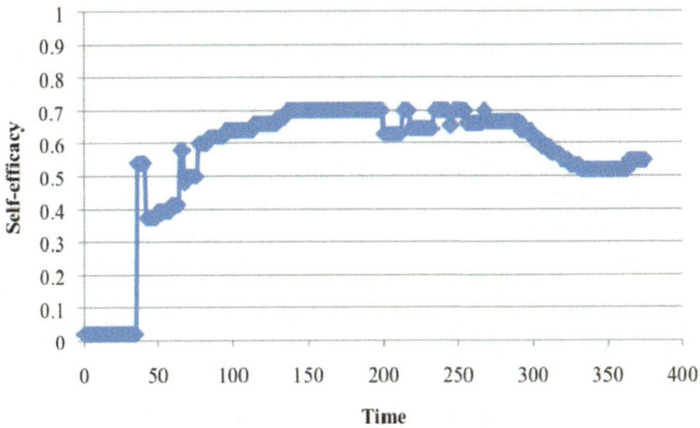

(c) Case 3

Fig. 10 (*continued*)

5 Summary

In this paper, we discussed the applicability of robots in language learning. First, we explained the robot-assisted language learning. Next, we discussed how to interact and communicate with students in the language education. We proposed learning conversation system of physical robot partners. The essence of the proposed method is how humans and robots will improve each other.

As future work, we will inspire the students to learn English. We need to clarify how long the robot can interact with humans. To do long-term communication, we will take gaming element to our system, and enable it to do long-term communication. For example, the value of self-efficacy decides victory or defeat.

Also we will examine whether children can enhance conversation and vocabulary capacity. This time we tried to make elementary school students use this system though, more than junior high school students would be able to use it effectively.

References

1. Kanda, T., Hirano, T., Eaton, D., Ishiguro, H.: Interactive Robots as Social Partners and Peer Tutors for Children: A Field Trial. Human-Computer Interaction 19, 61–84 (2004)
2. Han, J., Jo, M., Jones, V., Jo, J.H.: Comparative Study on the Educational Use of Home Robots for Children. Journal of Information Processing Systems 4(4), 159–168 (2008)
3. Kwon, O.-H., Koo, S.-Y., Kim, Y.-G., Kwon, D.-S.: Telepresence robot system for English tutoring. IEEE Workshop on Advanced Robotics and its Social Impacts (ARSO), 152–155 (2010)
4. Lee, S., Noh, H., Lee, J., Lee, K., Lee, G.G.: Cognitive effects of robot-assisted language learning on oral skills. In: Proceedings of the Interspeech 2010 Workshop on Second Language Studies: Acquisition, Learning, Education and Technology (2010)
5. Chang, C.-W., Lee, J.-H., Chao, P.-Y., Wang, C.-Y., Chen, G.-D.: Exploring the Possibility of Using Humanoid Robots as Instructional Tools for Teaching a Second Language in Primary School. Educational Technology & Society 13(2), 13–24 (2010)
6. Kanda, T., Sato, R., Saiwaki, N., Ishiguro, H.: A two-month Field Trial in an Elementary School for Long-term Human-robot Interaction. IEEE Transactions on Robotics (Special Issue on Human-Robot Interaction) 23(5), 962–971 (2007)
7. Nakano, M., et al.: Grounding New Words on the Physical World in Multi-Domain Human-Robot Dialogues. In: Dialog with Robots: Papers from the AAAI Fall Symposium (FS-10-05), pp. 74–79 (2010)
8. Uchida, Y., Araki, K.: Evaluation of a System for Noun Concepts Acquisition from Utterances about Images (SINCA) Using Daily Conversation Data. In: Proceeding of the North American Chapter of the Association for Computational Linguistics - Human Language Technologies (NAACL HLT) 2009 Conference, pp. 65–68 (2009)
9. Kubota, N., Wagatsuma, Y., Ozawa, S.: Intelligent Technologies for Edutainment using Multiple Robots. In: Proc. (CD-ROM) of the 5th International Symposium on Autonomous Minirobots for Research and Educatinment (AMiRE 2009), pp. 195–203 (2009)
10. Fiene, J.: The Robockey Cup. IEEE USA Robotics & Automation Magazine 17(3), 78–82 (2010)
11. Billard, A.: Robota: Clever Toy and Educational Tool. Robotics & Autonomous Systems 42, 259–269 (2003)
12. Yorita, A., Hashimoto, T., Kobayashi, H., Kubota, N.: Remote Education based on Robot Edutainment. In: Proc (CD-ROM) of the 5th International Symposium on Autonomous Minirobots for Research and Educatinment (AMiRE 2009), pp. 204–213 (2009)
13. Yorita, A., Kubota, N.: Essential Technology for Team Teaching Using Robot Partners. In: Liu, H., Ding, H., Xiong, Z., Zhu, X. (eds.) ICIRA 2010, Part II. LNCS, vol. 6425, pp. 517–528. Springer, Heidelberg (2010)

14. Kubota, N., Nishida, K.: Cooperative Perceptual Systems for Partner Robots Based on Sensor Network. International Journal of Computer Science and Network Security (IJCSNS) 6(11), 19–28 (2006)
15. Kubota, N., Yorita, A.: Structured Learning for Partner Robots based on Natural Communication. In: Proc. (CD-ROM) of 2008 IEEE Conference on Soft Computing in Industrial Applications (SMCia), pp. 303–308 (2008)
16. Bandura, A.: Self-efficacy: Toward a Unifying Theory of Behavioral Change. Psychological Review 84(2), 191–215 (1977)
17. Yashima, T.: Willingness to communicate in a second language: The Japanese EFL context. The Modern Language Journal 86(1), 54–66 (2002)
18. DTalker for Mac OSX Ver3.0, `http://www.createsystem.co.jp/dtalkerMacOSX.html`
19. Hofstadter, Douglas: Fluid Concepts and Creative Analogies: Computer Models of the Fundamental Mechanisms of Thought. Basic Books, New York (1995)
20. Wada, K., Shibata, T., Saito, T., Tanie, K.: Effects of Robot Assisted Activity to Elderly People who Stay at a Health Service Facility for the Aged. In: Proceedings of the 2003 IEEE/RSJ Intl. Conference on Intelligent Robots and Systems, pp. 2847–2852 (2003)
21. Kanoh, M., Kato, S., Itoh, H.: Facial Expressions Using Emotional Space in Sensitivity Communication Robot "ifbot". In: IEEE/RSJ International Conference on Intelligent Robots and Systems, pp. 1586–1591 (2004)
22. Fujita, M.: AIBO: towards the era of digital creatures. International Journal of Robotics Research 20(10), 781–794 (2001)
23. Yaguchi, A., Kubota, N.: The Style of Information Service by Robot Partners. In: Liu, H., Ding, H., Xiong, Z., Zhu, X. (eds.) ICIRA 2010, Part II. LNCS, vol. 6425, pp. 529–540. Springer, Heidelberg (2010)
24. Gonzalez-Gomez, J., Valero-Gomez, A., Prieto-Moreno, A., Abderrahim, M.: A New Open Source 3D-printable Mobile Robotic Platform for Education. In: Proceedings of the 6th International Symposium on Autonomous Minirobots for Research and Edutainment, p. S22 (2011)
25. Riedo, F., Rétornaz, P., Bergeron, L., Nyffeler, N., Mondada, F.: A two years informal learning experience using the Thymio robot. In: Proceedings of the 6th International Symposium on Autonomous Minirobots for Research and Edutainment, p. S11 (2011)
26. Salvini, P., Macrì, G., Cecchi, F., Orofino, S., Coppedè, S., Sacchini, S., Guiggi, P., Spadoni, E., Dario, P.: Teaching with minirobots: The Local Educational Laboratory on Robotics. In: Proceedings of the 6th International Symposium on Autonomous Minirobots for Research and Edutainment, p. S12 (2011)
27. `http://en.wikipedia.org/wiki/Doraemon`

The CoaX Micro-helicopter: A Flying Platform for Education and Research

Cédric Pradalier, Samir Bouabdallah, Pascal Gohl, Matthias Egli,
Gilles Caprari, and Roland Siegwart

Abstract. CoaX is a micro-helicopter designed for the research and education markets by Skybotix AG in Switzerland. It is a unique robotic coaxial helicopter equipped with state of the art sensors and processors: an integrated Inertial Measurement Unit (IMU), a pressure sensor, a down-looking sonar, three side looking range sensors and a color camera. To communicate with a ground station, the robot has a Bluetooth (or XBee) module and an optional WiFi module. Additionally, the CoaX supports the Overo series of tiny computers from Gumstix and is ready to fly out of the box with a set of attitude and altitude control functions. One can also control the system through an open-source API to give high-level commands for taking-off, landing or any other type of motion. In addition to presenting the CoaX, this paper reports on three experiments conducted to demonstrate the system's motto: "simple to fly, simple to program, simple to extend".

1 Introduction

From the point of view of physics, a helicopter is an unstable system as described by Castillo in [1]. Its main advantage against fixed-wing aircrafts is being able to hover. This ability allows it to perform complex tasks in a confined space, such as inspecting location unreachable by humans or entering hazardous areas. To become such a powerful tool, a micro-helicopter needs not only to be easily controllable but also needs to be designed so as to simplify the development of applications and the extension of the sensor set.

Developing such a micro-helicopter is a very challenging task that has taken the Autonomous Systems Lab from ETH Zürich and Skybotix AG close to 5 years. The outcome is the CoaX, a small helicopter designed with the needs of education

Cédric Pradalier · Samir Bouabdallah · Pascal Gohl · Matthias Egli ·
Gilles Caprari · Roland Siegwart
Autonomous Systems Lab, ETH Zürich
e-mail: `firstname.lastname@mavt.ethz.ch`

and research in mind: being easy to program, easy to extend, easy to control. This is possible because the CoaX comes with a very rich set of sensors and a powerful embedded computing system (see below for more details). To this we added an extensive and open application programming interface (API) and a large set of development tools using state-of-the-art languages and the ROS middleware.

This article is organized as follows: we first introduce the CoaX, its sensor set and its development tools. We then illustrate how this can be used to develop a simple wall-following application. After presenting the integration of a speed module built from the sensor of an optical mouse, we demonstrate how it can be used to develop position and speed controllers. Finally, we present results of the CoaX integration in the ROS middleware environment.

2 System Overview

The CoaX helicopter is shown in figure 1 with a short overview of its main dimensions. As a consequence of the two rotors turning in opposite directions, we obtain a very stable, robust and compact flying vehicle with only few mechanical parts. The standard electronics consist in a board with two microcontrollers and the basic IMU sensors. It is optimized as a research and education platform and therefore offers many free interfaces and a programmable firmware. The main electronic features are listed below:

- 2x dsPIC33 Microcontrollers
- Inertial Measurement Unit (IMU)
- Sonar and Pressure sensor (altitude)
- Infra-Red range sensor (horizontal)
- Bluetooth module or Xbee module
- Remote control (2.4 GHz)
- Optional Gumstix Overo computer
- Optional USB WIFI dongle and camera

Span:	0.34m
Height:	0.274m
Weight:	280g
Autonomy up to:	20min

Fig. 1 The CoaX helicopter and its specification

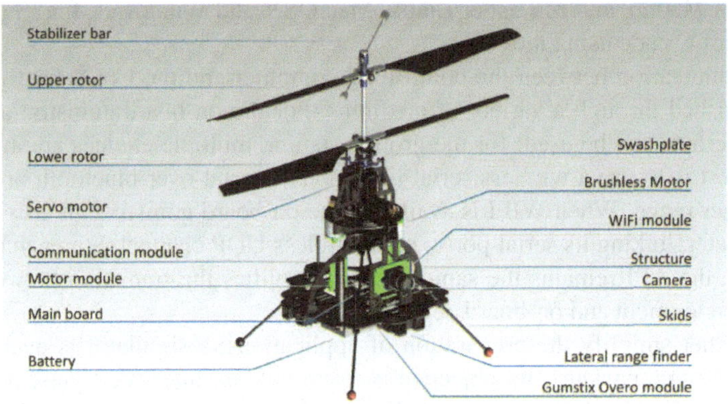

Fig. 2 Organisation of the CoaX hardware

2.1 Development Tools

As a platform for research and education, the CoaX is designed to allow programmers to modify its firmware and to develop control programs running either onboard on the gumstix computer or off-board on a host PC. All the required tools, including the firmware, are available under an open-source license through a SVN server.

When working on the firmware, the development is simplified by a custom-made bluetooth firmware updater. However, for most of the applications, the users have the possibility to use a programming interface (API) to communicate with all the functions of the firmware. Using this API, programs can be written in C/C++, python

```
// Configure and initialize
SBApiSimpleContext api;
sbSimpleDefaultContext(&api);
// Check the type of connection
sbSimpleParseChannel(&api,argv[1]);
sbSimpleInitialise(&api);
// Take off to 0.4 m
sbSimpleReachNavState(&api,SB_NAV_CTRLLED,30.0);
while (api.state.zrange < 0.35) {
    sbSimpleControl(&api,0.0,0.0,0.0,0.4);
    usleep(20000);
}
// Land and shut down
sbSimpleReachNavState(&api,SB_NAV_IDLE,30.0);
sbSimpleTerminate(&api);
```

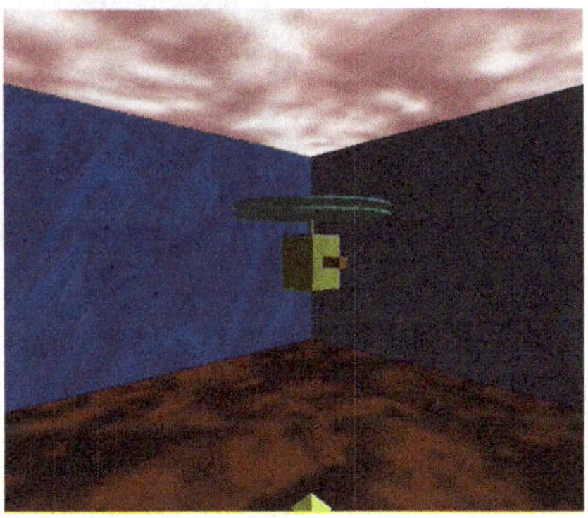

Fig. 3 Example of C code using the API (left) and the simplified simulator (right)

or even in Matlab, and run under Linux, Mac OSX and Windows[1]. Fig. 3 gives an example of C code using the API.

Communication between the controlling computer and the CoaX firmware can be established through a variety of mediums. For the on-board gumstix, a simple serial interface can be used; for the ground station, multiple choices are available. The easiest is to use a wireless serial link such as serial-over-bluetooth or a Xbee for a larger range. When WIFI is available, the on-board gumstix can also be used as a repeater, linking its serial port with a wireless UDP channel. For each of these channels, the API remains the same, which simplifies the transition between off-board development and on-board deployment.

To further simplify the preparation of applications, a simulator is available in the open-source package. Its objective is to provide an interface to test programs developed using the API, as well as a limited dynamic model. A screenshot of the simulator can be found in fig. 3.

2.2 *Optical Flow Sensor*

To measure the horizontal displacement of the CoaX, we implemented a speed module from a common optical flow sensor designed for computer mice. These sensors are independent from external components and easy to read out. To measure the optical flow on the ground, the speed module is equipped with a lens in focus from 10cm to infinity. The sensor is attached on the bottom side of the helicopter. In order to obtain good measurements without depending on ambient light, a bright LED module is integrated around the sensor lens. The speed module, mounted on the CoaX helicopter is shown in figures 4 and 5.

Fig. 4 Sensor with lights mounted on CoaX

[1] Even if Linux is the primary development OS.

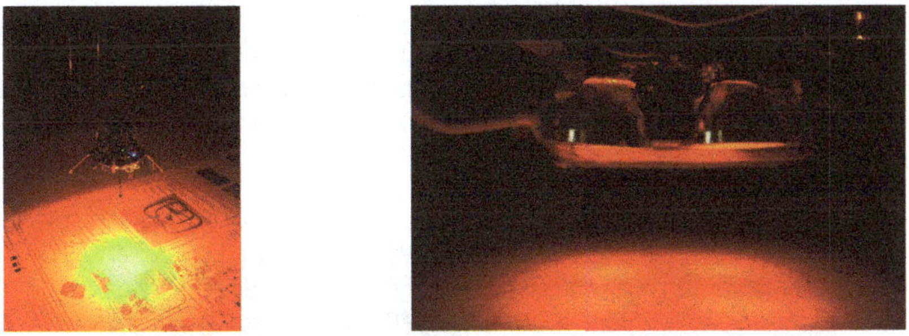

Fig. 5 The CoaX with the sensor light at full power

3 First Experiment: Wall Following

In this first experiment, we would like to describe how a simple control task can be implemented on the CoaX. To this end, we have chosen to implement a wall-following application, which uses the horizontal range sensors (infra-red) to detect the distance and orientation of a wall located in front of the flying CoaX. From this information, two simple control laws are implemented to control the observed angle to zero and the distance to a reference value (typically in $[0.6m\,;\,1.0m]$).

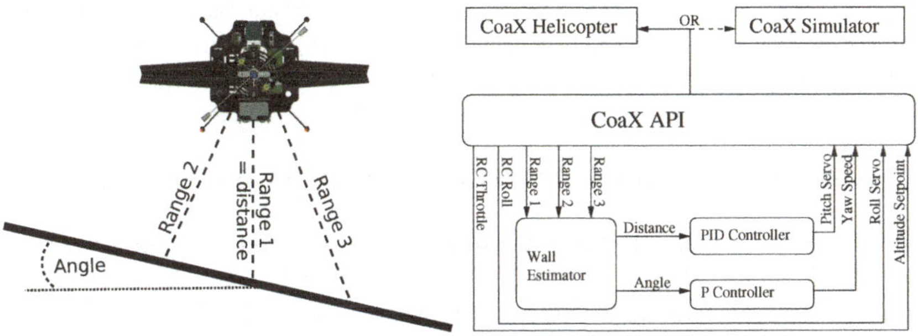

Fig. 6 Sensor arrangement and control law for the wall-following task (left, seen from top) and structure of the control law (right).

For this experiment, the CoaX three range sensors were moved to the front as illustrated in fig. 6 and visible on fig. 7. Controlling the distance to a set-point is achieved by setting the CoaX pitch servo[2] using a PID controller. Controlling the orientation with respect to the wall is achieved by setting the yawing speed using a P controller. Because the information obtained from the wall can only constrain two degrees of freedom, the lateral position along the wall and the CoaX altitude are directly connected to the remote control output provided by the API.

[2] Responsible for forward/backward movement.

The control law and logging system were implemented on the on-board gumstix using the API described in the previous function. The control could also have been implemented on a host computer using the bluetooth interface. On-board deployment was chosen to obtain a stable communication bandwidth of 100 messages/second.

Fig. 7 CoaX helicopter while running the wall-following task

Videos of this experiment can be found on the Skybotix youtube channel: `http://www.youtube.com/user/skybotix`. As can be seen on the video, the CoaX is very stable and shows a great robustness to perturbation, such as pushing it by hand, passing in front of the sensor beam or moving the target wall.

4 Second Experiment: Velocity and Position Control

Using the speed module, it is now possible to measure the helicopter velocity in longitudinal and lateral direction. Several software modules can take advantage of this information: integrating the velocity allows to estimate the CoaX odometry, to implement a velocity controller and finally develop a position control by combining the position estimation and velocity control.

4.1 Odometry

The speed module is based on the sensor of an optical mouse. This sensor is essentially a small camera looking at the floor through a lens and computing the displacement using inter-frame block matching. As a result, the measurement of the displacement is measured in pixels. This speed in pixels Δ is converted to a velocity v in meter/second using the focal length of the camera f and the height to the ground h:

$$v = h \times \frac{\Delta}{f} \tag{1}$$

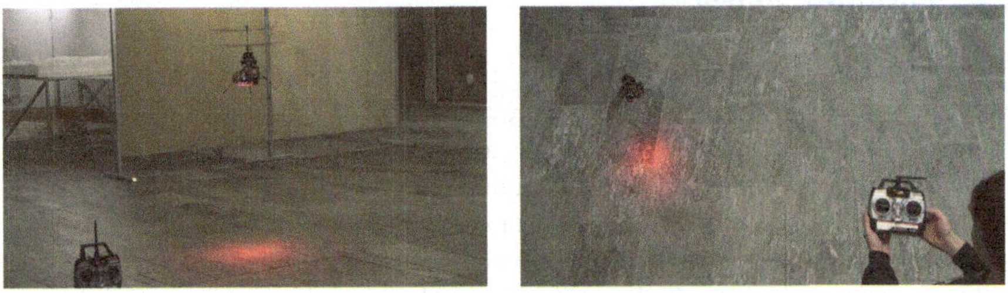

Fig. 8 CoaX helicopter in a velocity control experiment: hovering (left) and following a square path (right)

Note that this assumes a globally flat ground in the field of view of the camera.

The optical mouse sensor measures translations in longitudinal and lateral directions. Once these displacements have been converted to metrics, odometry is simply achieved by integration. See fig. 9 and 10 for paths recorded while hovering and following a square trajectory.

4.2 Speed Control

Knowing the lateral and longitudinal speeds, it is natural to think of implementing a velocity controller. The simple approach currently developed on the CoaX consists in using two decoupled controllers: the pitch servo is used to control the longitudinal direction while the roll servo controls the lateral direction. Each of these servo is mechanically linked to the swash-plate and act by slightly tilting the rotation plane of the lower rotor; as a result it creates a small horizontal force. Hence, velocity error and servo angle could theoretically be linked directly by a P controller. In practice, small mounting errors and uncertainties in the position of the center of gravity create static horizontal forces that must be compensated by an integral term in the controller. This is illustrated in fig. 9: the helicopter takes-off in the lower right corner with a velocity set-point of zero. The integral term slowly gets filled while the helicopter drifts by $25cm$. Once the static forces are compensated, the CoaX can hover in a $15cm$ radius, which roughly corresponds to the rotor size. Note that the path displayed is the recorded odometry. It is nonetheless consistent with a visual estimation of the hovering accuracy.

It is important to be aware that decoupled PI controllers are a theoretically sub-optimal solution for the CoaX as the real system dynamics exhibits a significant coupling between the longitudinal and lateral directions, especially for constant roll or pitch servo set-point. In practice, this solution has demonstrated very satisfying performances for this platform, and will provide users with a baseline system that can be improved upon.

4.3 Position Control

Position Control was implemented using a nested controller approach. The speed controller which was introduced in the previous section is fed with a desired speed computed using a P-controller to bring the actual position to a given set-point. This approach allows to follow trajectories defined as a sequence of set-points. Some results can be found in fig. 10. Significant oscilations can be noted in comparison with a perfect square. This can be attributed to sub-optimal tuning of the controller and to the absence of controlled acceleration and braking phases. However, it is also important to note the limited drift accumulated by the odometry while performing this trajectory 5 times. This stability is consistent with what was visually observed during the experiment.

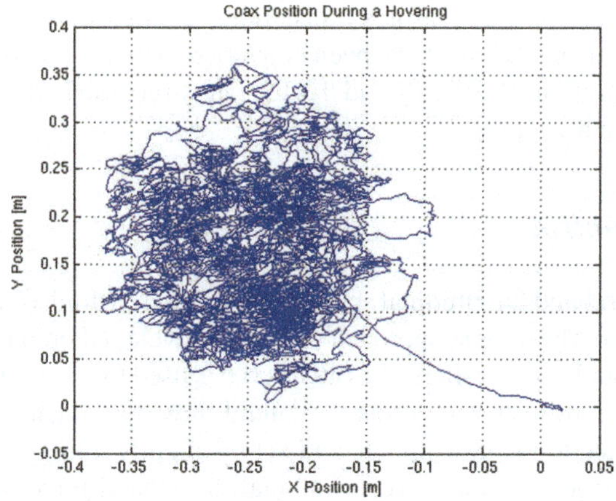

Fig. 9 Path of the CoaX helicopter in a hovering experiment

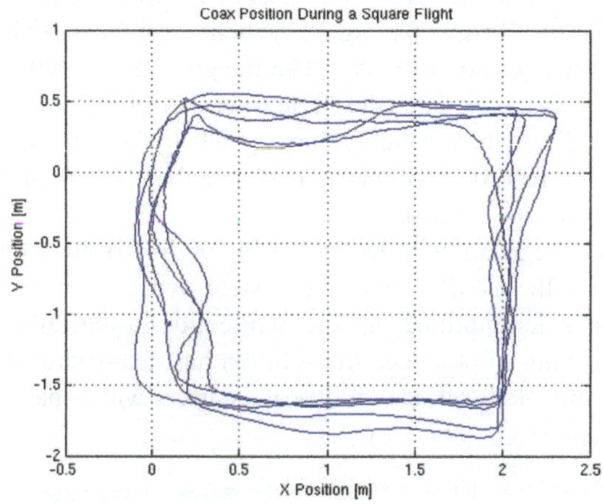

Fig. 10 Path of the CoaX helicopter in a velocity control experiment following a square path

5 Third Experiment: ROS Integration

ROS, or the Robotic Operating System, is a middleware developed by the company Willow Garage and released under an open-source license at `www.ros.org`. ROS is gaining an enormous momentum in the robotics community and many robotic systems now come with a pre-packaged ROS interface. This middleware allows multiple processes running on a multiple computers to exchange data. It also provides a lot of basic tools for developing robotic applications.

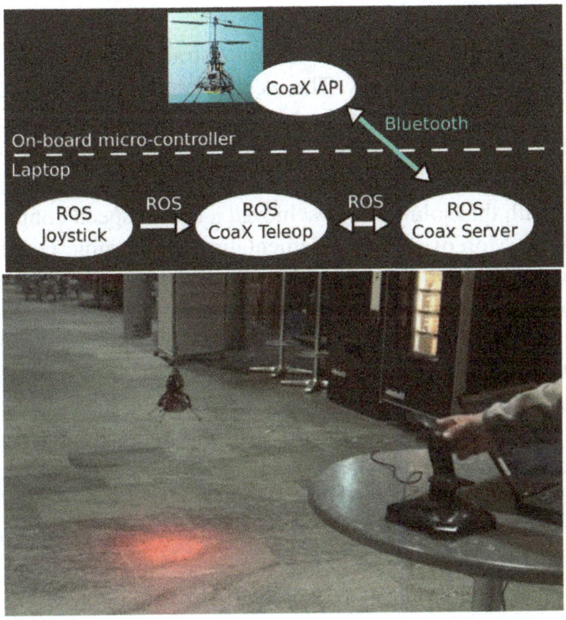

Fig. 11 CoaX helicopter remote controlled using ROS

The CoaX takes advantage of being equipped with an on-board computer running a full Linux system and with a WIFI connection. This allows it to offer a ROS interface either running on-board or off-board using one of the wireless communication mediums mentioned earlier. This interface gives access to all the functions of the API through a set of ROS services[3] and topics (data streams).

Once a ROS server is available, implementing a teleoperation application is a simple matter of connecting the ROS joystick node (available by default) with a teleoperation module converting the joystick controls into CoaX controls: yaw speed, altitude set-point, roll and pitch servos and take-off/landing requests. When the speed module is available, a simple service call switches between speed control and direct servo control, and allows the joystick to directly control the speed of the CoaX.

[3] Equivalent of remote procedure call (RPC).

The structure of the application is illustrated in fig. 11 and videos can be found on the Skybotix youtube channel mentioned earlier.

6 Conclusion and Outlook

This paper has presented the CoaX platform: a small autonomous helicopter for research and education purposes. The CoaX has been specifically designed to be programmed and extended by the end-users. For this reason, all the software including the firmware, API and ROS interfaces, is made available under an open-source license.

A second objective of this paper was to demonstrate how easy it is to design and run new experiments with the CoaX. The mechanics and electronics are stable and robust, and the CoaX is inherently safe due to its small weight and rotor diameter. The second and third experiments show that it is possible to stabilize a coaxial helicopter by measuring the speed over ground with an optical flow sensor designed for computer mice. With this solution, we achieved a robust speed control in both horizontal dimensions. Moreover the implementation of a simple position controller showed good results except for the small drift in the yaw orientation. The lighting module provides enough light to fly on adequate featured ground up to 2.5m in dark environment, while still being able to measure movements. Thanks to the speed module, the CoaX is nearly as easy to use as a wheeled robot: it has smooth dynamics, velocity control and a form of odometry.

We identify a strong limitation in the reaction to wind gusts and other fast disturbances. This is due to the fact that our controller only handles the speed over ground and does no use the attitude of the helicopter. M. Fässler showed in [2] that it is possible to counter-steer fast on wind gusts by taking advantage of the accelerometer data. The implementation of a fusion of Fässler's and our algorithm, improved by calculating a model-based controller, is a subject for future work, or for developments by CoaX users.

Our programmed square flight shows that it is possible to navigate using the speed module information for dead-reckoning, as long as we can assume a known orientation. This could be measured using a compass for instance, but compasses happens to be very unreliable in indoor environments. Future works will address the addition of other optical sensors, possibly with different orientations, or even the fusion of the speed module data with video processing results from the on-board camera.

References

1. Castillo, P.: Modelling and Control of Mini-Flying Machines. Springer, London (2005)
2. Fässler, M.: Inertial-based hovering with a micro helicopter. Bachelor thesis, ETH Zürich (2009)
3. Franklin, G.: Feedback Control of Dynamic Systems. Pearson Prentice Hall, Upper Saddle River (2006)

4. Wangler, C.: Optical flow based velocity estimation for miniature rotorcrafts. Semesterthesis, ETH Zürich (2009)
5. Haag, S.: Design of an optical flow sensor for off-road application. Bachelor thesis, ETH Zürich (2009)
6. Philips. Luxeon® k2. Technical Datasheet DS51
7. Philips. Thermal design using luxeon® power light sources. Application Brief AB05
8. Frank, A.C.(Dipl.-Ing.): Kühlkörper. ETH Zürich.
9. Avago. Adns-5050 optical mouse sensor. Data Sheet
10. Khatod. Kclp series reflectors for lambertian. Reflectors Test Report
11. Griffiths, S., et al.: Advances in Unmanned Aerial Vehicles. Springer, Netherlands (2007)
12. Agence des Médias Numérique. Thermal properties for pcbs (2009), http://www.frigprim.com/online/cond_pcb.html
13. Microchip Technology Inc. dspic® language tools libraries (2004)

AMiRo – Autonomous Mini Robot for Research and Education

Stefan Herbrechtsmeier, Ulrich Rückert, and Joaquin Sitte

Abstract. This paper describes the motivation, system architecture and design details of a mini robot for research and education. The main objective is to produce a set of electronic modules for sensor processing, actuator control and cognitive processing that fully utilise currently available electronics technology for the construction of mini robots capable of rich autonomous behaviours. These modules are used for the two wheeled AMiRo mini robot that meets the size requirements for participation in the AMiRESoT robot soccer league. All mechanical parts for the robot are off-the-shelf components or can be fabricated with common drilling, turning and milling machines. The connection between the modules is well defined and supports standard interfaces from parallel camera capture interfaces down to simple serial interfaces.

1 Introduction

Mini robots are small autonomous systems which need little space, can be used on a table or the floor of a small laboratory and are easy to transport. With the resources offered by today's microelectronic technology small robots can be given the sensing, information processing and actuation capabilities necessary for executing behaviours as complex as those of bigger robots. This makes mini robots a powerful tool for research and education [9, 10].

With the Khepera 1 mini robot Francesco Mondada and his colleagues [8] pioneered the concept of a small robot that teachers and researchers could buy at an

Stefan Herbrechtsmeier · Ulrich Rückert
Center of Excellence Cognitive Interaction Technology, Bielefeld University,
Universitätsstr. 21–23, 33615 Bielefeld, Germany
e-mail: sherbrec@cit-ec.uni-bielefeld.de

Joaquin Sitte
Queensland University of Technology,
Brisbane, Queensland, Australia
e-mail: j.sitte@qut.edu.au

affordable price, and that was ready for experimentation straight out of the box. The Khepera mini robot was used for nearly a decade, which is an extraordinary achievement. Several other mini robots followed from the same team of designers at EPFL (Ecole Polytechnique Fédérale de Lausanne): the larger s-bot and the larger but modular marXbot, both were never available commercially. The current commercial off-springs from these robots are the e-puck, the very cheap Thymio II and the substantially bigger Khepera 3 . Another team at EPFL produced the tiny Alice micro robot [4, 5].

At the same time other low cost commercial robots were developed for the research, educational and entertainment market but none of them was suitable for a desktop laboratory. Among the better known ones are the Sony Aibo, the AmigoBot, and lately the iRobot Create.

One can buy many components for building low cost hobby, educational and research robots. The capabilities of such low cost robots are mainly determined by the computing resources with which such robots can be equipped. The two main low cost computing modules available to the robot builder are the Arduino™and the Gumstix™set of processor and extension boards. The Arduino boards are very low cost and are built around the Atmel ATmega 8-bit microcontroller series and therefore offer rather limited computing capabilities. The Gumstix boards based on the 32-bit ARM Cortex A8 processor and are therefore one level up in performance from the Arduino boards. They are somewhat more expensive but in turn the Gumstix computer-on-module boards can run a full operating system like Linux. The Korebot from K-team offers similar performance as the Gumstix boards at a substantially higher cost but has the advantage that it fits the Khepera 3 robot base with little integration effort.

The significant integration work required for using either the Arduino or the Gumstix modules as a computing component for a small robot constitutes a barrier for their use in educational and research robotics.

With the AMiRo we aim at creating a set of affordable modules with state-of-the-art electronics that can be easily combined for building a variety of small autonomous robots. Specifically the design of AMiRo responds to the following demands:

- Must have a small size to be usable on a desktop or in a small laboratory area
- Must be capable of completely autonomous operation
- Must have sufficient processing capability for on board real time vision processing
- The information processing architecture must be modular and extensible for easy adaptation to different tasks
- Must cost less than EUR 1000.-
- Must be able to run current software frameworks for autonomous robots

To provide a concrete design focus we added the demand that the robot must be able to act as a player in a robot soccer game according to the 2008 AMiRoSoT rules [3].

The design for the AMiRo robot described in this paper draws from the experience gained with the BeBot mini robot [6] and numerous related student projects.

Furthermore some of the architectural ideas derive from a mini robot developed at Queensland University of Technology that had a modular distributed computing architecture based on Transputers [7]. Section 2 describes the modular system architecture for the AMiRo robot. Section 3 describes in detail the function and architecture of each of the modules in the current set. In Section 4 we describe the software architecture and the required robot programming tools after which we conclude in Section 5 with a brief evaluation and point to further work required.

2 System Architecture

2.1 Physical Structure

The rules for AMiRESoT robot soccer league require that the soccer robots fit in a vertical cylinder of 110 mm inner diameter [3]. Therefore a cylindrical body is an obvious choice for AMiRo. Cylindrical robot bodies are easy to fabricate from plastic tube material and provide strong structural support with low weight. The top of the cylindrical body can be closed with lids of various shapes. For example a hemispherical lid gives the robot *friendly* appearance.

The inside of this shell (figure 1) has to accommodate the power source (batteries), sensors, actuators (motors and wheels) and the computing hardware. Following

Fig. 1 See through view of the AMiRo robot showing the stackable modules (CAD drawing)

the principle of functional modularisation the computing hardware consists of several AMiRo modules (AMs) with a prescribed common electronic interface. Each AM is hosted on its own circuit board and contains its own processing unit that can be a microcontroller, a powerful processor or a programmable device. For best use of space the circuit boards are round and stacked vertically. The size of the circuit boards is such as to fit into cylindrical shell of the robot. We chose to make the body from clear or satinated plexiglas[1] tube with an outer diameter of 100 mm and a wall thickness of 3 mm. To provide dimensional tolerance we have set the maximum diameter of the circuit boards to 92 mm.

The modules interconnect through two 60 pin connector pairs mounted on the circuit boards. Each pair has the female connector on the top of the board and the male connector on the underside, such that boards can be plugged into each other to make a stack of boards. Corresponding connector pins on the top and bottom of the boards connect through the board creating a common signal bus for the modules.

2.2 *Electrical Interface*

Figure 2 shows a diagram of a stack of modules with their electrical interfaces. The electrical interface between the modules contains a $6-12$ V system supply (V_{SYS}), a 1.8 V ($V_{IO1.8}$) and 3.3 V ($V_{IO3.3}$) regulated power supply, control signals

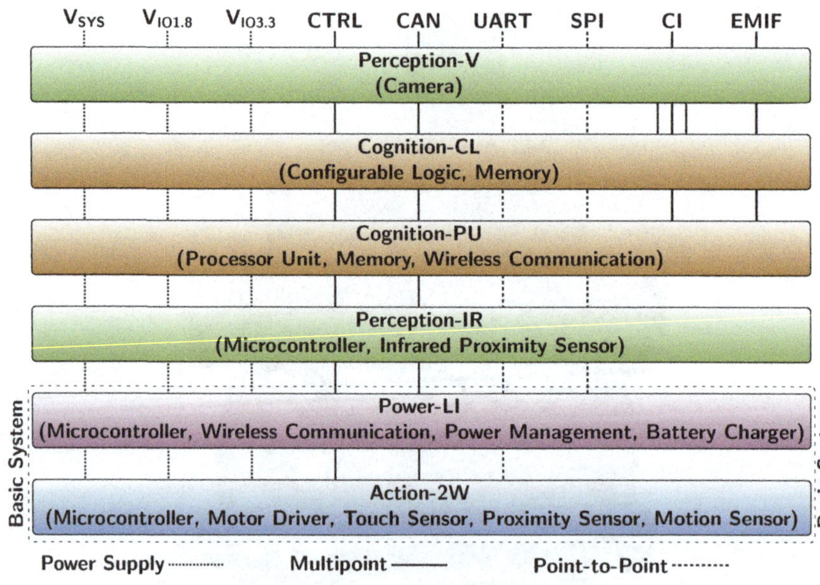

Fig. 2 System architecture

[1] Poly methyl methacrylate, also marketed under various other names such as Perspex.

(CTRL), serial communication standards like CAN, UART and SPI (on all modules except the action module) as well as a parallel camera capture interface (CI) and an external memory interface (EMIF). The CAN (Controller Area Network) bus is the main inter-module communication interface and must exist on each module. Therefore each AM has a CAN controller and transceiver, and consequently also a host processor. This is quite easy to provide as microcontrollers with integrated CAN controllers are available. Unless otherwise indicated the default host processor is an ARM 32 bit Cortex-M3 microcontroller from STMicroelectronics with 64 kbit main memory and 512 kbit flash memory running at 72 MHz. Each module also has an EEPROM to permit software based hardware detection. The UART ports for every module are accessible over the UART port on the power module. Therefore the UART TX (RX) line of all modules connects to the RX (TX) line of the UART port on the Power module. The control signals have the functions 'Power supply enable', 'Cold reset', 'Warm reset' and 'Power down request'. In addition each module has a serial programming connector (UART) for programing and debugging the microcontroller via the serial port of a computer or with a separate USB to serial converter cable.

2.3 Communicating with the Robot

The reasons for wanting to communicate with the robot are:

1. Interacting with the behaviour currently active on the robot. For example, sending steering commands when the robot is in remote control mode, or requesting sensor readings when the behaviour includes some diagnostic capabilities, or telling the robot to carry out a task.
2. Loading behaviours (programs) onto the robot.

Although there are communication interfaces on both the cognition processor module and the power module they can be used for both purposes on each module. On the cognition processor module there are Wi-Fi, Bluetooth and also a USB host and a USB On-The-Go port. The facilities are more limited on the power module, which has a Bluetooth module and low power wireless transceivers. The Bluetooth module can be used among other functions for a wireless serial communication which can be routed from the power AM over the bus to the UART of any other AM in the robot. As already described in the previous section each module has an additional UART programming connector.

3 AMiRo Module Descriptions

This section describes the AMs currently under development. According to their function there are four types of modules: power, perception, cognition and action. The following modules have been developed until now:

- Power-LI AM
- Action-2W AM (2 wheel motion)

- Cognition-PU AM (processor unit)
- Cognition-CL AM (configurable logic)
- Perception-IR AM (infrared)
- Perception-V AM (vision)

3.1 The Basis System

There are two modules that must always be present for a mobile robot, namely the action module and the power module. We have used these circuit boards as parts in a simple mechanical construction of the robot's drive train. The power and action AMs are structurally joined by two U-profiles. Each profile carries a gear motor with a wheel direct attached to its shaft. The action module forms the base plate of the robot. It has a diameter of 100 mm so that the cylindrical shell can stand on it. The U-profiles are screwed on the action module facing each other creating a box. The power module is screwed on top of the U-profiles giving the assembly the required mechanical stability. The two batteries fit in the space between the action and power modules, one in front of the motor assembly and one behind. Two plastic sliders are fitted underneath the action module, one at the front and one at the rear, as a substitute for caster wheels. The base system is the supporting structure for all other modules. Alone it provides wireless programming of the system and can run simple behaviours or can be remote controlled from a PC or Bluetooth controller.

3.2 Power-LI AM

The function of the module is to provide the power management, wireless communication and additional multi-color lighting. Figure 3 shows a block diagram of the power module.

The power management consists of two battery fuel gauges, two battery chargers, multiple power path controllers, three synchronous step-down converters thereof two with additional power monitor, a low power step-down converter with additional power monitor, a low dropout voltage regulator and a power connector. The fuel gauge provides information such as remaining battery capacity, state-of-charge, run-time to empty, battery voltage and temperature. Each of the battery chargers provides built-in 1.2 A charge management for two cell lithium ion battery packs. The synchronous step-down converters supply the lightning with a voltage of 5 V and the whole robot with a voltage of 1.8 V and 3.3 V. Each provides a maximal current of 1.5 A. The converters can be disabled in order to save power during off mode of the system. The low power step-down converter provides a voltage of 3.3 V with a current capability of 75 mA. The last converter is always active and powers only the microcontroller and a few components to keep the power management running during off mode of the system. The low dropout voltage regulator powers the two fuel gauges with a voltage of 2.5 V and a maximal current of 50 mA. The module receives power from two battery packs in parallel. Each pack consists of two lithium ion cells in serial and has a nominal voltage of 7.4 V and a capacity of 1950 mAh.

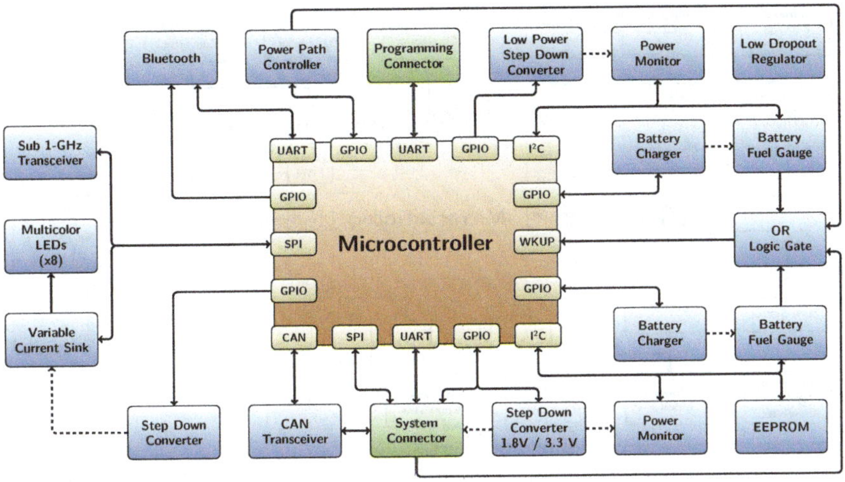

Fig. 3 Power-LI module architecture

Eight separate controllable multi-color light emitting diodes build a colored ring around the module which can be used to visualise the system status. A dedicated microcontroller controls the power management and the lighting and runs user applications. A Bluetooth interface allows wireless communication with a computer and can be used for programming and control of the robot. An additional low power wireless transceiver offers inter robot communication.

3.3 Action-2W AM

The main function of this action module is to separately control the voltage to each of the DC motors in the two-wheel differential drive train. The module can drive two 1 W DC gear motors with integrated optical encoders. Figure 4 shows a block diagram of the Action-2W module.

Besides the increments of the optical encoders the motor current can be monitored via two measuring shunts. The module is equipped with additional sensors that further assist in motion control. These sensors are accelerometer, gyroscope, compass, infrared proximity sensors and capacitive touch sensors. Three axes accelerometer, gyroscope and compass allow inertial navigation and realisation of a balancing mini robot. Seven infrared proximity sensors enable the robot to follow predefined lines on the ground and to detect obstacles in front and back of the robot. With the ground detection the robot can avoid falling off from a cliff (edge of a table, step of a stair). Six capacitive touch sensors on the bottom side of the module provide a touch button user interface to control the status and mode of the robot. An audio transducer allows the generation of various sounds. The module provides two power converters: A synchronous step-down converter supplies the motors with a voltage of 6 V and a maximal current of 1.5 A and a low power step-down

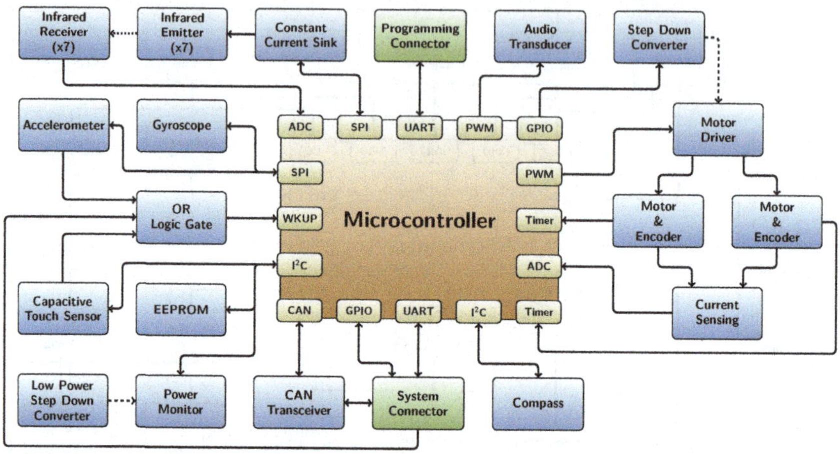

Fig. 4 Action-2W module architecture

converter provides a voltage of 3.3 V with a current capability of 75 mA to power the accelerometer, touch sensors and microcontroller. The converter for the motors can be disabled for saving power during off mode of the system. The 3.3 V converter is always active to allow a touch or motion based startup of the system. Two dedicated contacts allows the autonomous charging of the system in the environment. The microcontroller implements the user interface, feedback motor control and inertial navigation.

3.4 Cognition-PU AM

The main function of the processor cognition module is to run the robot behaviour programs and serve their communication needs: wired, wireless and audio. Figure 5 shows a block diagram of the Cognition-PU module.

This module hosts a Gumstix Overo Tide computer-on-module. This comes with a 720 MHz ARM Cortex-A8 processor, a 520 MHz Texas Instruments C64x+ digital signal processor and 512 MB low power main memory. A microSD memory card slot provides maximal 16 GB of flash memory. A USB host and On-The-Go interface allow the plug in of USB devices into the system and the attachment of the system to a computer. Power is supplied to the computer-on-module by a synchronous step-down converter with a voltage of 4.2 V and a maximal current of 1.5 A. A second synchronous step-down convert supplies the USB host with a voltage of 5 V and a maximal current of 500 mA. The converters can be supervised with a power monitor and disabled in order to save power during off mode of the system. A dedicated CAN controller connects the processor to the remaining system. Additional microphones and speakers on the module allow audio processing and user interaction. Wi-Fi and Bluetooth enable the module to be part of distributed wireless communication networks.

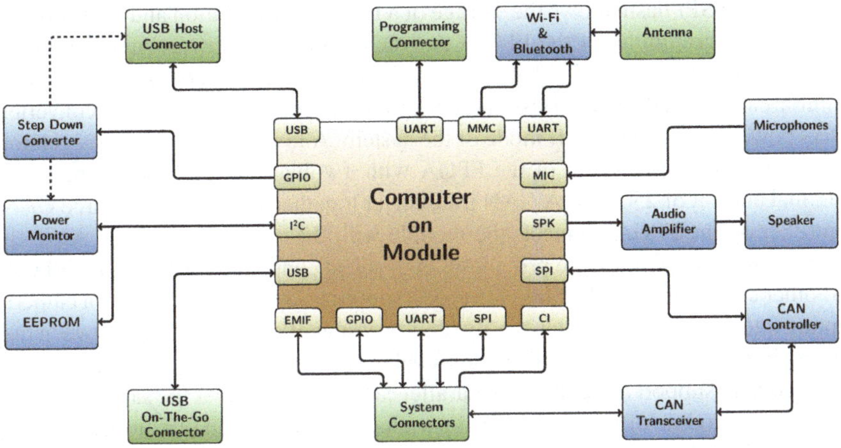

Fig. 5 Cognition-PU processor module architecture

3.5 *Cognition-CL AM*

The main function of this cognition module is to serve as a hardware accelerator for the cognition processor module, as preprocessing unit for the camera data or as platform for parallel vision algorithms. Figure 6 shows a block diagram of the cognition configurable logic module.

The module includes a Xilinx Spartan 6 field programmable gate array (FPGA) with 101261 logic cells, 256 MB low power main memory and a microSD card slot for flash memory. The digital clock management is driven by a 100 MHz oscillator.

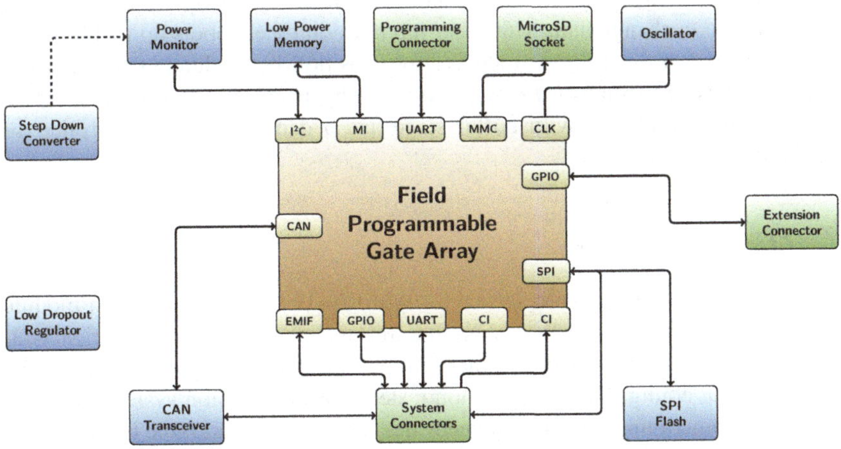

Fig. 6 Cognition-CL configurable logic module architecture

An additional SPI flash allows the storage of a FPGA configuration and supports the automatic programming of the FPGA. A synchronous step-down convert supplies the FPGA with a voltage of 1.2 V and provides a maximal current of 3 A. A power monitor can be used to supervise the converter and the disabling of the converter offers power saving during off mode of the system. A low dropout voltage regulator powers the auxiliary system of the FPGA with a voltage of 2.5 V and supports a maximal current of 250 mA. A CAN transceiver together with an open source CAN controller on the FPGA connects the module with the remaining system. A parallel external memory interface provides a high speed connection between the FPGA and other cognition modules, like the processor unit. Separate up and down camera capture interfaces allows the on-the-fly preprocessing of camera images between the sensor above and other cognition modules below. An additional extension connector can be used to support more than one parallel camera interfaces at the same time. Altogether the module offers fine grained parallel implementation of computationally intensive algorithms. Furthermore it is possible to implement a complete system on a chip on the module and to use the module without the cognition processor module.

3.6 Perception-IR AM

The function of the infrared perception module is to provide obstacle detection capability in the near vicinity of the robot. This is required for the soccer game where the ball may be too close to the robot to be in the field of view of the camera. Besides proximity sensing, infrared transmission can also be used for wireless communication with other robots. Combining the proximity sensing and communication into one sensor system reduces the resource requirements. Figure 7 shows a block diagram of the infrared perception module.

The infrared perception module supports eight enhanced infrared proximity sensors around the body. The module is able to detect objects up to a distance of 300 mm and to exchange data with other modules on other robots up to 1100 mm. Furthermore it can determinate their relative position by evaluating the received infrared signals. The improvement is achieved by using constant current sinks, transimpedance amplifiers with buffer stage and a dedicated microcontroller for processing.

3.7 Perception-V AM

Vision is meant to be the main sensor for AMiRo. The perception vision module adds one or more cameras to the system. It builds the mechanical and electrical connection between an off-the-shelf camera module like the Gumstix Caspa and the robot. Different configuration like a mono, stereo or omni vision systems are possible. Multi-camera implementation needs the cognition configurable logic module for image preprocessing and multiplexing or fusion of the camera streams to one stream which can be processed by the cognition processor module.

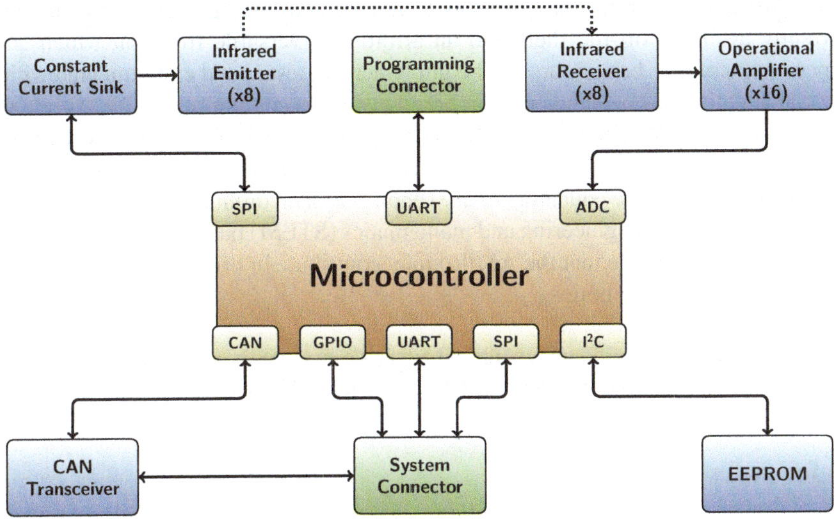

Fig. 7 Perception-IR infrared module architecture

4 Software

There are three parts of the AMiRo software. The embedded software for the microcontrollers on the perception, power and action AMs, the behaviour software on the Cognition-PU AM, and the FPGA configuration on the Cognition-CL module. The embedded software is written to the on-chip flash memory of the 32 bit ARM Cortex-M3 microcontrollers either via a serial cable connection to the programming connector or from the power or cognition module over the CAN bus. Currently, there is no operating system kernel on the microcontroller boards. Low level services code is purpose build for every module. Any suitable microprocessor program development environment and appropriate cross compilers can be used. We use Eclipse as the software development environment for the embedded code with cross compiler from CodeSourcery. The Cognition-PU module runs Linux stored on the microSD card and therefore any of the Linux based software frameworks for robotics such as Player [1] or ROS [2] can be used. Drivers for a Player server were developed for the BeBot and will be ported to the AMiRo. Nodes for ROS have not been written yet.

5 Conclusion

The AMiRo design described in this paper draws on the knowledge gained in over 15 years of designing mini robots and extension modules for mini robots. With regards to circuit technology the immediate predecessor of the AMiRo is the BeBot robot developed at the Heinz Nixdorf Institute, University of Paderborn, Germany.

The AMiRo incorporates the advances in microelectronic and sensor technology currently available off-the-shelf into an affordable, versatile robot on which advanced autonomous behaviours can be implemented, tested and experimented with. Special attention has been given to on-board vision processing capability. Although we have designed AMiRo as a robot for soccer playing, the modules provide the necessary services for robots of other shapes and sizes. We hope that with the AMiRo we can encourage more educators to use robotics as a medium for teaching not only science, technology, engineering and mathematics (STEM) but also humanistic subjects. Likewise we hope that the AMiRo may contribute in broadening the research base for autonomous robotics.

Acknowledgements. This work was funded by the German Research Foundation (DFG) in the course of the Excellence Cluster 277 (CITEC, Bielefeld University).

References

1. Player robot device interface, `http://playerstage.sourceforge.net/`
2. ROS (Robot Operating System), `http://www.ros.org/`
3. AMiRE. AMiRESot, Robot Soccer Rules (2008), `http://amiresymposia.org/AmiresotRules2008.pdf` (last accessed May 15, 2011)
4. Caprari, G., Arras, K.O., Siegwart, R.: Robot Navigation in Centimeter Range Labyrinths. In: 1st International Symposium on Autonomous Minirobots for Research and Edutainment (AMIRE 2001), Paderborn, Germany, pp. 83–92 (2001)
5. Caprari, G., Siegwart, R.: Design and Control of the Mobile Micro Robot Alice. In: 2nd International Symposium on Autonomous Minirobots for Research and Edutainment, Brisbane, Australia, pp. 23–32 (2003)
6. Herbrechtsmeier, S., Witkowski, U., Rückert, U.: BeBot: A Modular Mobile Miniature Robot Platform Supporting Hardware Reconfiguration and Multi-standard Communication. In: Kim, J.-H., Ge, S.S., Vadakkepat, P., Jesse, N., Al Manum, A., Puthusserypady, S., Rückert, U., Sitte, J., Witkowski, U., Nakatsu, R., Braunl, T., Baltes, J., Anderson, J., Wong, C.-C., Verner, I., Ahlgren, D. (eds.) Progress in Robotics. CCIS, vol. 44, pp. 346–356. Springer, Heidelberg (2009)
7. Malmstrom, K., Sitte, J.: Continuous Action Space Reinforcement Learning on a Real Autonomous Robot. In: Sitte, J., Rueckert, U., Witkowski, U. (eds.) Proceedings of the 5th International Heinz Nixdorf Symposium: Autonomous Minirobots for Research and Edutainment (AMiRE 2001), vol. 97, pp. 151–159. HNI-Verlagsschriftenreihe, Paderborn (2001)
8. Mondada, F., Franzi, E., Ienne, P.: Mobile Robot Miniaturisation: A Tool for Investigation in Control Algorithms. In: Yoshikawa, T.Y., Miyazaki, F. (eds.) Experimental Robotics III. LNCIS, vol. 200, pp. 501–513. Springer, Heidelberg (1994)
9. Rückert, U., Sitte, J., Witkowski, U. (eds.): Proceedings of the 5th International Heinz Nixdorf Symposium: Autonomous Minirobots for Research and Edutainment (AMiRE 2001), vol. 97, HNI-Verlagsschriftenreihe, Paderborn, Paderborn, Germany, Heinz Nixdorf Institut, Universität Paderborn (2001)
10. Rückert, U., Sitte, J., Witkowski, U. (eds.): Proceedings of the 4th International AMiRE Symposium: Autonomous Minirobots for Research and Edutainment (AMiRE 2007), vol. 216. Buenos Aires, Argentina, Heinz Nixdorf Institut, Universität Paderborn (2007)

Modular Robot Platform for Teaching Digital Hardware Engineering and for Playing Robot Soccer in the AMiREsot League

Thomas Tetzlaff and Ulf Witkowski

Abstract. The design of digital hardware is nowadays very well supported by software tools, which allow hardware description, hardware synthesis, and complex simulations ranging from device to system simulations. For teaching at universities often a bunch of complex software tool chains is available and is used in classes and for lab work, but practical experience with real hardware is often neglected. Our approach is a combination of teaching hardware design theoretically in a large group and to implement hardware in several small project groups to enable the students to learn about real problems of hardware design and integration. Besides the definition of several small projects defined for one semester lab work we have the objective to design a robot soccer platform that is consistent with the rules of the AMiREsot robot soccer league. This complex platform is a very good tool for making advanced experience in the area of hardware and software design.

1 Introduction

The efficient design of digital, analogue, and mixed digital-analogue hardware components and systems usually requires a lot of practical experience. Our approach for teaching design of hardware systems is a combination of teaching hardware design theoretically in a large group and as a second pillar to realize small hardware systems in several mini projects to enable the students to learn about real problems in hardware design and integration. By working on the mini projects the students usually gain relevant skills to be able to work efficiently on more complex projects like the robot soccer platform. Our objective is to develop and program a robot soccer platform, which can be used in the AMiREsot league that requires small autonomous robots with team sizes of 1, 3, or five robots [1].

Thomas Tetzlaff · Ulf Witkowski
South Westphalia University of Applied Sciences, Luebecker Ring 2,
59494 Soest, Germany
e-mail: witkowski@fh-swf.de

In section 2 a few mini projects are introduced to give an idea of the typical tasks and the complexity of the projects. Section 3 introduces our AMiREsot robot platform. Two experiments are reported in section 4. Section 5 concludes the paper.

2 Mini Projects

In this section a brief overview of selected mini projects is given. Within a class individual students or small groups of up to three students have to solve a hardware, software, or often combined hardware-software development task. In the beginning of a semester several topics are suggested, but the students are also encouraged to suggest design projects. Students decide on their projects they want to work on. On a weekly basis the progress and problems are discussed with presence of all students. After a half semester an intermediate presentation has to be given by each group. At the end of the semester a final presentation typically including a demo is foreseen.

Advantage of these projects is that the students get a deep practical inside into design challenges. They make the experience what does it mean to implement a real hardware being able to work as desired that usually requires several iteration of implementation, debugging, and testing. The motivation of the students is usually quite high, because they want to have a system running at the end. In addition, the final presentations at the end of the class help the students not only to share experience, but also to get fruitful insights into several design tasks. The following sections briefly introduce a selection of mini projects.

2.1 VHDL Camera Controller

Task in this project was to develop a camera controller, which is located in the Xilinx Spartan FPGA (Type XC3S100E) [2]. The camera is from Omnivision, model OV9655, offering 1.3M pixel. The controller was completely coded in VHDL and has been synthesized for the FPGA using the ISE tool chain from Xilinx [3]. Due to unavailability of additional RAM on the FPGA-uC-Board the internal block RAM of the FPGA has been used for storing down scaled images. Configuration of the FPGA and testing has been done via a microcontroller (LPC2136 by NXP). The same microcontroller could be used to wirelessly transmit the captured images to a PC via a Bluetooth link using SPP.

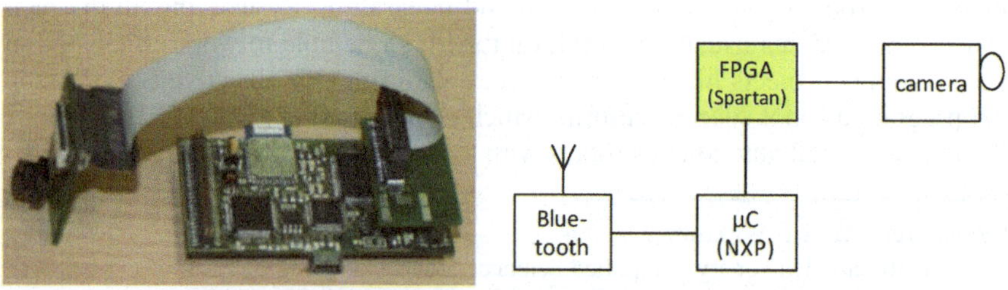

Fig. 1 Small image sensor connected to an FPGA with µC and Bluetooth support

2.2 Digital Audio Output with DAC and Amplifier

Basis for several projects is a small microcontroller-FPGA board (cf. also fig. 1). It integrates an FPGA (Spartan XC3S100E), a microcontroller (NXP LPC2136) and a Bluetooth module (Mitsumi WML-C46). Programming of the microcontroller as well as configuration of the FPGA is done via an USB link to a PC running a GUI to access the board. Task of this project was to extend the board by an audio output module integrating a digital-to-analogue converter (DAC), a filter, an amplifier, and a speaker. The student has designed the PCB integrating mentioned components including connectors and he has coded the controller for the DAC using VHDL. A played back sound demonstrated the well working board.

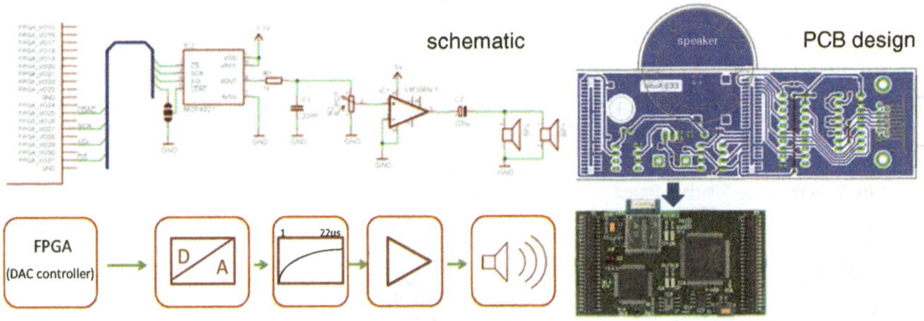

Fig. 2 Schematic and PCB design of an audio output unit

2.3 LED Hardware Clock

In the LED hardware clock project a PCB has been designed with a round shape and a diameter of 10 cm. 60 LEDs are integrated to indicate the minutes (or seconds) of the current time. 12 LEDs indicate the hour, see fig. 3.

The clock is controlled by a CPLD (Xilinx XC2C256). To configure the CPLD a VHDL design has been synthesized by using the Xilinx ISE tool chain. Three

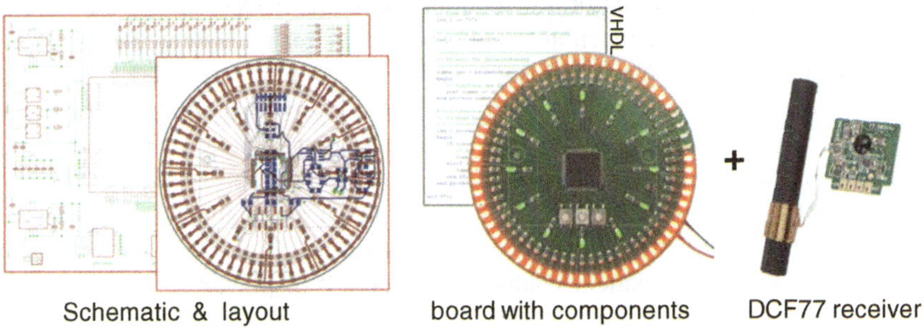

Fig. 3 LED clock design with control via a CPLD configured using VHDL / ISE

buttons can be used to set the clock or to activate a LED blinking demo. Optionally it is possible to connect a DCF77 receiver to automatically get the current time (and date) to initialize the correct time.

2.4 Image Processing on a Mobile Computer under MeeGo

Aim of this project was to port the operating system MeeGo [4] onto the mobile computing platform "Overo Fire" that has been developed by Gumstix [5]. MeeGo is an open source operating system with focus on usage on smartphones. The operating system has been initiated by Intel and Nokia. As an example application under MeeGo we have realized an image processing routine that processes images being captured by a connected USB camera. Via a touch screen TFT programs can be started and the images or video streams are being displayed. The developed system integration can be used as a starting point for mobile processing hardware system of a mobile robot platform, see also section 3.

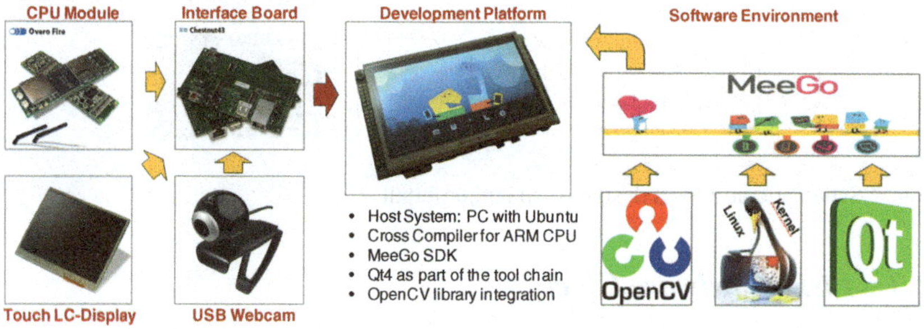

Fig. 4 Integration of display, camera and CPU module running MeeGo

2.5 Remote Control of a Mobile Robot Platform

A wireless connection to a robot realized via Bluetooth enables a designer to easily debug a mobile robot platform without negative effects caused by a cable connection. The Bluetooth connection can also be used to realize a remote control of a vehicle including the display of sensor signals on the remote control device. Basis of the remote control is standard smartphone (here Palm Treo running Windows). For connecting to the robot platform a Bluetooth point-to-point connection (SPP profile) is used. On the mobile robot a Bluetooth devices based on a BTM-182 (Rayson [6]) has been integrated. With the depicted application running on the phone (see fig. 5) the robot can be controlled forward, backward, left and right. Additionally, the current battery voltage of the robot is displayed.

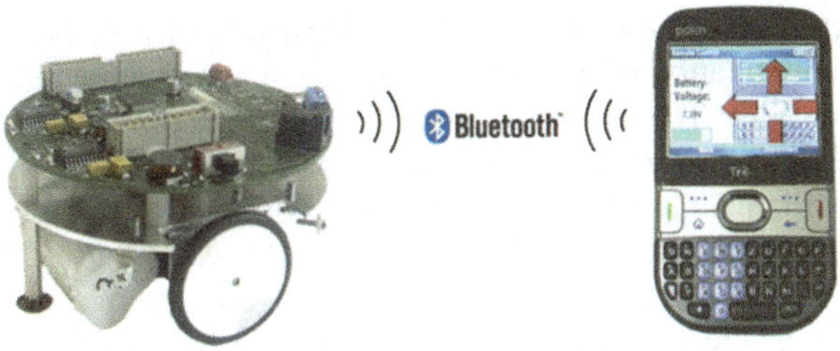

Fig. 5 Remote control and wireless debugging via Bluetooth

3 Mobile Robot Platform

In this section a mobile robot platform is introduced that benefits from the mini projects by integrating already evaluated hardware components and software. The mobile robot platform has a modular architecture that eases adaptation of the platform with respect to technical demands of a project. An important aim of the development of the platform is its use for playing robot soccer in the AMiREsot league. AMiREsot is one of the leagues that are played at national and international robot soccer tournaments under the umbrella of the FIRA organization [7]. AMiREsot robots have a small size with a maximum diameter of 110 mm, cf. fig. 6. Matches can be played with team sizes of one, three, or five robots. All robots in the pitch have to act autonomously, i.e., they have to sense, decide, and act based on integrated sensors and information processing hardware. A wireless communication system can be used to control the robots during kick-off.

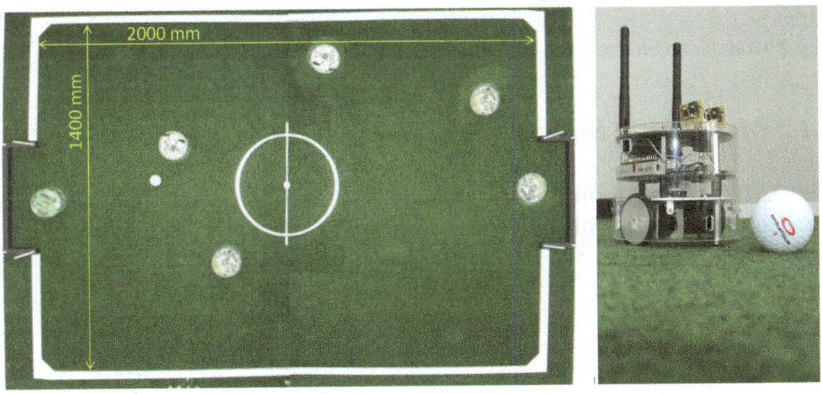

Fig. 6 AMiREsot robot soccer pitch (left), robot in front of goal with ball (right)

This paper introduces an AMiREsot robot platform. It is a modular design integrating the drive system, power supply, necessary electronics, and IR sensors in the base board. An extension board offers computing resources by a mobile processor and an FPGA. For perception a camera can be integrated by attaching it to the processor via USB or to the FPGA to do parallel image pre-processing.

The architecture of our AMiREsot robot is depicted in fig. 7. The minimal configuration is a combination of the mechanical base and a board containing sensors and a microcontroller. The extension of the robot's functionalities is realized by additional modules that can be stacked on top of each other. For communication between the boards mainly I^2C is used. But depending on the requirements of the communication other busses or high speed serial IOs can be used. For example, the top module integrating a touch screen is connected to the subjacent board via a standardized touch screen interface. Main board used for information processing and behaviour execution is the extension board with a mobile computing stick and an FPGA. The FPGA is used to speed up image processing, e.g., to find blobs.

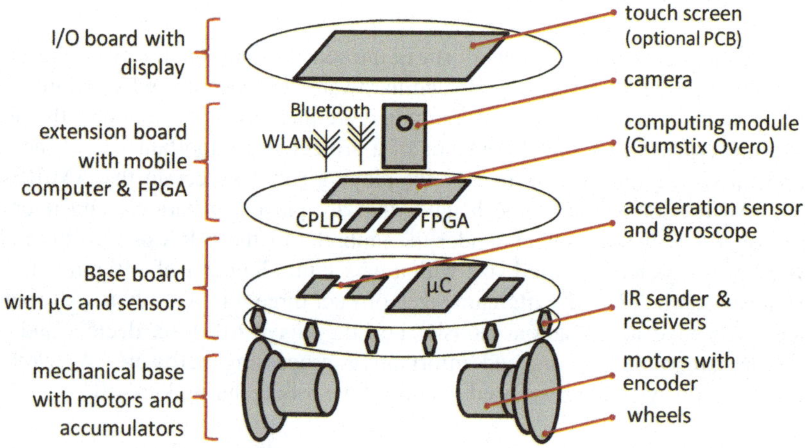

Fig. 7 Modular architecture of the AMiREsot robot soccer platform

3.1 *Mechanical Structure and Chassis*

Our AMiREsot platform follows a modular approach, i.e., the base module is a mechanical platform with an aluminium board at which motors and accumulators are mounted, see figure 8.

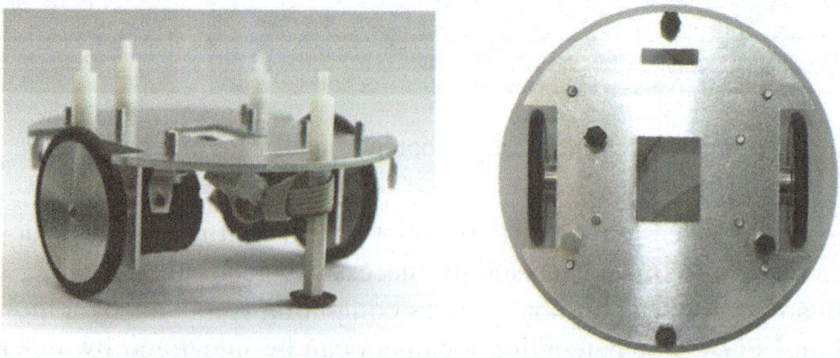

Fig. 8 Mechanical base with mounted motors and wheels; side view and top view

Diameter of the platform is 100 mm. Considering an additional plastic (e.g. PMMA) cover the total diameter is within the range of 110 mm specified in the AMiREsot soccer rules. Depending on the specific needs of the platform for performing experiments additional modules can be mounted on top of the mechanical base via plastic spacers. The motors are from Faulhaber type 2619S006SR 33:1 IE2-16 with max. power of 1,1W each. The accumulators are two packs of three AA NiMh accumulators with a capacity 2400 mAh at 7.2 Volts. Battery life time for a fully equipped robot driving at medium speed is about 4 hours.

3.2 Base Board

The minimal configuration of the AMiREsot platform is a combination of the mechanical base and a PCB integrating a microcontroller and sensors. The microcontroller (STM32F103 from ST Microelectronics [8]) controls the motor speed and reads sensor signals from 12 IR sensors symmetrically arranged at the cover of the robot. By connecting a Bluetooth module to a serial line of the microcontroller a cable replacement can be realized and used for debugging. An extension board is connected via 34 pin connector. Here a relatively large pin spacing of 2,54 mm has been used to ease interfacing and development of boards (PCBs) that are fabricated manually by students as part of a mini project.

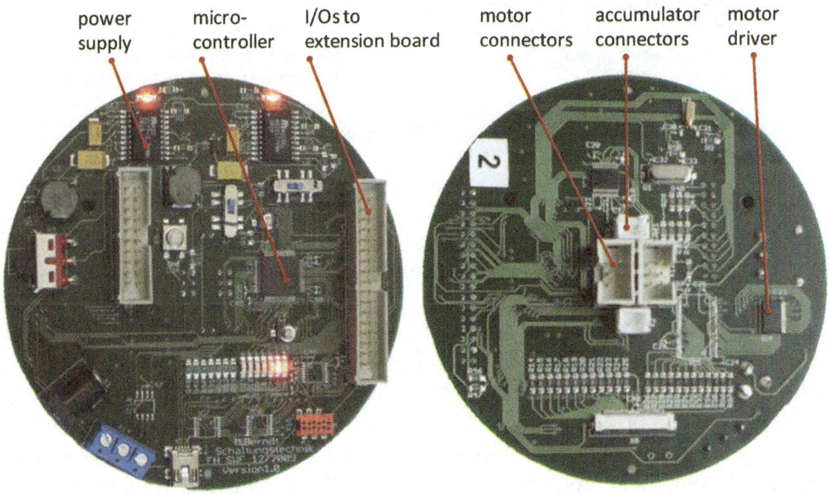

Fig. 9 Base board with microcontroller and interfaces; top view and bottom view

3.3 Extension Board

By using the base board only on the robot platform a simple behaviour like the Braitenberg behaviour can be realized using the IR sensors. Or it is possible to remotely control the robot. For advanced behaviour especially playing robot soccer a vision sensor and a more powerful processor is mandatory. The extension board for the robot which has been designed by a student integrates a processing module offering up to date mobile computing power. Core device is the computing module "Overo" from the Gumstix company [5]. This computing module has a small footprint and it can be easily exchanged by latest hardware if desired. On the robot platform we are using Linux that eases programming of the robot by using C as a programming language. We have added functions into the operating system which provide access to the sensor data of the base module and also access to the motor controller to set the speed of both motors.

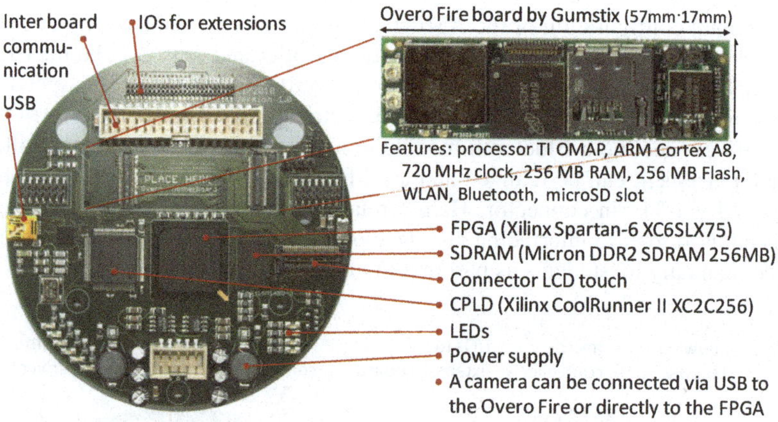

Fig. 10 Extension board with computing module and FPGA

The mobile processor is running Linux. It can be programmed in C, C++, via shell-script or Phyton. It is also possible to install other compilers and then to use other programming languages. The minimal functionality of the processor board is to realize a simple behavior on the robot, i.e. to set motor speed of both motors by writing the speed data into the microcontroller of the base board via I^2C. The mobile processor can be accessed by SSH. It is possible to log in via WLAN or Bluetooth in order to program or to remotely control the robot.

The board supports the usage of an FPGA, i.e. it is possible to (re-)configure the FPGA from the operating system that eases hardware access and speeds up software development. As configuration controller a CPLD is used. Via I^2C the configuration data is transferred from the processor to the CPLD. The FPGA is programmed in VHLD or Verilog with use of the XILINX ISE Design Suite. Data captured form the camera can be stored in 256MB DDR2 SDRAM. From this the FPGA can calculate the position of the robot, the ball and the goals to name an application example in the robot soccer context. Symbolic information as extracted positions can be sent to the mobile processor via the I^2C bus.

4 Experiments

Main objective of the development of the presented robot platform is to perform robot soccer matches with varying team sizes starting with one robot per team. Currently we are realizing the image processing to be able to detect opponents, the ball, the goals, and the lines in the pitch. This part has not been finished yet, therefore it still not possible to play matches. Currently the robot can be remote controlled by a PDA via the Bluetooth link, see figure 11. We can test ball pushing behavior, dynamics of the robot and are able to collect sensor data to implement and optimize the behavior of the robot.

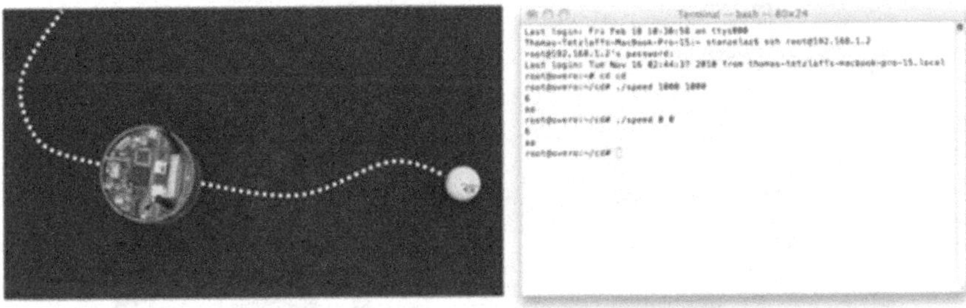

Fig. 11 AMiREsot robot in the pitch following and pushing the ball

A fully equipped AMiREsot robot, i.e., with integrated base board and extension board, is shown in fig. 12 left. Besides playing robot soccer the platform can be used to teach students in several engineering disciplines. This is processor and microcontroller programming, FPGA hardware design, classic control, behavior design including cooperative behavior schemes, perception, and computational intelligence. As one example in control a student has used the robot platform to realize a pole balancing behavior, see figure 12 right. Sensor signals as input for the controller are the incremental encoders of both motors, gyroscope data, and acceleration data from a 3d accelerometer. To get appropriate signals the gyroscope has been mounted on top in an upright position.

5 Conclusions

The design of microelectronic systems is a complex task. Our approach to teach the students in design of these systems is to define mini projects which have been identified to be very useful for the students in learning to design, to program, to commission, and to test real systems. Students make several practical experiences and by presenting the achieved results they get a very good overview of all projects that have been developed within the class. In total, approximately 10 new projects are defined every semester. Our experience is that students like this kind of project work. They like the proposed topics, they are requested to work self-contained and they like to show a running demonstrator at the end. The amount of time of the students spent for the projects is about two hours per person per week plus about three additional days at the end of the semester to complete the final demo and the presentation.

Students who are motivated to spend more time for a project work are welcome to work on the AMiREsot robot soccer platform. Currently they are working on the image processing to robustly detect the ball, the opponents, the goals, and the robot's position. Our objective is to have within the next 6 months a robot that is able to play matches 1 vs 1. Afterwards we focus on team play with 3 robots per team. Students who are interested have the option to attend international robot soccer tournaments as organized by the FIRA.

Fig. 12 AMiREsot robot (equipped with base board, extension board) in the pitch (left); self-balancing robot with additional gyroscope and accelerometer (right)

References

1. Witkowski, U., Sitte, J., Herbrechtsmeier, S., Rückert, U.: AMiRESot – A new robot soccer league with autonomous miniature robots. In: Proceedings of the FIRA RoboWorld Congress 2009, Korea, vol. 44 (2009)
2. Xilinx FPGAs, `http://www.xilinx.com/products/silicon-devices/fpga/index.htm`
3. Xilinx Tools, `http://www.xilinx.com/products/design-tools/ise-design-suite/index.htm`
4. MeeGo project, `http://meego.com`
5. Gumstix Overo Fire (2011), `http://www.gumstix.com`
6. Bluetooth module (2011), `http://www.rayson.com/btm180.html`
7. FIRA (Federation of International Robot-soccer Association), `http://www.fira.net`
8. STMicroelectronics, `http://www.st.com/internet/mcu/product/216824.jsp`

Radar Sensor Implementation into a Small Autonomous Vehicle

Ivan Ricardo Silva Ruiz, Dominik Aufderheide, and Ulf Witkowski

Abstract. The local behavior and local navigation of a mobile robot depends on the availability and quality of environment data, e.g., the information on the presence of obstacles or free space area as a minimum. In this work we focus on a sensor system that can be used under low and no visibility conditions as they may appear in dense smoke or dust. The related tasks are to characterize and to implement a small static radar sensor into a mobile robot platform with a size of about 40 cm by 30 cm. The sensor is able to determine ranges and angles between the robot and obstacles for close and medium distances. In this paper we focus on the implementation of the radar senor into a vehicle performing local navigation autonomously. Indoor and outdoor tests show the applicability of a radar sensor at a small vehicle.

1 Introduction

The development and use of mobile robots have become lately one of the biggest scientific areas where research and development efforts have increased significantly. Especially the sector of autonomous range navigation vehicles is in focus of attention, because such mobile robots can navigate very robustly and reliable in applications that require indoor or outdoor courses in considerable hazardous environments like fire situations, chemical exposures or any other situation that may harm a human being. The presented scientific approach of this investigation was derived from an idea of a previous research work founded by the European Union called Guardians [1]. The project covers a main disaster scenario including a large warehouse on fire with a dense cloud of black smoke which makes it difficult to orientate for the firefighters and thus complicate the task of finding and locating the source of the fire or even for the rescuing of trapped people from smoke or flames. Optical sensors that are often used in mobile robot systems have signal

Ivan Ricardo Silva Ruiz · Dominik Aufderheide · Ulf Witkowski
South Westphalia University of Applied Sciences, Luebecker Ring 2,
59494 Soest, Germany
e-mail: witkowski@fh-swf.de

degradation affronting low visibility conditions like smoke. The innovation presented in this work is that the used sensor employs high resolution radar which is integrated into a small mobile robot. The design of the mobile robot starts with the implementation of a CAN interface for controlling and manipulating the high resolution radar sensor as shown in section 2 of this paper. Once the radar sensor is implemented and the results from the sensor are obtained, the design continues with the identification or construction of a vehicle structure that supports all necessary system components and provides the possibility to move in flat and slightly rough terrains. Afterwards the design continues with the implementation of the electronic hardware and software which is responsible for controlling the movements of the mobile robot. A microcontroller acts as a decision unit for the mobile robot in terms of directivity depending on the information received from the radar sensor. Details of the integration are presented in section 3. Indoor and outdoor navigation tests are summarized in section 4. Section 5 concludes the work and identifies possibilities for future implementations.

2 Radar Sensor Unit

A radar electronic device operates by transmitting a particular type of electromagnetic wave to detect objects or materials in the atmosphere from the nature of the echo signal that the radar device receives. "The radar is used to extend the capability of man's senses for observing his environment, especially the sense of vision. The value of radar lies not in being a substitute for the eye, but in doing what the eye cannot do. Radar cannot resolve detail as well as the eye, nor is it yet capable of recognizing the color of objects to the degree of sophistication of which the eye is capable. However, radar can be designed to see through those conditions impervious to normal human vision, such as darkness, haze, fog, rain and snow. In addition, radar has the advantage of being able to measure the distance or range to the object", see [2]. For the radar system there are a few fundamental modules, as shown in Figure 1, that can be found in any radar system.

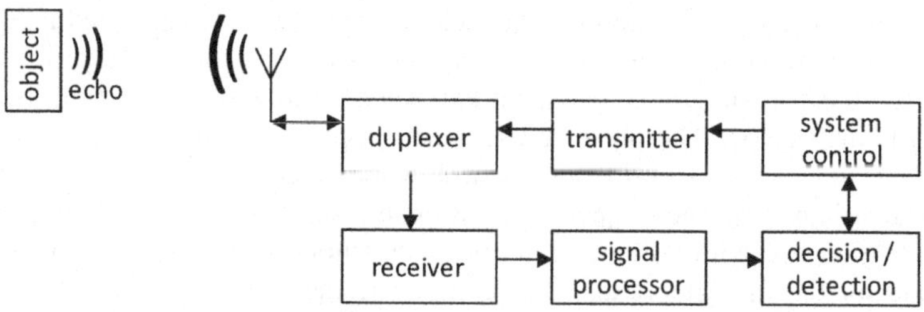

Fig. 1 Block Diagram of a Radar System

The transmitter module is responsible for the generation of electromagnetic radiation through the atmosphere by means of an oscillator; the receiver makes the detection, demodulation and amplification of the returning echo signal. The processor interprets the signal coming from the receiver, extracting the necessary information for the specific purpose in which the radar has been set up. The decision unit identifies objects with respect to specific features like distance, angle, speed, and size.

Radar has been employed on the ground, in the air, and on the sea and undoubtedly will be used in space. Ground space radar has been applied chiefly for detection and location of aircrafts or space targets. Shipboard radar may observe other ships or aircrafts, or it may be used as a navigation aid to locate shore lines or buoys. Airborne radar is mainly used to detect other aircrafts, ships or land vehicles, or it may be used for storm avoidance and navigation purposes. One application in the automotive area is the use of radar to detect obstacles in the vicinity of a car as part of an accident avoidance system. In this context the Adaptive Cruise Control (ACC) is a function for longitudinal control of vehicles. The ACC controls the speed of a car to guarantee a specific distance between the controlled car and a vehicle travelling ahead. Thus the relative speed to preceding vehicles is controlled by employing suitable actuator systems [3]. The used radar technology for such an ACC operates in both long range (up to 150m) and short range (up to 20m) to achieve additional safety for the automobile. The short range radar devices are used for creating a virtual safety belt around the car. The partial combination of long range and short range radar sensors provides the possibility to implement additional functions of the ACC such as stop & go, pre-cash and parking assistance. Figure 2 gives an overview of the different detection zones of different surround-sensing technologies being used for such driver assistance systems.

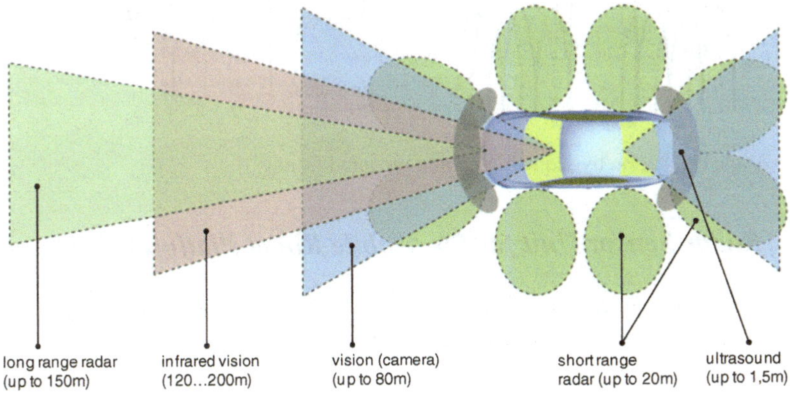

Fig. 2 Sensor surrounding system for cars

2.1 Range and Angle Detection

The ability to determine the range between two objects by measuring the time for the radar signal to propagate to the target and back is probably the distinguishing and most important characteristic of conventional radar. No other sensor can measure ranges with the accuracy achievable with radar technology even for long distances and under adverse weather conditions [4]. The typical radar wave form used for range measurements is a short pulse. "The shorter the pulse, the more precise can be the range measurement. A short pulse (compressed pulse) has a wide spectral width (bandwidth). The effect of a short pulse can be obtained with a long pulse (uncompressed pulse) whose spectral width has been increased by frequency modulation. The angle or angular direction of a target is determined by sensing the angle at which the returning wave front arrives to the radar." [4] Latest UWB (ultra wide band) devices as the used sensor are based on the same short pulse principle [5]. The direction in which the antenna points when a maximum of the received signal can be determined indicates the direction of the target, neglecting atmospheric disturbance. Angle estimation uses different measurement techniques at different beam positions for greater accuracy but particularly for the sensor used in this application sequential lobing is used. The general idea of sequential lobing is depicted in Figure 3. Basis for sequential lobing is a switching of the beam between two angular positions. By measuring the amplitude of the target return the angle to an object (target) can be calculated [6].

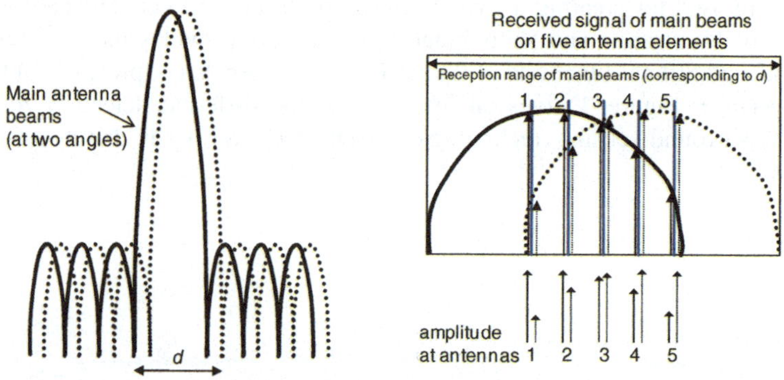

Fig. 3 Angle estimation by sequential lobing (adapted from [7])

2.2 *Radar Sensor Unit for the Mobile Robot Platform*

The radar device used in this paper is a 24.125 GHz radar sensor developed by COBHAM industries. The main features of this device are shown in Table 1.

Table 1 Selected features of the integrated radar sensor

Weight	130 grams
Frequency	24.125 GHz
Bandwidth	3 – 4 GHz
Detection Cycle Time	40 ms
Number of Objects	10
Detection Range	0.2 – 30 m

A graphical user interface (GUI) in LabVIEW was developed for evaluation purposes during the testing period whereat the communication with the sensor was realized by using a CAN interface. The GUI, as shown in Figure 4, realises the visualization and archiving of range and angle estimates provided by the sensor. The final implementation was realized in C and is executed on a microcontroller that offers a CAN interface which allows direct sensor integration.

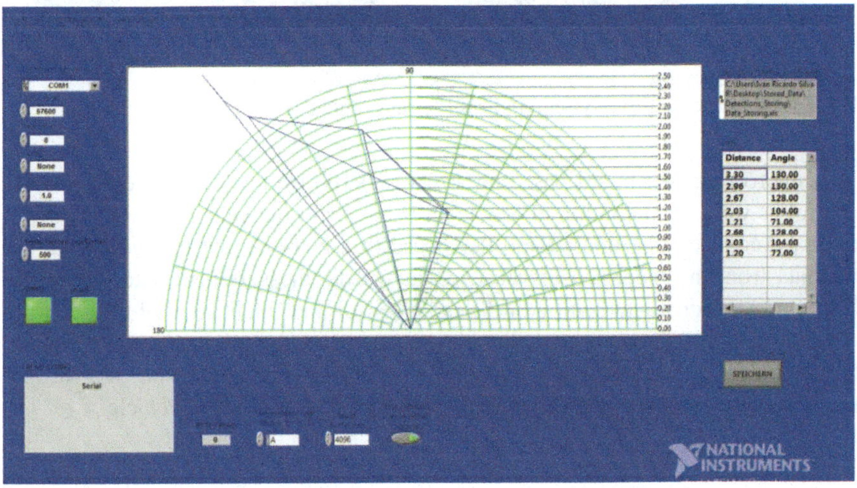

Fig. 4 GUI in LabVIEW to display received data (distance and angle to objects)

3 Radar Sensor Unit

Basis for the implementation of the radar senor is the vehicle depicted in figure 5. The vehicle has a very solid structure which allows the integration of additional devices such as sensors or processing units on top without strict limitations regarding size and weight. This is actually a big advantage for the particular application because the vehicle should carry objects such as the radar sensor, laptop computer, battery pack, and the corresponding CAN interfaces. The robust design of this vehicle provides a very good stability for different types of terrain, such as rough or slippery terrains. Thus the mobile platform can be applied in a wide range of applications. Two DC motors each with 25W are integrated to drive the system. The power supply is realized via a 7.2 V / 4500 mAh accumulator. For information processing a modular architecture has been chosen. The engine power card for the vehicle was designed to carry out the power transmission from the batteries to the motors and also for the motion control of the vehicle. The power transmission of the batteries uses the capabilities of an electrically operated switch to control the vehicle's ignition and also to protect the electrical circuits from overloads or faults.

Fig. 5 Base including drive system of the all-terrain vehicle (lateral and front)

A microcontroller board based on a STM32F103 [8] device realizes the motor control as well as integration of low level behavior of the vehicle. A RS232 serial interface connects the microcontroller board to a netbook that is used to capture the data from the radar sensor and to visualize the measurement data during debugging. After characterization of the radar sensor and finalization of the software development the sensor can be directly connected to the microcontroller that offers a CAN interface. This is part of future work. Figure 6 gives an overview of the hardware components and their connections of the completed vehicle.

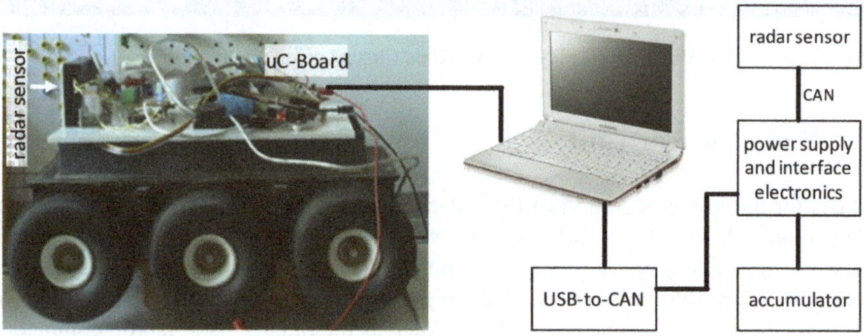

Fig. 6 Mobile robot block diagram. The radar sensor is mounted onto the vehicle.

It should be pointed out that the current configuration of the robot is just considered for evaluating the performance of the proposed radar based local navigation. Nowadays multi-modal approaches are state-of-the-art for robot navigation and localization and even for low level behavioral algorithms, such as obstacle avoidance (OA), the application of multi-sensor data fusion (MSDF) techniques as L-E-U Kalman filters (see [9]) or Monte Carlo particle filters is mandatory. Adequate sensor systems for local robot orientation include ultrasound, tactile perception, ultra wide band (UWB) devices, etc.

4 Indoor and Outdoor Tests

Several tests have been performed to analyze the characteristic and performance of the radar sensor in terms of resolution in range and angle as well as the dependence of the object detection with respect to object size. In addition several indoor and outdoor tests have been defined to get the overall vehicle's performance. For range measurements experiments were made in a room with three different objects of similar material but different shapes. The objects are (A) metal cylinder (18 cm long and 7.5 cm diameter), (B) iron cylinder (24 cm long and 7.5 cm diameter), and (C) rectangular metal plate (27 cm by 7 cm). The experiment consists in the placing of the objects (A, B or C) in 5 different positions (0.25, 0.50, 1, 1.5, 2 m distance to the robot) and the comparison between ground truth and measured distances as shown in Figure 7.

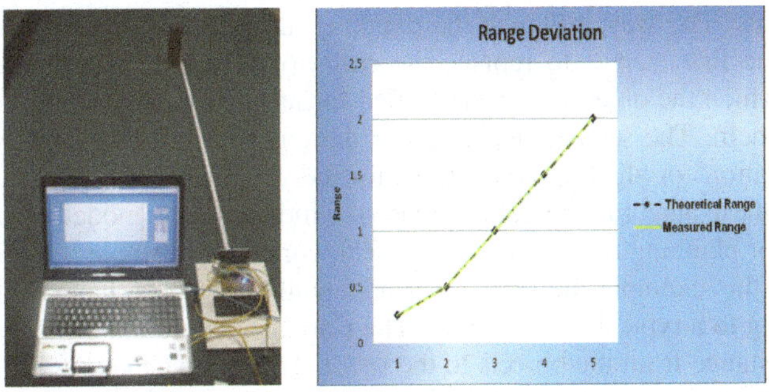

Fig. 7 Measurement setup and achieved range deviation

For angle measurements the objects have been placed at different distances in front of the sensor under a range of different angles. Results are depicted in Fig. 8.

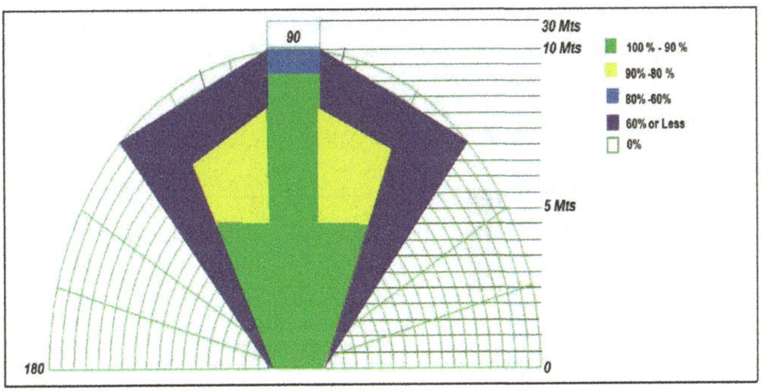

Fig. 8 Accuracy for angle measurements

Regarding the measures of the angle the results depend on the distances between sensor and object. This situation was expected because in the datasheet of the sensor the accuracy for the angle detection is reported to ±5° with increase for bigger angles. Thus at 90 degrees the deviation is minimal compared to the other angles 60 and 120 degrees, i.e., maximal accuracy is achieved for objects located in a central position.

4.1 Indoor Testing

The radar sensor is used for object detection as part of the low level behavior implementation. Therefore the performed tests focus mainly on obstacle avoidance. The approach for realizing the low level control is based on the virtual spring method as suggested in [10]. The virtual spring method is a theory in which a potential field is set up to attract the links of a robot towards the goal whilst repelling them from obstacles. The way in which the planning algorithm proceeds is divided into two steps, the first step is to replace the real world by a Dynamic World Model (DWM) in which the objects have simplified forms and specific charges (forces) associated to them. The second step contains the computation of DWM with mathematical operations or algebraic operations for the vectors used to direct a vehicle. A solution for these algebraic operations maps the motion of the modelled world which is the base for planning the motions in the real world. An example setup is shown in Figure 9. In this example the objects, or apparently obstacles, have been modelled corresponding to a typical real scenario. The Cartesian coordinates in the figure represent the distance from the objects to the robot. This allows the usage of vector operations for the virtual spring method. The results of the algorithms are estimates for the magnitude (distance) and direction of the objects (angle).

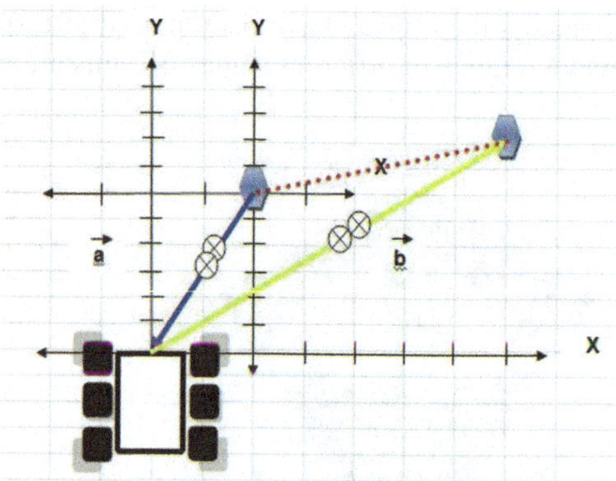

Fig. 9 Virtual spring model setup with two arbitrarily positioned objects

The virtual spring method has been combined with a rule based system to generate the robot's behavior. In order to have a refinement where the concept of virtual springs prevail the identification of the proximity of the targets by using the benefits of the radar sensor was implemented. Here the detection of a considerable number of targets and the evaluation in a rule based system helps to coordinate the actions of the robot (movements) towards its goal to drive autonomously while at the same time it can avoid obstacles.

For the first indoor trial a test was made in a narrow indoor scenario in a basement location of a four level building as depicted in the diagram of the Figure 10. The results of this first test where favourable because with it the mobile robot was able to drive through the corridor autonomously and besides that the ability of the system to avoid targets could be perceived in some occasions where the mobile robot was approaching to the walls of the corridor.

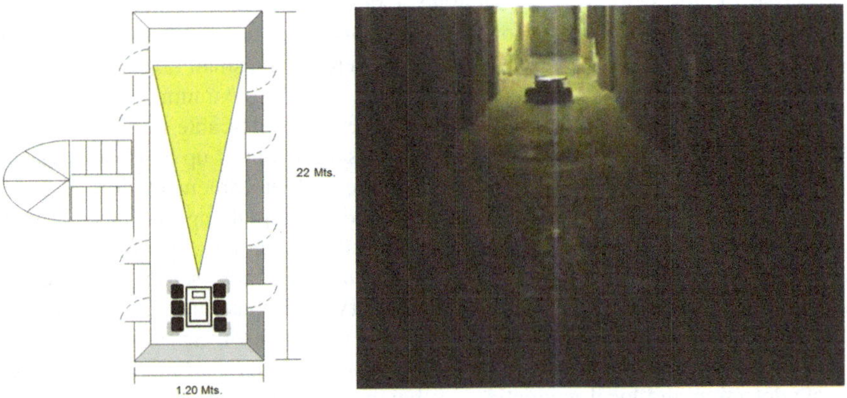

Fig. 10 Indoor testing diagram scenario and photo with real robot in corridor

4.2 *Outdoor Testing*

Outdoor experiments have been performed on narrow roads limited with trees or dense grass. The task for the vehicle was to follow the road autonomously. The mobile robot is able cover even distances of more than 80 m and beyond in continuous driving and without leaving the marks of the road, cf. figure 11.

Fig. 11 Outdoor testing scenarios (one robot had been used for experiments only; to show trajectory several photos have been overlaid)

The road following behaviour could be achieved because the height of the vegetation on both sides of the road has made that the mobile robot detect such vegetation as obstacles which have to be avoided. Nevertheless with the result obtained, the mobile robot still needs to improve its performance at least in relation of its own speed. In the experiment the speed was approximately 0.5 m/s. For higher speed new adjustments of the robot control unit are necessary and the integration of additional sensor system is mandatory.

5 Conclusions

Objective of this work was to integrate a radar sensor into a small vehicle to evaluate a small radar sensor for obstacle sensing. In several tests it has been shown that the selected radar sensor is able to detect even small obstacles within a range up to 30 m. For local path planning of the robot a maximum detection range of 5 to 10 m is usually sufficient. The radar device is also able to detect objects that are not directly in front of the sensor. A misalignment of up to 45 degrees can be detected and used for robot navigation. In the current implementation the radar sensor has been integrated using a notebook and corresponding interface devices. This eases program development, debugging and fine tuning of parameters in the control algorithms. Next step is to directly connect the radar sensor with the microcontroller that is equipped with the necessary CAN bus interface. The indoor and outdoor experiments have shown that the radar sensor can also be used on small vehicles to perform medium and long distance range measurements for obstacle detection and local navigation not being based on vision or laser. For more complex robot behavior the integration of additional sensors and a powerful information processing unit based on recently suggested methods from MSDF is foreseen.

References

1. Penders, J., Alboul, L., Witkowski, U., Naghsh, A., Saez-Pons, J., Herbrechtsmeier, S., El-Habbal, M.: A robot swarm assisting a human fire-fighter. Advanced Robotics 25, 93–117 (2011)
2. Skolnik, M.: Introduction to Radar Systems. McGraw-Hill (2000)
3. Marek, J., Trah, H.P.: Sensors Application - Sensors for Automotive Technology. VCH 4 (2003)
4. Skolnik, M.: Radar Handbook. McGraw-Hill (2008)
5. Aiello, R., Batra, A.: Ultra Wideband Systems. Technologies and Applications. Newnes (2006)
6. Knott, E.F., Shaeffer, J.F., Tuley, M.T.: Radar Cross Section. Scitech Pub. Inc. (2004)
7. MIT Lincoln Laboratory (2011), `http://www.ll.mit.edu/workshops/education/videocourses/introradar/index.html`
8. STMicroelectronics, `http://www.st.com/internet/mcu/product/216824.jsp`
9. Carelli, R., Oliveira Freire, E.: Corridor navigation and wall-following stable control for sonar-based mobile robots. Robotics and Autonomous Systems 45, 235–247 (2003)
10. LaValle, S.M.: Planning Algorithms. Cambridge University Press (2006)

The Wanda Robot and Its Development System for Swarm Algorithms

Alexander Kettler, Marc Szymanski, and Heinz Wörn

Abstract. We introduce a new development system for swarm algorithms to be used in research and education, composed of a swarm of Wanda miniature robots, a beamer assisted arena for robot experiments and a new framework for the accurate simulation of robotic swarms. The Wanda robot is easy manufacturable and was especially designed to be used as a swarm robot. It is accurately implemented in the new simulation framework which provides powerful methods for the efficient and exact simulation of sensor data for all kinds of vision based sensors, such as rgb-sensors, camera sensors, infrared communication and ranging sensors by utilizing graphics hardware. Utilizing the behavior based controller language MDL2ε which is implemented on the real robot and in the simulation, the complete system aims to allow for a quick and easy development, testing and demonstration of swarm algorithms.

1 Introduction

In a robotic swarm, large groups of simple autonomous mobile robots perform tasks in a cooperative and decentralized manner, that otherwise would be impossible to

Alexander Kettler
Institute for Process Control and Robotics, Karlsruhe Institute of Technology,
Engler-Bunte-Ring 8, D-76131 Karlsruhe,
e-mail: alexander.kettler@kit.edu

Marc Szymanski
Institute for Process Control and Robotics, Karlsruhe Institute of Technology,
Engler-Bunte-Ring 8, D-76131 Karlsruhe,
e-mail: marc.szymanski@kit.edu

Heinz Wörn
Institute for Process Control and Robotics, Karlsruhe Institute of Technology,
Engler-Bunte-Ring 8, D-76131 Karlsruhe,
e-mail: heinz.woern@kit.edu

achieve for a single robot of the group. The main advantages of such a decentralized and self-organizing system are its robustness, scalability and flexibility. Due to the autonomous mobility of the single individuals, such systems are able to easily adapt to environmental changes. Removal or adding of additional individuals allows to provide for changes in the size of a given task or to cope with the failure of single individuals. The development of control algorithms for swarm robots is a challenging task as the global behavior that emerges from the many interactions between the robots is often hard to predict and experiments with swarm robots are often expensive and time consuming. In the present paper, we will introduce a new compound system for research and education in swarm robotics that tries to simplify the simulation, experimental testing, and demonstration of algorithms for multi-robot-systems. The paper is organized as follows: In section 2 we will give an overview on the Wanda robot and the arena that is used in the development system. Section 3 describes the swarm robot simulation framework that is used for, but not limited to the simulation of the Wanda robot. Section 4 demonstrates the usability of the development system by means of two example scenarios.

2 The Swarm Robot Wanda

Designing small autonomous mobile robots for research in evolutionary robotics, swarm robotics, artificial intelligence, and artificial life has a long history and dates back to the mid 90's with the design of the Khepera robot [10]. Since then several robots, like the Alice [5], Jasmine [12] or the ePuck robot [9], to give just a few of them, have been designed each especially strong in parts of those topics. However, sometimes it seems that there is still much effort to be invested by the researcher to perform experiments with those robots and to properly shape the environment for the robot as most of the existing platforms only feature a limited set of sensors or require expensive extension boards. With the design of the Wanda robot we wanted to cope with these difficulties by providing an easy usable platform with a large number of different sensors already included to allow for a wide variety of experiments

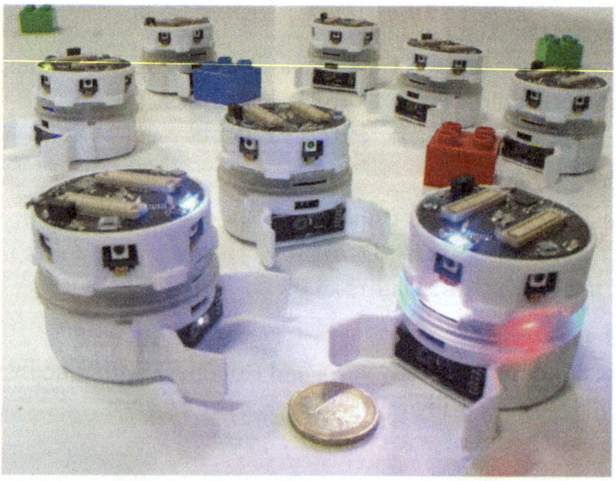

Fig. 1 A small group of Wanda miniature robots for research and education

while also making the robot easy to manufacture at a low price in order to be able to conduct experiments with a large (> 100) number of robots. This also implies a powerful infrared communication system that is especially designed to be used in a swarm of robots. As we wanted the robot to be small enough to conduct basic experiments on a desktop or in small arenas, we ended up with the robot featuring a size of 45 mm in height and a diameter of 51 mm which seems to be a reasonable trade-off between small size and still providing enough functionality and accuracy. In the following we will give an overview on the basic design of the robot and the arena system used in conjunction with it (for a more complete description, see [7],[8]).

2.1 *Basic Design*

One key aspect during the development of the Wanda robot was to give it a modular structure and to make it easily extensible. Therefore, the basic structure of the robot is made up by a variable number of stacked printed circuit boards, which are almost completely manufacturable via automated pick-and-place and can be easily sticked together. Manual steps are only necessary for the assembly of the differential drive, the installation of the batteries and the fixation of the front-sensor.

2.1.1 CPU

As it's central processing unit, the robot features a LM3S1960 [15] micro controller (μC) from Texas Instruments (TI) with 50 MHz clock frequency, 256 Kb of Flash memory and 64 Kb of RAM, which in our opinion is a good trade-off between power consumption and computational power. Additionally, the robot can be extended with a CM-BF-561 Dual Core Blackfin board running ucLinux.

2.1.2 Sensors

The Wanda robot has a rich set of sensors that can be used to fulfill multiple purposes depending on the given task. These sensors will be shortly described in the following.

Floor and Touch Sensor: Three active infrared sensors arranged in one line parallel to the axis at the bottom of the robot serve as floor sensors that allow for simple line following or to mark areas for different purposes. One additional sensor of the same type, located on the front of the robot between its grippers serves as a near distance proximity sensor.

Color Sensor: Next to the near distance proximity sensor, a color sensor is located. It allows for passive color detection, ambient light detection, and active color detection by emitting white light from an LED which gets triggered by the near distance sensor.

Accelerometer: The accelerometer can be used for artificial intelligence and evolutionary robotics. It should help to detect if the robot hit any obstacles and thereby serves as a direct and important feedback to online evolutionary robotics.

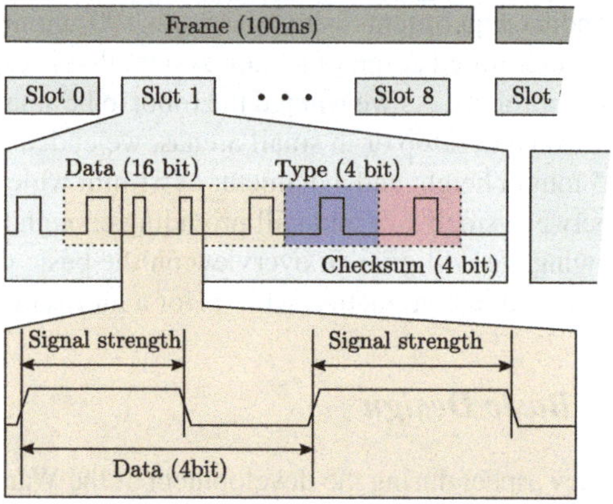

Fig. 2 The infrared protocol used by the Wanda robot: Each robot occupies one out of eight time-slots. During one time-slot, the robot sends one starting pulse and six subsequent data pulses. Each data pulse contains four bits of data, encoded in the distance between itself and its predecessor. This gives 16 bit of user data plus four additional bits for control data and another four bits used for error detection.

Communication and Ranging: The Wanda robot features a infrared communication system which enables it to locally communicate with multiple other Wanda robots over distances of up to 1 meter while measuring the distance, direction and orientation of each communication partner at the same time. Furthermore, the robot is able to measure the distance and bearing to passive obstacles by receiving its own reflected messages.

The system consists of six infrared diodes and six infrared photo transistors, aligned symmetrically around the center of the robot with an angular spacing of 60°. A simple but effective preprocessing of the sensor data that converts pulse intensities into pulse lengths eliminates the need for an analog digital converter and allows for simple interrupt driven processing of communication data.

In order to update its information about the environment, the Wanda robot has to send data at periodic intervals. To prevent collisions with packets from other robots a simple self-synchronizing time division multiplexing algorithm with eight time-slots is applied (see fig. 2). Upon sending, the robot checks for ongoing transmissions and changes its own slot if necessary. Eight time-slots have shown to be sufficient even for larger groups of robots (>40 robots) because the average number of robots within direct sight will usually be below eight robots.

By measuring the intensity of each single pulse of a packet, the Wanda robot is able not only to determine the angle from where it received a packet, but also to calculate the relative orientation of the sending robot with respect to itself (see fig. 3): Let $I_{m,n}$ denote the intensity of the n-th pulse as received by receiver R_m. The direction from receiver to sender α and the direction from sender to receiver β are estimated by

$$\alpha = \operatorname{atan2}(a_y, a_x), \quad \mathbf{a} = \sum_{m=0}^{5} \sum_{n=0}^{6} \begin{pmatrix} \cos(\frac{m}{6}\,2\pi) \\ \sin(\frac{m}{6}\,2\pi) \end{pmatrix} \cdot I_{m,n} \tag{1}$$

$$\beta = \operatorname{atan2}(b_y, b_x), \quad \mathbf{b} = \sum_{m=0}^{5} \sum_{n=0}^{5} \begin{pmatrix} \cos(\frac{n}{6}\,2\pi) \\ \sin(\frac{n}{6}\,2\pi) \end{pmatrix} \cdot I_{m,n+1} \tag{2}$$

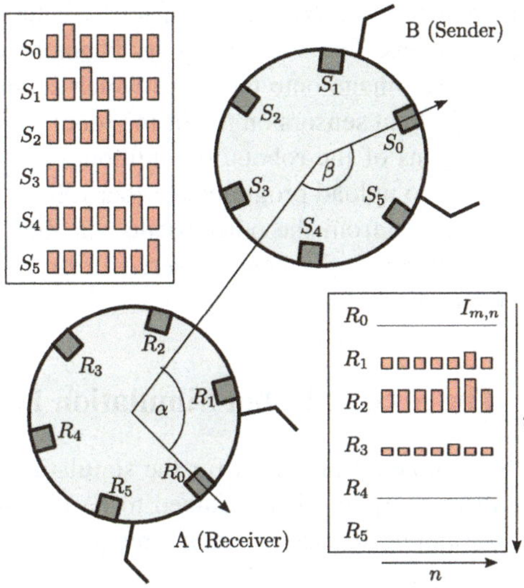

Fig. 3 Schematic illustration how the orientation information is encoded in a packet consisting of seven infrared pulses. The gray bars denote the intensity of the pulses as they are sent from each infrared-LED ($S_0,\ldots,S_5$) of robot B and received from robot A ($R_0,\ldots,R_5$). Robot A can calculate α by comparing the average intensity of the packet as it is received on its six receivers (see eq. 1). Additionally, it can calculate β by comparing the intensities of each individual pulse received (see eq. 2).

2.1.3 Actuators and Power System

The robot is equipped with two DC motors arranged as a differential drive with incremental encoder for both wheels. The wheels are fabricated as a printed circuit board in conjunction with the rest of the electronics but coated with silicon for increased friction. For moving small objects the chassis has a passive gripper with enough space for e.g. a 2×2 LEGO® DUPLO brick. Visual signaling is possible via five dimmable RGB LEDs that can be perceived by the RGB sensor up to a distance of 30 cm. Two 250 mAh LiPoly cells allow the robot to operate autonomously for up to 2.5 hours. Moreover, the robot is able to recharge itself using two sliding contacts at the bottom.

2.1.4 Interfaces, Software and Arena System

The robot features a JTAG debugging interface and a UART bus, both accessible via USB. Wanda's software is derived from the SymbricatorRTOS [13], which has been designed and implemented in the SYMBRION and REPLICATOR projects [6] and utilizes FreeRTOS [2], an open source embedded real-time operating system. All devices of the Wanda robot can be accessed using a high-level C++ API. A very generic implementation of a world model gives access to filtered and raw sensor data. A command shell allows for calibration, configuration, debugging or task control. Controllers for the robot can be written either as a custom task in C++ or as a behavior based controller using MDL2ε [14].

Setting up, conducting and evaluating experiments with a swarm of robots can be very time consuming. To improve this situation, the Wanda robot can be used in conjunction with a powerful interactive arena for swarm experiments. The system originates from the I-SWARM project [11, 4] and has been designed for the Jasmine robot [12]. It has been transferred to be used by the Wanda robot with slight improvements. The system utilizes beamers to mark special areas and to project a gray code image onto the arena which enables the robots to localize themselves using two light sensors on top of the robot. A camera at the ceiling is used to track the positions of the robots. In addition, each Wanda robot can be used as a host controller to upload programs to other robots, start and stop or setup experiments, and poll data from the other robots via ZigBee. Using this combination, we are able to automatically conduct series of experiments without the need for manual intervention.

3 The Swarm Robot Simulation Framework (SRSF)

SRSF is a new framework for the simulation of robotic swarms. Its development is still ongoing and it is planned to be released as an open-source project under general public license, soon. Its intention is to provide a simple, clear structured and highly customizable C++ library that can be used as a basis for various swarm robot simulations. One main aspect is the accurate simulation of more complex local communication systems like the one used in the Wanda robot or other types of vision based sensors. A screen-shot of the SRSF in its current state can be seen in fig. 4.

There exists a huge number of different powerful, both commercial and open-sourced robot simulators such as Stage[1], WeBots[2] or Robot3D[3] to mention just a few of them (A more complete overview can be found in [16]). Some of them are capable of simulating large numbers of robots. However these simulators often use software ray-casting for the calculation of sensor values which can be very costly in terms of computing time. When trying to obtain results of high accuracy, these simulators are often not fast enough for the simulation of large swarms. SRS uses graphics hardware to accelerate the calculation of sensor data. It aims for a compromise between an efficient and accurate simulation while still being fast enough for the simulation of reasonably large robotic swarms ($>$ 100 robots).

3.1 Basic Structure

SRSF uses Bullet [1], an open-source physics library for simulating rigid body dynamics. If it is not needed, rigid body physics can also be switched off to improve speed. Graphics are handled by OGRE [3], an open-source graphics rendering engine which is not only used for displaying the scene but also to generate sensor data. We chose these two libraries as they are to the best of our knowledge the most actively developed open-source projects in their field.

---

1 http://playerstage.sourceforge.net

2 http://www.cyberbotics.com

3 https://launchpad.net/robot3d

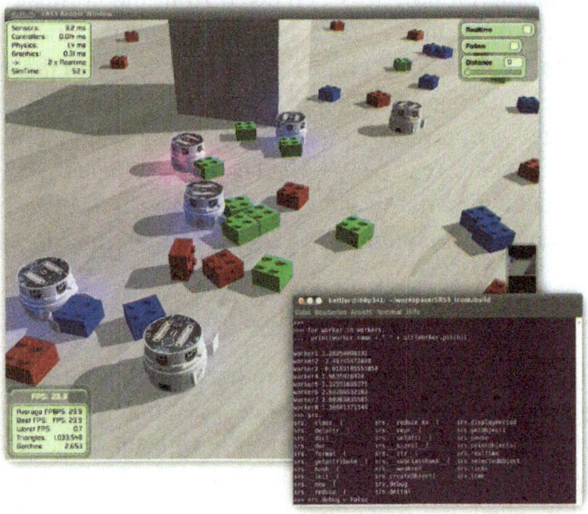

Fig. 4 Screen-shot of SRSF in its current state running a simulation of a swarm of Wanda robots pushing differently colored bricks

SRSF provides a very simple C++ API to instance combinations of various types of classes like visual objects, rigid bodies, sensors, actuators or controllers and to register them with the simulation. Collision shapes of rigid body objects and their corresponding visual objects can both be independently specified as complex OGRE mesh files or as a combination of simple geometric primitives. Additionally, SRSF features a Python interface which provides access to all objects of the simulation and to various simulation parameters. This can be used as a quick and easy way to script simulation runs with different parameters, implement controllers, or to interact with a running simulation through the console interface.

SRSF separates physical object representation from logical functionality. Each object of the simulation inherits from one or more of three device classes: sensor, controller and actuator and thereby gets registered with the global device list. A typical simulation loop consists of the sequential update of all sensors, controllers and actuators in this order. After that, one step of the rigid body dynamics is calculated and the scene graph gets updated.

3.2 Sensor Modeling

One main aspect of the framework is the efficient and accurate simulation of sensor data. When simulating a large number of robots, usually the simulation of infrared or sonar sensors makes up the biggest part of the used computing time and is usually done by ray-casting which can be quite expensive if one is interested in high fidelity. Therefore, SRSF utilizes graphics hardware to calculate sensor values for all types of sensors that can be seen as some sort of "visual" sensors, like infrared sensors, ultrasonic sensors, color or ambient light sensors. This works by maintaining multiple scene graphs, each one responsible for one specific type of visual sensors. Each visual object can have a different or no representation in each of the scene graphs depending on how it is perceived from the corresponding sensor type and on the amount of accuracy that is needed. Sensor data is then calculated by rendering the corresponding scene graph from the viewpoint of the visual sensor into an off-screen render buffer and evaluating the obtained image. Physical properties of how the object gets perceived by the sensor (like e.g. amount of reflection, angular sensitivity of receivers,...) can be specified using the OpenGL Shading Language GLSL and get evaluated directly on the graphics hardware. Therefore the amount of calculations that have to be conducted on the CPU are reduced to a minimum.

3.2.1 Simulation of Wanda's Infrared Communication System

The Wanda robot features a rather complex infrared communication and ranging system, which is implemented very accurately within the SRS framework. All properties of the emitter and receiver have been modeled in the GLSL shaders according to specifications available from their data-sheets. Each time a sensor needs to get updated, the scene gets rendered into a 64×64 pixel off-screen render buffer which corresponds to a ray-cast using 4096 rays. In order to obtain local communication data in addition to the intensity value, the color of a pixel in the rendered image contains not only an intensity but also a unique identifier encoded in the alpha value. This allows to reassign each pixel to its corresponding object in the simulation and therefore to easily determine which emitters can be perceived. This serves as a good example of how a rather complex sensor system can be easily implemented within the framework to be evaluated using graphics hardware. On a decent machine (2.66GHz Dualcore, GeForce™ 9500GT) the framework is still able to simulate rigid body physics and sensors of a number of up to 40 Wanda robots equipped with 6 infrared-sensors and 1 rgb-sensor in realtime.

4 Example Use-Cases

4.1 Aggregation

In the following, we will show a short comparison of results obtained from 20 simulation runs using SRSF with 20 experiments with real Wanda robots for an aggregation scenario. The experiments where performed almost automatically including the repositioning of the robots after each experiment. Using the optical positioning system of the arena, all robots continuously determined their global positions which where then transmitted to the host controller and logged. The program that was run on the robots during the experiment and in the simulation was given as a MDL2ε controller that was obtained using a genetic algorithm. It leads to a formation of larger clusters by using simple combinations of collision avoidance behavior and stopping depending on the number of perceived neighbors (see fig. 5).

Due to the high number of interactions between the robots, a comparison of simulation and experiment by the absolute positions of the robots is not very meaningful. Therefore, we compared the average number of the resulting clusters to quantify the

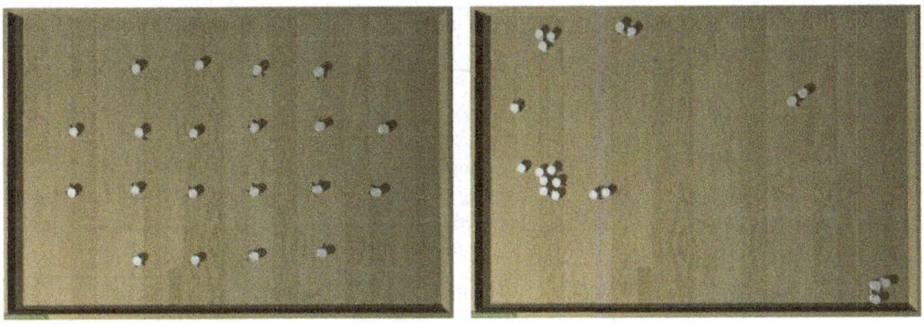

Fig. 5 The initial state (left image) and after 200 seconds (right image) as obtained from a simulation run. At the beginning, all robots are placed at a regular grid with random headings. During the experiments, the robots aggregate into clusters of different sizes.

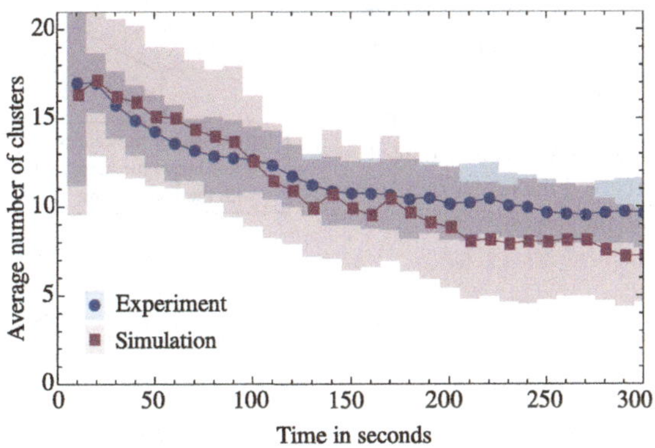

Fig. 6 Comparison of the average number of clusters over time. The pale bars denote the standard deviation with respect to the individual runs. Over time, the number of clusters decreases as more robots aggregate in fewer but bigger clusters.

correlation. As one can see from fig. 6, the results from simulations are in quite good agreement with the experiments. However, the results from simulation exhibit a slightly higher standard deviation. One explanation for this may be the strength of Wanda's differential drive which was assumed a little too high in the simulation. This enables individual robots to eventually push smaller clusters apart more easily, which leads to more fluctuations.

4.2 Collecting Boxes

The objective of the second scenario was to pick up small boxes that are initially located at random positions inside the arena and to carry them towards a designated spot. One robot sitting at this spot therefore acts as a beacon by continuously sending an infrared message of a certain type (target message) that differs from the standard message sent by all other robots. If a robot receives a target message from the beacon and is carrying a box it will drive towards the direction from where it perceived the message and increment a counter. If this counter is high enough it unloads its box,

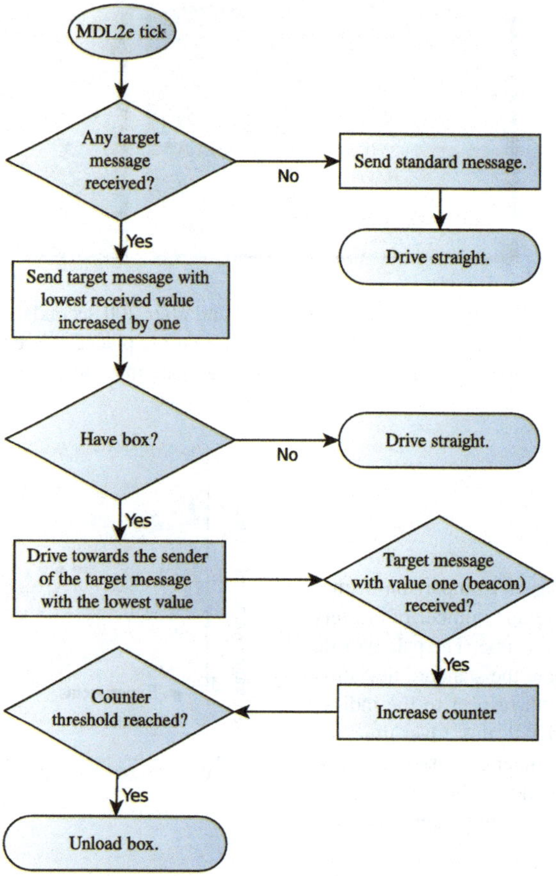

Fig. 7 Schematic illustration of the control algorithm including the hop-count mechanism used to propagate the direction of the beacon through the swarm. The beacon sends a target message with a value of one. Robots that perceive one or more target messages send a target message by themselves with a value of the lowest perceived value of all received target messages, increased by one. Robots that do not perceive any target message send a standard message. In order to find the beacon, a robot drives into the direction from where it perceived the target message with the lowest value.

turns and starts to search for the next box. A robot searching for a box just drives straight and avoids obstacles if needed.

In order to increase the efficency, in a second setup the direction towards the beacon was propagated through the swarm by utilizing a simple hop-count mechanism. A schematic illustration of the algorithm can be seen in figure 7.

The controller for the robots was developped manually using the simulation framework. The program was then transferred to the real robots without further modifications.

Figure 8 shows an example for the initial positions of the robots and the boxes inside the arena. Figure 9 shows the state of the experiment after 7 minutes for the two different setups.

A quantitative comparison of the average result obtained from 10 experimental runs for each setup versus the results obtained from 50 simulation runs can be seen in figure 10.

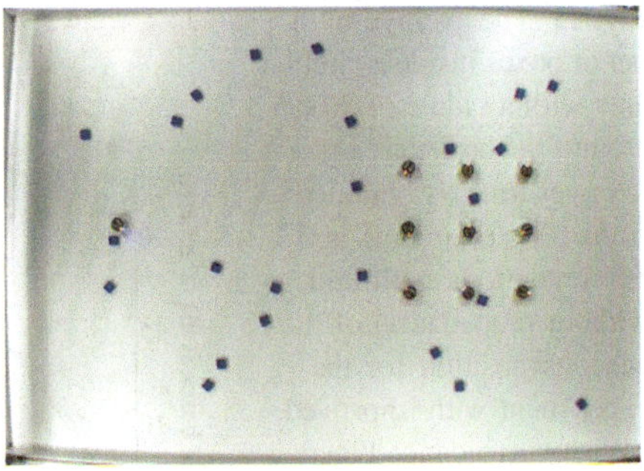

Fig. 8 Example configuration for the initial positions of the robots and the boxes at the beginning of the experiment. The robot on the left side plays the role of the beacon. All robots are starting from fixed positions in the right half of the arena.

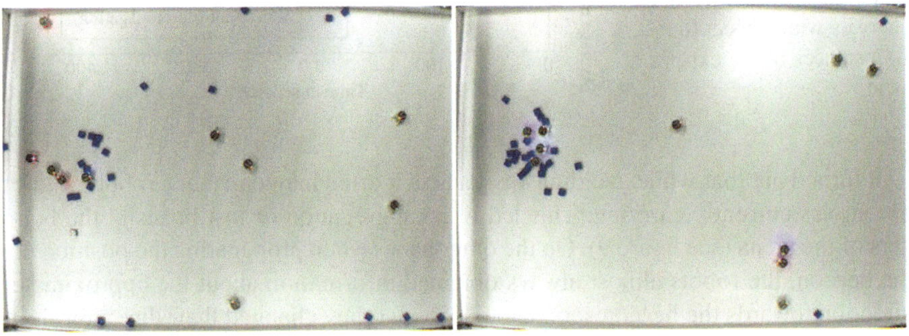

Fig. 9 Examples for the state at the end of the collecting experiment. The left picture was taken from an experiment without propagating the information about the beacon, the right one with propagating.

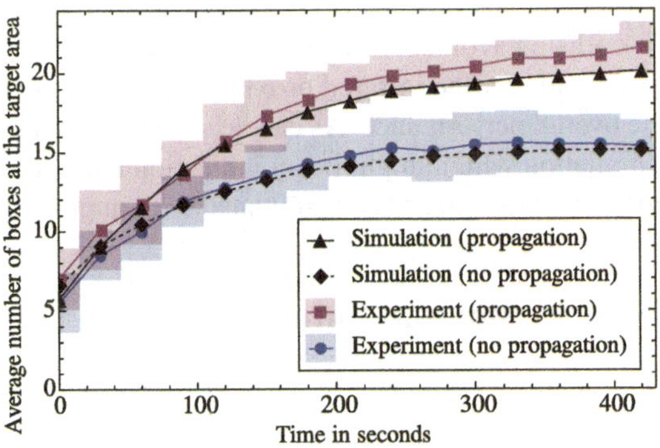

Fig. 10 The average number of boxes inside a circular area with a radius of 50 cm around the target area plotted over time as obtained from 10 experiments and 50 simulations with and without propagation of the beacons position. The pale bars denote the 95% confidence intervals with respect to the 10 instances of the experiment.

As can be seen, the effect of propagating the beacons position strongly increases the efficiency. What can also be observed is that the final number of collected boxes lies below the total number of all boxes (which was 24), especially in the case where the position of the beacon was not propagated through the swarm.

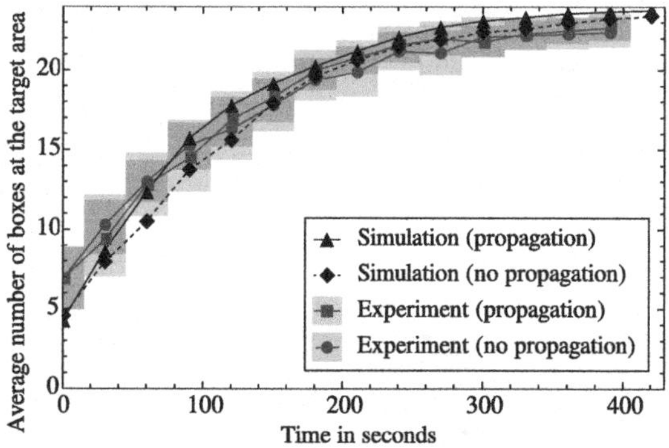

Fig. 11 The average number of boxes inside a circular area with a radius of 50 cm around the target area plotted over time as obtained from 10 experiments and 50 simulations with and without propagation of the beacons position for the experiment with improved collision avoidance behavior. The pale bars again denote the 95% confidence intervals with respect to the 10 instances of the experiment.

It turned out that while avoiding an obstacle it often happend that a robot lost the box it was currently carrying. This led to an aggregation of lost boxes at the borders of the arena (see figure 9). On the other hand, when propagating the position of the beacon, the robots almost always obtained information about the approximate direction towards the beacon soon enough so that they changed their directions before they reached the borders of the arena. Therefore the number of lost boxes at the arena boundaries is much lower in this case which leads to a much improved performance.

We conducted the same experiments a second time but this time with an improved collision avoidance behavior. Figure 11 shows the results. As can be seen, the effect of propagating the position of the beacon is no longer noticeable. The improvement in the collision avoidance behavior was obtained mainly due to a slight variation of the thresholds for the infrared proximity values which nevertheless had a big effect on the overall performance. As can be seen this change was in good agreement with the simulation which might be an indication that the communication system of the wanda robot has indeed been simulated with quite good accuracy.

5 Conclusions and Future Work

For the future development of new swarm robot systems, it is essential to provide fast, accurate and reliable predictions of their behavior. The work presented in this paper aims to improve this by providing tools that may allow for the systematic identification of critical parameters in swarm robot simulations. All experiments conducted so far have shown an excellent usability of the automated arena system. The new swarm robot simulation framework promises to be applicable to a wide area of swarm robotic scenarios due to its flexible and open structure. Future work will now concentrate on methods for the automatic fitting of simulation parameters to data from experiments to further increase the accuracy of the simulation and to simplify the adaptation of the simulation to new types of robots and sensors.

Acknowledgements. The authors are supported by the DFG Research Training Group 1194: "Self-organizing Sensor-Actuator-Networks" and the SYMBRION and REPLICATOR projects. REPLICATOR and SYMBRION projects are funded by European Commission within the 7th framework program. The Wanda robot has been funded by the KIT Collective Robotics Lab (KITCoRoL).

References

1. Bullet physics library, `http://www.bulletphysics.org`
2. Freertos - the standard solution for small embedded systems, `http://www.freertos.org`
3. Ogre: Object-oriented graphics rendering engine, `http://www.ogre3d.org/`
4. Boletis, A., Brunete, A., Driesen, W., Breguet, J.M.: Solar Cell Powering with Integrated Global Positioning System for mm3 Size Robots. In: Proceedings of the 2006 IEEE/RSJ Int. Conf. on Intelligent Robots and Systems (IROS 2006), pp. 5528–5533 (2006)
5. Caprari, G., Estier, T., Siegwart, R.: Fascination of Down Scaling - Alice the Sugar Cube Robot. Journal of Micro-Mechatronics 1(3), 177–189 (2002)
6. Kernbach, S., Meister, E., Schlachter, F., Jebens, K., Szymanski, M., Liedke, J., Laneri, D., Winkler, L., Schmickl, T., Thenius, R., Corradi, P., Ricotti, L.: Symbiotic robot organisms: Replicator and symbrion projects. In: PerMIS 2008 (2008)
7. Kettler, A., Szymanski, M., Liedke, J., Wörn, H.: Introducing wanda - a new robot for research, education, and arts. In: 2010 IEEE/RSJ International Conference on Intelligent Robots and Systems (IROS), pp. 4181–4186 (2010), doi:10.1109/IROS.2010.5649564
8. Winkler, L., Kettler, A., Szymanski, M., Wörn, H.: The Robot Formation Language - A Formal Descriptions of Formations for Collective Robots. In: Proceedings of IEEE Symposium on Swarm Intelligence 2011 (SIS 2011), pp. 102–109 (2011)
9. Mondada, F., Bonani, M., Raemy, X., Pugh, J., Cianci, C., Klaptocz, A., Magnenat, S., Zufferey, J.C., Floreano, D., Martinoli, A.: The e-puck, a Robot Designed for Education in Engineering. In: Gonalves, P., Torres, P., Alves, C. (eds.) Proceedings of the 9th Conference on Autonomous Robot Systems and Competitions, Portugal, vol. 1, pp. 59–65 (2009) ISBN 978-972-99143-8-6
10. Mondada, F., Franzi, E., Guignard, A.: The Development of Khepera. In: Experiments with the Mini-Robot Khepera, pp. 7–14. HNI-Verlagsschriftenreihe, Heinz Nixdorf Institut. (1999)
11. Seyfried, J., Szymanski, M., Bender, N., Estana, R., Thiel, M., Wörn, H.: The I-SWARM project: Intelligent Small World Autonomous Robots for Micro-manipulation. In: From Animals to Animats 8: The Eighth International Conference on the Simulation of Adaptive Behavior (SAB 2004) Workshop on Swarm Robotics (2004)
12. Szymanski, M., Kernbach, S.: Jasmine - open source swarm robot (2007), `http://www.swarmrobot.org`

13. Szymanski, M., Winkler, L., Laneri, D., Schlachter, F., Van Rossum, A.C., Schmickl, T., Thenius, R.: Symbricatorrtos: a flexible and dynamic framework for bio-inspired robot control systems and evolution. In: CEC 2009: Proceedings of the Eleventh Conference on Congress on Evolutionary Computation, pp. 3314–3321. IEEE Press, Piscataway (2009)
14. Szymanski, M., Wörn, H.: JaMOS - A MDL2ε based Operating System for Swarm Micro Robotics. In: IEEE Swarm Intelligence Symposium, pp. 324–331 (2007)
15. Texas Instruments, Austin, TX, USA: Stellaris®LM3S1960 Microcontroller - Data Sheet (2009)
16. Vaughan, R.: Massively multi-robot simulation in stage. Swarm Intelligence 2(2), 189–208 (2008), doi:10.1007/s11721-008-0014-4

Multi-Robot System Validation: From Simulation to Prototyping with Mini Robots in the Teleworkbench

Andry Tanoto, Felix Werner, and Ulrich Rückert

Abstract. One challenging aspect in the development of multi-robot systems is their validation in a real environment. However, experiments with real robots are considerably tedious as experimenting is repetitive and consists of several steps: setup, execution, data logging, monitoring, and analysis. Moreover, experiments also require many resources especially in the case when involving many robots. This paper describes the role of the Teleworkbench as a platform for conducting experiments involving mini robots. The Teleworkbench offers functionality that can help users in validating their robot software from simulation to prototyping using mini robots. A traffic management system is used as a scenario for demonstrating the support of the Teleworkbench for validating multi-robot systems.

1 Introduction

There are several problems in developing software for intelligent autonomous robots and multi-robot systems that must be solved for robots to become viable commercial products. For example, it is arguably impossible to specify the complete system requirements in advance because the variety of application scenarios is seemingly unlimited and the undefined nature of a task means that an operator or a robot itself needs to be able to fine tune the task specifications during operation. Thus, robot software development is not yet well defined and established as a structured formal process.

On the contrary, the process of developing large software systems is well understood and several models such as the Spiral Model [5] or the V-Model [15] for such

Andry Tanoto
Heinz Nixdorf Institute, University of Paderborn, Fürstenallee 11, Paderborn, Germany
e-mail: tanoto@hni.upb.de

Felix Werner · Ulrich Rückert
Cognitronics and Sensor Systems Group, Cognitive Interaction Technology Centre of Excellence (CITEC) Bielefeld University, 33615 Bielefeld, Germany
e-mail: {fwerner,rueckert}@cit-ec.uni-bielefeld.de

processes exist, each describes approaches to a variety of tasks or activities that take place during the process. A particular activity that is part of all models is software validation. Validation is the process of checking that a software system is compliant with the specified requirements.

In general, there are two groups of validation methods: formal and informal validation. Formal validation methods, such as model checking, suffer from state space explosion which occurs because of the difficulty in formalizing real-world scenarios and is almost impractical for multi-robot systems [12]. Hence, informal validation methods, such as simulation or testing, are used widely in robot software development. However, results from simulation cannot guarantee the completeness of the validation, as it is difficult to model the physical environment of a robot, interaction or environment dynamics correctly for testing of all possible states a robot system can be in.

The most common method to validate robot system is testing in the desired application domain. However, conducting experiments has its own drawbacks. Experimenting is considerably tedious, especially when many robots are involved. It is also resource intensive and consists of several steps: setup, execution, data logging, monitoring, and analysis. Moreover, the experiments must be repeatable and reproducible to ensure the validity of the results.

To help tackling these issues in simulation and experimentation, prototyping is commonly employed in software development process. Prototyping can provide a way to prove a concept, to gather information through experimentation, or to identify and to validate requirements which are otherwise difficult to define or specify. Additionally, prototyping can also be used for providing a basis for the system development, e.g. in determining software architecture and algorithms. Prototypes can be applied in different aspects of robotics: appearance [14], mechanical parts [1], electronics [4], software [13, 17, 21], and interaction [7].

In this paper, we demonstrate the assistance of the Teleworkbench [19, 18, 20] in developing and validating software of multiple robots used for prototyping a complex traffic management system. Several aspects of such systems including communication, cooperation and path planning can be validated in the development process using mini robots in real environments.

The Teleworkbench (TWB) offers a standardized environment in which geographically distributed users can test, validate and compare their algorithms and programs using real robots. Moreover, the TWB provides functionality which assist researchers and developers in several aspects: (i) integration with a robot simulator, (ii) remote-download of user-designed robot programs, (iii) automatic environment building, (iv) information logging, (v) position tracking of up to sixty-four robots, and (vi) a visualization tool for experiment analysis. As experiments run in a standardized environment, researchers can reproduce and compare the results of the experiments.

The remainder of this paper is organized as follows: In the next section, we present a short literature review of prototyping in robotics. In Section 3 we present an overview of the Teleworkbench. Section 4 explains the role of the Teleworkbench in bridging the two validation processes and describes the scenario used to demonstrate the functionality. The paper is concluded in Section 5 and future work of the Teleworkbench is discussed.

2 Prototyping in Robotics

Robot development is a multi-disciplinary effort which requires cooperation from different domains. Glesner et al. [9] present some possible problems faced in the development of a complex mechatronic systems such as robots: (i) difficulty in describing formally different concepts in different domains, (ii) in turn, it is very difficult to model and thus to simulate such as a system, and (iii) difficulty in defining requirements and specification in an ambiguous manner for engineers from different domains. One possible way to solve the aforementioned problems is the application of prototyping.

In software engineering, a prototype can have different functions. It can be used as a proof of concept, for information gathering through experimentation, or for identification and validation of requirements which are otherwise difficult to define or specify. Additionally, a prototype can be used as well for providing a basis for the system development, e.g. in determining software architecture and algorithms.

In robotics, prototypes can be applied in different aspects of robot system: appearance [14], mechanical parts [1], electronics [4], software [13, 17, 21], and interaction [7]. Bartney and Hu give an overview of rapid prototyping for building interactive robots [3]. They argue that robotics faces similar problems in software engineering and accordingly can take advantages of rapid prototyping for building robotic systems. They also point out the difference between robot and software prototyping due to the fact that robotics deals not only with the software but also with diverse types of components, physical existence, and also physical appearance and behavior. Another difference is the lifetime of the prototypes. In robotics, a prototype is usually intended to quickly elicit the user requirements in the early development stage, and will be no longer used in the next stage of development, e.g. industrial reproduction. Shen et al. present a virtual prototyping system that can be used by a team of developers from different domains in the development of mechatronic systems [16]. They argue that the multi-disciplinary aspect of mechatronic systems can lead to communication problems among engineers from different domains. Zalzal et al. present an open source middleware robotic framework called *Acropolis* that can be used for fast software prototyping [21]. The framework supports the reuse of program codes and extensibility through extension modules called plugins. A robot behavior can be designed by interconnecting the needed modules in a graphical user interface. Smuda describes a methodology for automatic generation of software wrappers to simplify prototyping and development of robotic systems [17]. The role of software wrappers is to enable insertion of modules into a visual prototyping environment. The methodology promises reduced work and consistent behavior of module interfaces during the system development.

In comparison to the other works, the Teleworkbench offers prototyping for multi-robot systems. In multi-robot systems, some aspects, such as communication,

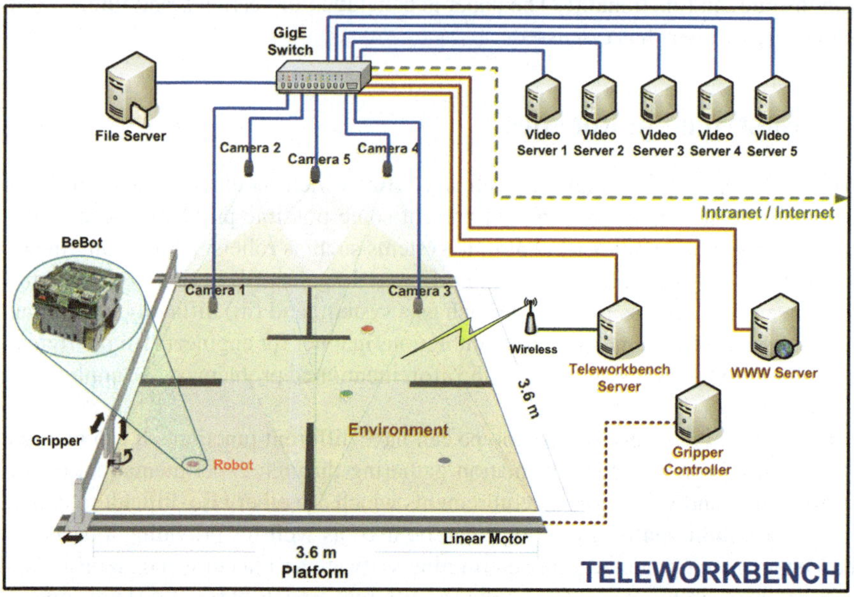

Fig. 1 The diagram showing the general system architecture of the Teleworkbench system

cooperation, path planning, etc., can early be developed and validated on mini robots before they are implemented on the actual hardware platform. Additionally, the environment where the experiments take place can be rebuilt and modified during runtime so that different algorithms or hardware can be tested and compared with each other to explore the design requirements and specifications. As the Teleworkbench supports interoperability with Stage robot simulator [8, 6], code portability can be achieved. Hence, development and deployment of multi-robot systems can be made easier.

3 The Teleworkbench

The distributed system architecture of the Teleworkbench is shown in Figure 1. Earlier papers [18, 20] describe the Teleworkbench in more details. In this paper, we will briefly describe the system and its main components.

The Teleworkbench comprises a main experiment field of 3.6×3.6m that is partitionable into four sub-fields, each of which can independently be used for an experiment. A gripper module with four degrees of freedom (3 translational and one rotational) enables automatic environment building by using plastic blocks. Additionally, it can also place robots at predefined locations and orientations. Three different robotic platforms are currently used on the Teleworkbench: *Khepera II*, *Khepera III* [11], and the *BeBot* [10]. Five 1-megapixel Gigabit-Ethernet cameras are mounted above the experiment field, four of which are assigned to the sub-fields and the other one monitors the entire experiment field. Each camera is connected

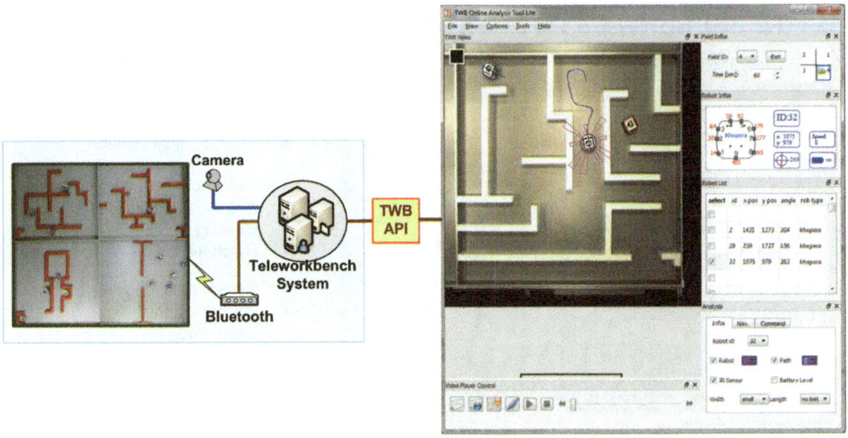

Fig. 2 The GUI of the online analysis tool. The API is used to communicate with the Teleworkbench as well as with the robots.

to a video server that processes the video data to provide the GPS-like position and orientation information of the robots as well as to record and stream the video. Currently, up to 64 robots can be identified and tracked by the help of barcode-like markers that are placed on top of the robots. One server is responsible for scheduling, queuing and execution of experiments. Additionally, the server handles messages among robots that are transmitted wirelessly, e.g. with Bluetooth or WLAN. Another server is assigned for intermediating users and the Teleworkbench. A website is provided to support users in performing different activities, e.g. setup and execution of an experiment, live-monitoring and data acquisition during the experiment. A file server is deployed to store all data that accumulates during experiments.

Additionally, an application programming interface (API) is provided to support interoperability with other systems and to help users in developing programs interacting with the robots or the Teleworkbench System. As an example, the API can be used by a graphical user interface, as shown in Figure 2, for monitoring and analyzing the experiments.

4 Validating Multi-Robot System with the Teleworkbench

This section presents the use of the Teleworkbench in providing a tool for validating a multi-robot system. A simplified robot software development process is depicted in Figure 3.

The Teleworkbench aims to provide a seamless transition from simulation to experiments with real robots. Users first login to the Teleworkbench website and setup experiments by specifying parameters such as the model of the environment, the experiment duration, number of robots, robot programs, and the initial positions. The same environment model that is used in the simulator can be used in the experiment

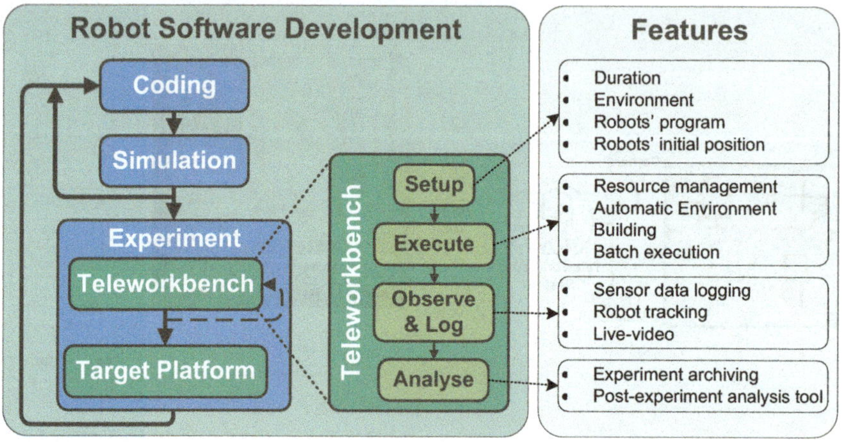

Fig. 3 The process flow of the robot software validation using the Teleworkbench

with mini robots. When the experiment is set and ready, the defined environment model is realized by using plastic blocks arranged by the gripper module. Afterwards, the uploaded programs are deployed and executed. There are two possible deployment platforms for the robot programs: PCs or robots. On PCs, the robot programs control the robots over the wireless communication channel. For programs running on the robot, the code must be firstly compiled with the robot-platform-specific cross-compiler before it is copied to and executed on the robot.

During experiments, the communicated messages among agents are logged and can be retrieved after the end of the experiment. At the same time, users can also observe the experiment using the developed graphical user interface (GUI) that can display the streamed live-video overlaid by some robot information such as robot symbol, robot path, sensor information, and exchanged messages (see Figure 2).

4.1 Demo Scenario: Traffic Management System

The scenario used in the demo is a traffic management system (see Figure 4). The goal of such a system is to enhance the transport efficiency and the traffic safety. In this scenario, vehicles posses a certain level of autonomy and are able to follow the route autonomously. Moreover, they are independent of each other and do not exchange any message in performing their tasks. The main objective of the system is to manage these autonomous agents so that they can move along their route in a fast and safe manner. We identify four functional requirements of the system: (i) *a traffic light controller must be able to manage one crossing correctly without causing any collision among vehicles*, (ii) *every vehicle has to obey the traffic lights*, (iii) *a vehicle can always avoid collision with the other vehicle in front of it or any other object (wall)*, and (iv) *the vehicles must be able to follow the route properly.*

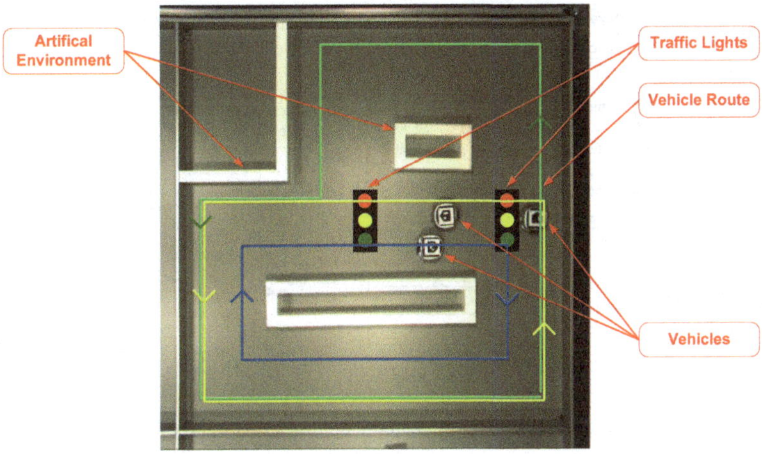

Fig. 4 The traffic management scenario. Some vehicles follow their specified routes. In some places, they have to cross intersections which are managed by traffic light controllers.

The system is composed of many agents with different roles: *Trafficlight Controller* (TC), *Blackboard* (BB), and *Robot Controller* (RC) (see Figure 5). The TC agent is responsible to control a set of traffic lights at one location that requires traffic management, namely a *crossing*. In the current scenario, only one direction at a crossing can have a green light and there is no communication among TC agents. The TC agents update the status of all traffic lights through a *topic* managed by the BB agent in a *publish-subscribe* fashion. Any agent which needs the status of a specific traffic light can subscribe to that particular topic. The RC agent is responsible for controlling a vehicle. Each vehicle has its specific route that may go through one or more crossings. The RC agent periodically receives the position of the vehicle and inquires the BB agent for the status of the nearby traffic light depending on the route. Accordingly, it controls the vehicle to follow the traffic rules, e.g. stop if the traffic light turns red or continue following the route when it turns green.

4.2 Implementation

Each agent described earlier is implemented as an independent component that runs on its own process. They are developed using the Player/Stage framework [8] and written in C++. They can be flexibly deployed: all in one machine or distributively across several machines. As the vehicles, Khepera III mini robots are employed. Each robot is equipped with a *Korebot* extension module with embedded Linux operating system and *Player Server* running on it.

Two crossings are defined (see Figure 4) in the environment. One TC agent is deployed to control each crossroad. At each crossing, a vehicle can go in six different directions. For each direction, one traffic light is assigned. Thus, in total one TC agent manages six traffic lights.

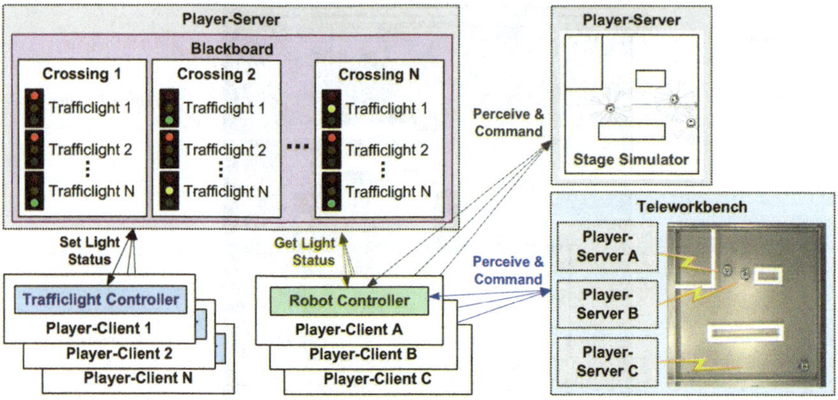

Fig. 5 The block diagram of the traffic management system validated both on the simulator and the Teleworkbench

Three Khepera III mini robots are used in the scenario. Accordingly three homogeneous RC agents are deployed, each uses a route different from each other. The routes are defined as a set of positive points that are within the area of the defined environment. The RC agent has three main behaviors: *follow the route*, *obey the traffic lights*, and *collision avoidance*. These behaviors are implemented as a schema based on the schema-based architecture [2]. Figure 6 shows the architecture of the RC agent. One of the unique features of the architecture is that it is component based which supports the exchangeability, extensibility, expandibility, or reusability of the system components.

4.3 Test

The aim of validating the traffic management system is to ensure that it meets its requirements. Besides validating the system, we report (i) *waiting time, which is the time spent by a robot at the crossing*, (ii) *the crossing frequency*, (iii) *the collision frequency*, and (iv) *the vehicle average speed* to demonstrate the system performance.

The first step of system validation is done through simulation using the Stage simulator. The environment is an area of size 1.8m$\times$1.8m that is defined and modeled as a bitmap loadable by the simulator. The Trafficlight Controller agents are deployed on a PC and work independently from each other. They broadcast the states of each traffic light to the Blackboard agent that runs on the same machine as the simulator. The Robot Controller agents are executed on a PC. They control the virtual robots running in the Stage simulator. During the simulation, some system parameters are empirically adjusted, e.g. the route by taking into consideration the robot and the road size, the distance threshold for the robot to stop when it detects another robot ahead, and the time for changing a traffic light from one state to the

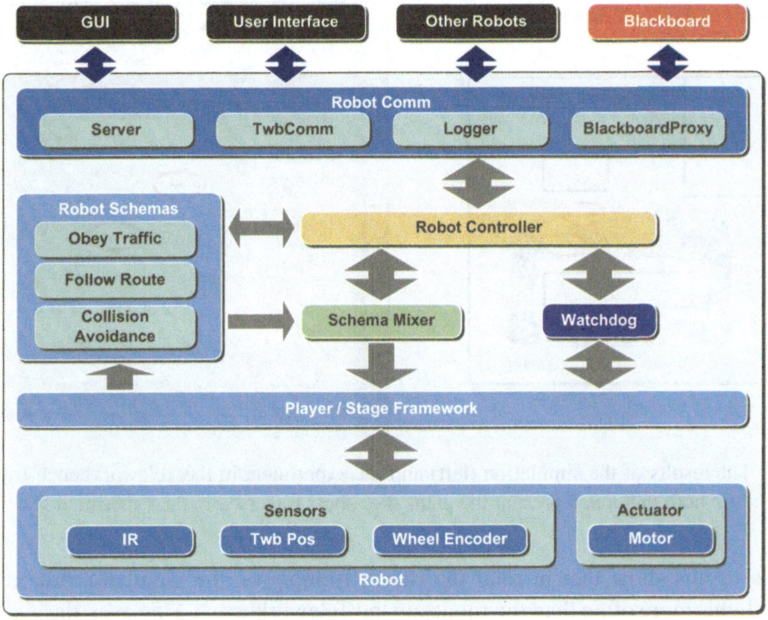

Fig. 6 The software architecture of the robot controller

other. Figure 7 shows a snapshot of the Stage simulation environment with the paths of the robots overlaid on it.

In the next step, we change the test environment from the Stage simulator to the Teleworkbench. The environment for the experiment is firstly built using the gripper module. The input of the module is the same bitmap file used in the simulator. Figure 8 shows a snapshot of the environment building. In the tests we undertake, the Robot Controller agents are again executed on a PC. However, in this case they are set to control the real Khepera III mini robots. We use the same parameters as used in the simulator.

We run the system three times on simulator and three times on the Teleworkbench to measure the performance of the system. Each simulation and experiment is set to run in a specific runtime (ten to twelve minutes). Figure 7 shows the paths traversed by the robots in two different contexts, i.e. the simulation and experiment with real robots. From the figures we can see that the robots can follow the specified routes without colliding to the walls. Four performance metrics are used: *waiting frequency*, *waiting time*, *number of collision*, and *average speed*. Figure 9 shows the system performance which is extracted from the log file. The waiting frequency measures how often the robots stop at the crossing. The waiting time measures how long the robots stop at the crossing. The number of collision represents the number of times the robots stop to avoid collision with other robots. The average speed is the total distance the robots travels divided by the runtime, which can be used to represent the flow rate of the traffic management system.

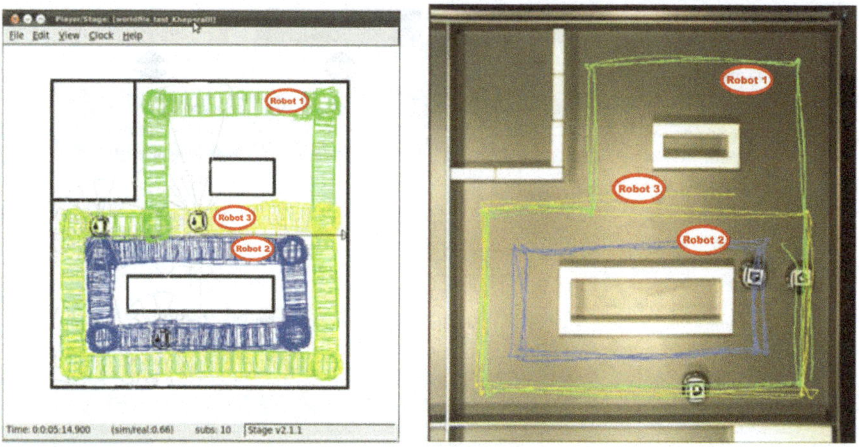

Fig. 7 The results of the simulation (left) and the experiment in the Teleworkbench (right). The lines in both pictures represent the paths the robots traversed in the experiments.

The results show that in total the robots running on the simulator stop at the traffic lights more often than the robots on the Teleworkbench. However, the average time spent at the traffic light is lower in the case of simulation compared to the real experiment although the traffic light controllers are the same. In general, the robots on the simulator run faster than the ones on the Teleworkbench although the Robot Controller agents are set with the same parameters. This can be explained by the difference between the real Khepera III mini robot and its model on the Stage simulator, in terms of actual movement speed given an input speed value. The last graph of Figure 9 shows that two robots (Robot 1 and Robot 3) that are running on the simulator stop due to collision avoidance more often than the ones on the Teleworkbench.

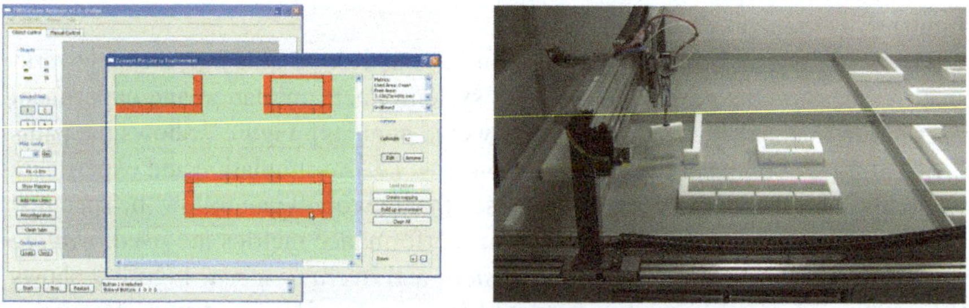

Fig. 8 The gripper module building the environment based on the bitmap files used in the simulator. On the left is mapping done by the gripper software to match the available plastic blocks with the input bitmap file. The right picture shows a snapshot of the gripper in action.

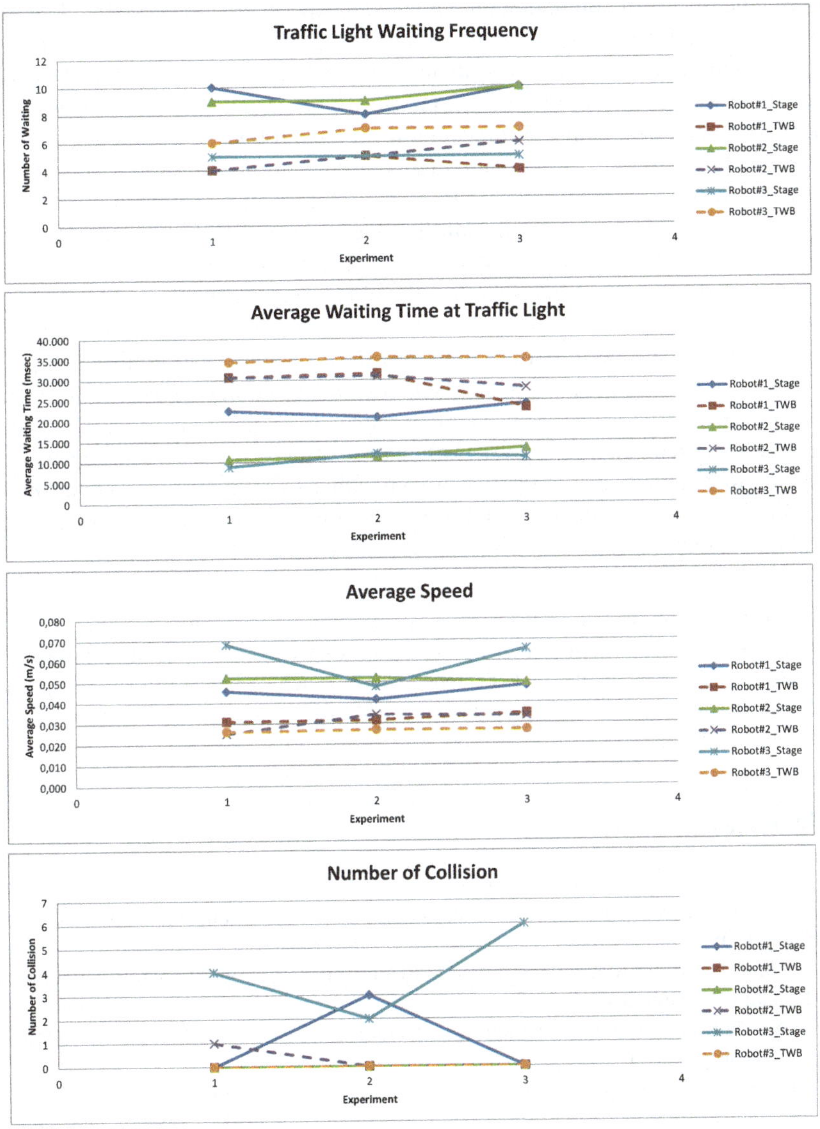

Fig. 9 The system performance based on the results of three runs using the Stage robot simulator and the Teleworkbench

5 Discussion

This paper demonstrates the functionality of the Teleworkbench in multi-robot system validation by bridging the two validation processes, i.e. simulation and experimentation using a system prototype. The presented results show discrepancies between the results of simulation on the PC and experimentation on the Teleworkbench, which is mainly due to the difference between the robot model used in the simulator and the actual robot. Nevertheless, simulation is still important in the early stage of the software development as it provides a faster and less tedious validation process in comparison with experimentation with real robots.

Seamless transition from simulation to experimentation with mini robots on the Teleworkbench by using the Player/Stage framework is also presented. The seamless transition can be necessary in fulfilling the repeatability requirement of robot software validation. Another advantageous feature of the Teleworkbench is batch experiment execution that offers the functionality to automatically conduct experiment in batch, which is enabled by the automatic environment building capability.

We have noticed one important issue based on our observation during experimentation on the Teleworkbench, i.e. the *automatic experiment termination*. At the moment, experiments are terminated either manually by the users or automatically when the specified runtime is over. Although it suffices in many cases, it is also necessary to be able to detect failures or some other specific conditions that can be used for experiment termination. In this way, the total experiment time and the user workload for experiment monitoring can be reduced.

Acknowledgements. This work has been supported within the Excellence Initiative of the German Research Foundation (DFG) by the Center of Excellence "Cognitive Interaction Technology" (EXC277), by the DFG Colloborative Research Center "Self-Optimizing concepts and structures in mechanical engineering" (SFB614), and by the DFG grant (INST 214/47-1 FUGG).

References

1. Ambrose, R.: Interactive robot joint design, analysis and prototyping. In: Proceedings of 1995 IEEE International Conference on Robotics and Automation, vol. 2, pp. 2119–2124 (May 1995)
2. Arkin, R.C.: Motor schema-based mobile robot navigation. The International Journal of Robotics Research 8(4), 92–112 (1989)
3. Bartneck, C., Hu, J.: Rapid prototyping for interactive robots. In: Information Assurance and Security (2004)
4. Birk, A.: Fast robot prototyping with the cubesystem. In: IEEE International Conference on Robotics and Automation 2004 Proceedings. ICRA 2004, April-1, May, vol. 5, pp. 5177–5182 (2004)
5. Boehm, B.: A spiral model of software development and enhancement. SIGSOFT Softw. Eng. Notes 11, 14–24 (1986)
6. Collett, T.H., MacDonald, B.A., Gerkey, B.P.: Player 2.0: Toward a practical robot programming framework. In: Proc. of the Australasian Conf. on Robotics and Automation, ACRA (2005)
7. Dudenhoeffer, D.D., Bruemmer, D.J., Davis, M.L.: Modeling and simulation for exploring human-robot team interaction requirements. In: Proceedings of the 33nd Conference on Winter Simulation, WSC 2001, pp. 730–739. IEEE Computer Society, Washington, DC, USA (2001)

8. Gerkey, B.P., Vaughan, R.T., Howard, A.: The Player/Stage Project: Tools for Multi-Robot and Distributed Sensor Systems. In: Proc. of the ICAR 2003, pp. 317–323 (2003)
9. Glesner, M., Kirschbaum, A., Renner, F.-M., Voss, B.: State-of-the-art in rapid prototyping for mechatronic systems. Mechatronics 12(8), 987–998 (2002)
10. Herbrechtsmeier, S., Witkowski, U., Rückert, U.: BeBot: A modular mobile miniature robot platform supporting hardware reconfiguration and multi-standard communication. In: Kim, J.-H., Ge, S.S., Vadakkepat, P., Jesse, N., Al Manum, A., Puthusserypady, S., Rückert, U., Sitte, J., Witkowski, U., Nakatsu, R., Braunl, T., Baltes, J., Anderson, J., Wong, C.-C., Verner, I., Ahlgren, D. (eds.) Progress in Robotics. CCIS, vol. 44, pp. 346–356. Springer, Heidelberg (2009)
11. K-Team Corp. Khepera III (October 2010)
12. Karlsson, D., Eles, P., Peng, Z.: Model validation for embedded systems using formal method-aided simulation. Computers Digital Techniques, IET 2(6), 413–433 (2008)
13. Kenn, H., Carpin, S., Pfingsthorn, M., Hepes, B., Ciocov, C., Birk, A.: FAST-Robots: a rapid-prototyping framework for intelligent mobile robotics. In: Artificial Intelligence and Applications Conference (2003)
14. Kim, W.-S.: Advanced kinematic cardboard prototyping for robot development. In: IASDR 2009 International Association of Societies of Design Research, pp. 3075–3084. Seoul National University of Technology, Department of Industrial Design (2009)
15. Pressman, R.S.: Software engineering: a practitioner's approach, 2nd edn. McGraw-Hill, Inc., New York (1986)
16. Shen, Q., Gausemeier, J., Bauch, J., Radkowski, R.: A cooperative virtual prototyping system for mechatronic solution elements based assembly. Adv. Eng. Inform. 19, 169–177 (2005)
17. Smuda, B.: Software wrappers for rapid prototyping jaus-based systems. In: SPIE, vol. 5804, pp. 718–726 (2005)
18. Tanoto, A., Rückert, U., Witkowski, U.: Teleworkbench: A teleoperated platform for experiments in multi-robotics. In: Web-Based Control and Robotics Education, vol. 38, ch. 12, pp. 287–316. Springer (2009)
19. Tanoto, A., Witkowski, U., Rückert, U.: Teleworkbench: A Teleoperated Platform for Multi-Robot Experiments. In: Murase, K., Sekiyama, K., Kubota, N., Naniwa, T., Sitte, J. (eds.) Proc. of the 3rd International Symposium on Autonomous Minirobots for Research and Edutainment (AMiRE 2005), pp. 49–54 (September 2005)
20. Werner, F., Rückert, U., Tanoto, A., Welzel, J.: The Teleworkbench - a platform for performing and comparing experiments in robot navigation. In: Proc. of the Workshop on The Role of Experiments in Robotics Research, ICRA 2010 (May 2010)
21. Zalzal, V., Gava, R., Kelouwani, S., Cohen, P.: Acropolis: A fast protoyping robotic application. International Journal of Advanced Robotic Systems 6(1), 1–6 (2009) ISSN: 1729-8806

An Experimental Testbed for Robotic Network Applications

Donato Di Paola, Annalisa Milella, and Grazia Cicirelli

Abstract. In the last few years, multi-robot systems have augmented their complexity, due to the increased potential of novel sensors and actuators, and in order to satisfy the requirements of the applications they are involved into. For the development and testing of networked robotic systems, experimental testbeds are fundamental in order to verify the effectiveness of robot control methods in real contexts. In this paper, we present our Networked Robot Arena (NRA), which is a software/hardware framework for experimental testing of control and cooperation algorithms in the field of multi-robot systems. The main objective is to provide a user-friendly and flexible testbed that allows researchers and students to easily test their projects in a real-world multi-robot environment. We describe the software and hardware architecture of the NRA system and present an example of multi-mission control of a network of robots to demonstrate the effectiveness of the proposed framework.

1 Introduction

In these last years, multi-robot systems are being widely investigated, since they offer better performances than single robot systems in challenging applications, such as exploration of hostile environments, terrain mapping, space and rescue operations [1],[2],[3],[4].

Due to the intrinsic difficulty in developing and assessing the performances of multi-agent control algorithms in real contexts, much work has been carried out only analytically or in simulation [5],[6],[7]. Advanced simulation tools for multi-robot systems have been also developed, and are available as open source tools (e.g., Stage and Gazebo) or as commercial products (e.g., Webots, Microsoft Robotic Studio etc.).

Donato Di Paola · Annalisa Milella · Grazia Cicirelli
Institute of Intelligent Systems for Automation (ISSIA), National Research Council (CNR), via G.Amendola 122/D, 70126 Bari, Italy
e-mail: `{dipaola,milella,grace}@ba.issia.cnr.it`

Nevertheless, for complete analysis and testing, and to provide additional insights to theory, real world experiments are generally desirable. On the other hand, experimentation in robotics for comparison of different methods is often difficult due to the variety of robotic platforms and environments that may be encountered.

Therefore, the development of experimental testbeds is a key issue for the design of effective multi-robot control systems. These testbeds can support all stages of the design process, including algorithm development and testing, conventional simulation, robot-in-the-loop simulation, and real robot control, in an integrated way. An example of platform for conducting, evaluating and analyzing experiments in robotics, named the Teleworkbench, can be found in [8].

In this paper, we describe our experimental set-up, referred to as Networked Robot Arena (NRA) that allows us to monitor, control, and test realistic behaviours of relatively simple multi-robot systems. This set-up includes both hardware and software components. The main contribution of the work is the development of a low cost, robust, flexible, and scalable software/hardware system, which uses desktop mobile robots with a vision-based identification and localization system, and a multi-mission control framework. The control architecture is a hybrid distributed architecture in which the high-level multi-mission control is performed by a software module with a number of controllers, each one in charge of managing a sub-network, while the low-level operations are implemented on board each robot.

The proposed framework can serve as a "middle ground" between a totally software-based simulation tool and a full-scale real-world implementation. From the one hand, it allows one to encompass real-world problems difficult to model and manage in a virtual environment, such as errors in robot identification and localization, or interactions with other objects. On the other hand, it is based on user-friendly and flexible tools, which can be used by both researchers and students to easily test their projects.

2 Overview of the Networked Robot Arena

In this section, we provide an overview of the NRA. The main goal of the proposed system is to set up a hardware and software framework to assist the development and testing process of robotic network applications for educational and research purposes.

The NRA has the following characteristics, which are typically desired for education and research testbeds: open software and hardware platform; desktop mobile robots; low cost; robust, flexible and scalable software; user friendly software development and maintenance. In order to satisfy these requirements, the NRA features the logical structure shown in Figure 1. The main components of the system are the arena, i.e., the physical environment where the robots operate, and the workstation, i.e., the computer in charge of communicating with the robots and controlling most of software modules.

The arena is a bounded and controlled environment in which the robots, wireless connected, can move to perform the desired tasks. In order to satisfy the small size requirements, in our implementation, the arena is a desk with boundaries and movable objects (i.e., small wooden items), which can be arranged to create walls and divide the environment in different zones. A camera is placed high above the arena to monitor positions and actions of the robots over time. This camera sends images to the workstation for elaboration, through a USB port. The output of the image processing algorithm consists of a set of robot IDs and their positions in the arena, which are fed into the control system of the network. A short-range wireless connection is adopted to link the robot network with the workstation. The latter runs a software architecture, which is in charge of: managing the connections between the workstation and the network of robots, using appropriate communication libraries; processing the images acquired by the overhead camera; controlling the network with high-level decision making algorithms; communicating the control decisions to the robots. Indeed, control is not totally centralized in the control module of the architecture, but it is distributed between the workstation and the network of robots. Furthermore, no a priori knowledge about the environment configuration is required by the control system.

In order to facilitate the use of the testbed by researchers and students, MATLAB is employed as software development environment. In the next sections, each part of the system architecture is described in more detail.

2.1 Robots and Hardware Infrastructure

As discussed above, two of the design requirements of the NRA are the desktop-size dimensions of the robots and the possibility to have an open hardware platform.

Keeping in mind these requirements, we analyzed different solutions. First, we selected three robots as possible candidates for the NRA platform: the K-Team Khepera robot [9], the Lego Mindstorms NXT, and the E-Puck robot. Khepera is a small size robot, which can be equipped with various sensors, like infra-red proximity and ambient light sensors, cameras, and ultrasonic sensors, but it is relatively expensive. The Lego Mindstorms NXT is a modular platform with a rich set of sensors; however, it is not an open platform, and it is not suitable for desktop size

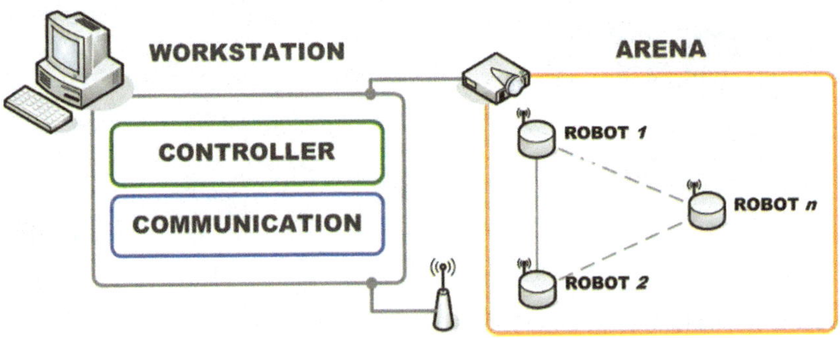

Fig. 1 Schematic representation of the Networked Robot Arena (NRA).

robotic networks. Finally, we chose the E-Puck robot (see Figure 2(a)), because of its small size, low cost and open hardware characteristics [10].

The E-puck is a "desktop size" (7.0 cm diameter) differential wheeled mobile robot. It was originally designed as an education tool for engineering and robotics. It is equipped with a powerful Microchip dsPIC (i.e., dsPIC 30F6014A) microcontroller, and the following sensors and actuators: eight infrared (IR) proximity sensors, placed around the body; a 3D accelerometer that can be used to measure the inclination and acceleration of the robot; three microphones, to localize source of sound by triangulation; a CMOS colour camera (640 × 480), mounted in the front of the body; two stepper motors for differential drive; a speaker, connected to the audio module; a set of LEDs placed on the body to provide a visual interface to the user; a set of communication interfaces (i.e., a Bluetooth radio link, a RS232 serial interface, and an infrared remote control receiver).

The hardware infrastructure of the NRA testbed consists of a bounded arena (60 cm × 110 cm) with a wireless communication network, a USB camera, and a high performance workstation (see Figure 2(b)). The camera consists of a USB colour device, placed at about 150 cm above the arena. It provides robot identity and position information for the control system, and can be also used as ground-truth basis for experimental validation of the algorithms. For the given arena and robot size, an image resolution of 640 × 480 is sufficient to detect and track all the robots. The workstation consists of a high performance computer, equipped with an Intel Core 2

(a)

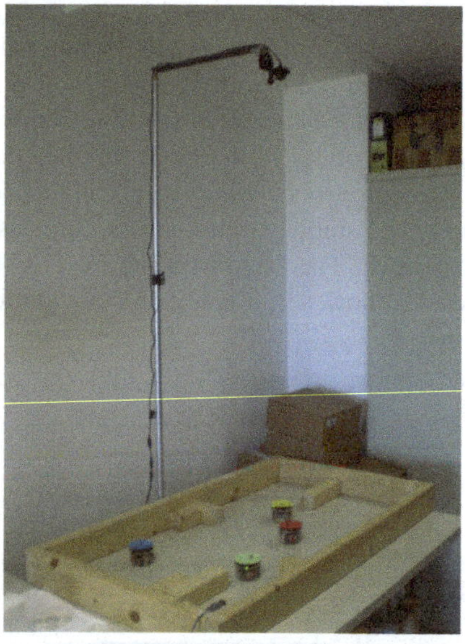

(b)

Fig. 2 Robots and Hardware Infrastructure: (a) close up of an E-Puck robot inside our arena; (b) the Networked Robot Arena.

vPro microprocessor and 4 Gb RAM, and is used for control, visual data processing, and connection with the network of robots.

Since we need short-range communication among the agents, and between the agents and the workstation, we use a Bluetooth wireless network. Due to the small amount of data passed through the network and the hardware and software module embedded within the E-Puck, the Bluetooth communication protocol revealed a good choice.

2.2 *Software*

The main goal of the NRA software architecture is to provide a distributed mission control system, able to manage all the activities of the robotic network. We assume that the network has to execute a number of missions. Each mission consists of a predefined sequence of tasks, connected by logical rules. Activation and completion of missions and tasks are triggered by input and output events, respectively. In order to execute each task, we assume that a set of behaviours, i.e. simple reactive actions, is properly activated on board each robot.

The layout of the software architecture is shown in Figure 3. It consists of the following main parts: the high-level control, communication, and localization and identification modules, running on the workstation, and the execution module

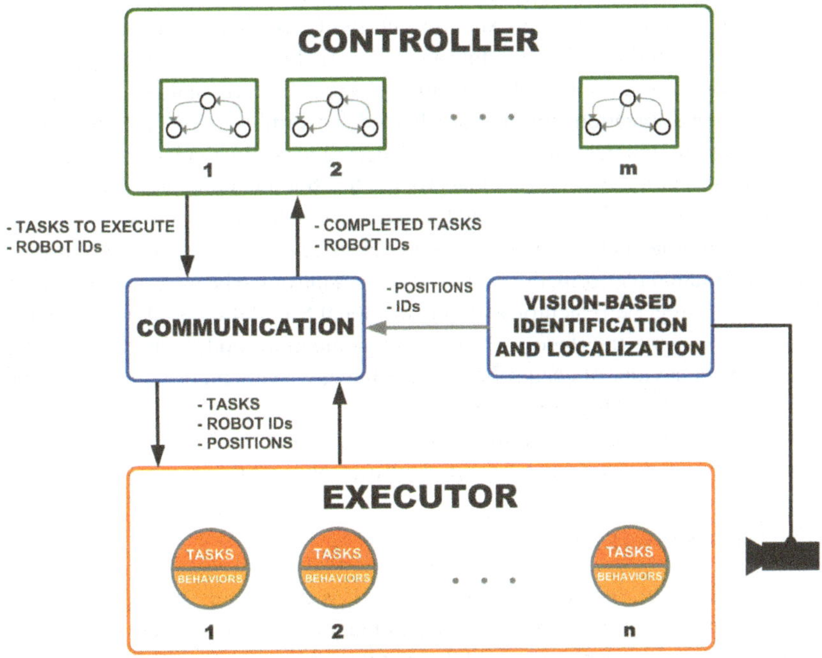

Fig. 3 Layout of the software architecture.

running on each robot. All components are executed on a discrete-time basis: state updates, event detection, and commands are all processed at the beginning of each sample time.

The Controller implements high-level control functionalities, monitoring mission execution and performing task assignment. In this module, one or more controller can be run. Each controller, called leader, is in charge of controlling a part of the network, in terms of missions and tasks. More specifically, each leader controls the execution of the missions in progress (e.g., guaranteeing the satisfaction of precedence constraints), sends the configuration information about the tasks that must be started on each robot, and receives information about the completed tasks. Each leader performs its work using a discrete-event model of the domain, a task execution controller, and a conflict resolution strategy, which will be described in the following section. Obviously, if only one leader exists, the control of the network can be viewed as a centralized system.

The Executor performs the control at the lower level. This component is implemented on each robot through a set of tasks and a set of behaviours needed to accomplish a given goal. For each task, multiple behaviours can be executed at the same time. Each behaviour computes an output and when multiple behaviours are active at the same time, a predefined behaviour arbitration algorithm (e.g., subsumption [11] or cooperative methods [12]) can be used to obtain the final control signal. Currently, in our system, a subsumption mechanism is implemented. At the end of a task, the completion flag and the results are sent to the upper level.

Communication between the Executor and the Controller is guaranteed by the Communication module. This module provides all high-level control commands from the leaders to the other robots in the network. In particular, in this module, the proper communication libraries are loaded, according to the specified robot. Moreover, the communication module is connected to the vision based localization and identification module that provides, for each robot, position and identity information, used to accomplish an assigned task and for behaviour control.

It is worth to note that, although the main objective of this work is that of providing an experimental platform, the proposed system may also be used as a simulation tool by substituting the Executor with an additional Simulation module. The latter, which uses the same interface with the Communication module as the Executor, simulates the behaviour of each robot in a virtual environment, providing robot positions and simulating task execution.

In the rest of this section, further details concerning the hybrid control of the network, as well as the localization and identification module, are provided.

2.2.1 Robot Network Control

As mentioned above, the control of the network is distributed among the components of the architecture. The overall management of the robotic network is based on a hybrid control framework: the global network activities are modeled and controlled as a Discrete Event System (DES) and the low-level behaviours, for each robot, are considered as continuous-time control algorithms trigged by the DES supervisor. Moreover, the global model of the network is distributed among a set of leaders. Each leader can control all activities or a sub-set of them in the network.

The activities domain of the network is modeled as follows: each robot must execute a set of missions; each mission is a set of atomic tasks, which may be subject to priority constraints (i.e., the end of one task may be a necessary prerequisite for the start of some other ones). To execute each task, the robot must perform a predetermined sequence of behaviours (e.g. *GoTo*, *AvoidObstacle*, *WallFollowing*, *Wandering*, etc.). Each behaviour runs until an event (either triggered by sensor data or by the supervisory controller) occurs. Two or more behaviours can be active simultaneously. In such a case, the behaviour arbitration algorithm, on board the robot, computes the appropriate command to be sent to robot actuators. Missions may be triggered at unknown times, while the number and type of tasks composing a mission is known a priori. Since different missions can run simultaneously, different tasks may require the same subset of robots. In these cases, conflicts between the robots have to be handled using an efficient task allocation policy in order to avoid deadlock situations.

The overall control scheme, proposed in this work, is based on the Matrix-based Discrete Event modeling and control Framework (MDEF) [13], a tool for modelling and analyzing complex interconnected DES with shared resources, for routing decisions, and managing dynamic resources. More precisely, in this work a partial decentralization of the MDEF is implemented. The set of leaders are considered as a group of centralized MDEFs working on separate parts of the network [14].

All the MDEF builds over a set of key matrices and vectors. Although this presentation is mainly intended for self-containment purposes, it is obviously not possible to provide all the details of the formalism in the limited space available here. Reader therefore is referred to [14] for further details. Suppose that the RN is composed of $\mathscr{V} = \{v_1, \ldots, v_n\}$ robots, which have to accomplish $\mathscr{M} = \{m_1, \ldots, m_l\}$ missions. Each mission is, in turn, viewable as a sequence of primitive tasks (e.g., reach the target, take a picture of the target, measure the temperature, collect the target, etc.), which requires some low-level behaviors and may be subject to precedence constraints (e.g., the target can be collected after it has been found). Thus, $\mathscr{T} = \{t_1, \ldots, t_m\}$ denotes the set of all possible tasks for the RN.

The execution of tasks is governed by a set of *rules*. More specifically, each rule specifies all the preconditions for tasks execution, and the resulting consequences (postconditions). All rules for all missions in $\mathscr{M}$ are defined by the set $\mathscr{X} = \{x_1, \ldots, x_q\}$. Thus a mission m_j is associated to a specific set of rules $\mathscr{X}_{m_j} \subseteq \mathscr{X}$. The set of all possible input events (a sensory input, a user command, etc.) is indicated as $\mathscr{U} = \{u_1, \ldots, u_p\}$. One or more inputs can be considered as preconditions for a generic mission m_j, thus these event are element of the set $\mathscr{U}_{m_j} \subseteq \mathscr{U}$. Besides tasks and rules, generic outputs can be considered as postconditions. All possible outputs in the network are indicated by the set $\mathscr{Y} = \{y_1, \ldots, y_o\}$.

Briefly, the MDEF controls the execution of tasks by checking the status of the rules at each time sample. Each rule whose preconditions are all true in the current time sample is fired, i.e., all the actions described in the consequent part of the rule are triggered. The result of the logical preconditions evaluation provide the information needed to properly trigger tasks. Since conflicts in the assignment of tasks can occur, before starting the execution of task a multi-agent task allocation algorithm is performed, such as the Look-Ahead strategy proposed in [15]. After the task allocation step, postconditions are fired properly and robots start assigned tasks and eventually output events are produced.

2.2.2 Robot Identification and Localization

Robot identification and localization is performed using a vision-based tracking system. Indeed, E-Puck robots are equipped with encoders, and therefore odometry could be adopted for robot location estimate. Nevertheless, we use an external camera, since it provides an absolute positioning system, thus avoiding error accumulation problems. In addition, based on visual information, the robot identification issue can be easily solved.

We developed an algorithm that is able to simultaneously estimate the global positions of all the robots in the arena, determining, at the same time, which location belongs to which robot. We call this module Simultaneous Localization and Identification (SLI). The output of the SLI module is fed into the control system. It is assumed that each robot is equipped with a circular coloured "hat" (see Figure 2(b)). Different colours are used for the different robots. Each hat serves as an artificial landmark for robot detection and identification. Specifically, the SLI includes a semi-automatic learning procedure that allows us to learn and store a model for each landmark; then, based on this model, it performs robot localization and identification through a model matching technique.

The learning phase consists of the following steps. First, various pictures of each robot in the arena are acquired by the overhead camera; then, in each picture a polygonal area containing the robot is selected manually, and shape and colour information are extracted by analyzing the pixels of the selected area. Shape information is represented by using the radius of the minimum enclosing circle for the selected polygonal area. Colour is defined, instead, by a three-dimensional vector containing the median of each color component in the HSV color space. The procedure is

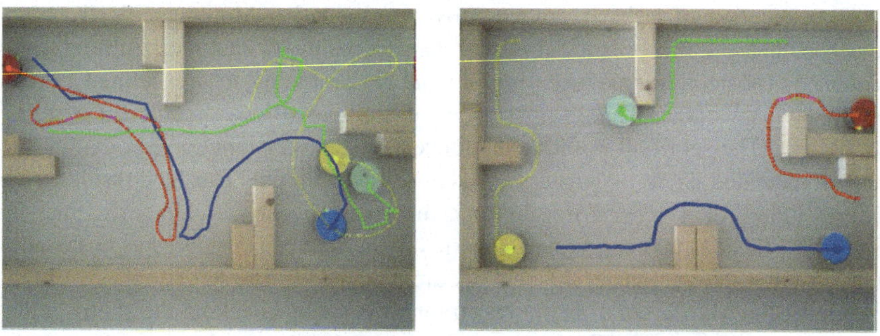

(a) (b)

Fig. 4 Results of the SLI module: (a) wandering behaviour; (b) wall-following behaviour.

repeated for all the pictures, and median values for both the radius and the color vector are computed and stored. Once the model of each robot has been learned, it can be used for detecting and identifying all the robots in the arena.

The detection phase is based on a Circular Hough Transform, which allows one to detect all the circular objects within a specified radius interval based on the results of the model learning phase. Then, the pixels internal to each detected circle are analyzed for robot identification, using colour information. Specifically, the median of hue, saturation and value of the pixels internal to each circle are computed and are stored in a 3D color vector, which is then compared with the available robot models. A circle is finally recognized as being one of the robots if the Euclidean distance between its associated color vector and the closest robot color model is less than an experimentally determined threshold.

At this step, the position of each robot is expressed as pixel coordinates of the centre of the circle in the image reference frame. A camera calibration procedure performed once, after camera installation, allows us to transform the pixel position in a real-world position. In order to improve the performance of the algorithm in terms of speed and accuracy of both detection and identification phase, we implemented a Kalman filter-based robot tracking. As an example, Figure 4 shows the result of the SLI module for four robots in two different experiments. Specifically, in Figure 4(a) the robots are tracked while wandering in the arena; in Figure 4(b), each robot carries out a wall-following task.

3 Experimental Results

In this section, we show the results of an experimental test performed using the setup described in the previous sections. In this experiment, the network evolves from a single-leader configuration to a multiple-leader configuration, in order for the robots to perform exploration tasks in different parts of the arena starting from a *rendez vous* position. In the various phases of the test, the leader robots can be distinguished from the other ones, as they are green-lighted. In Figure 5(a), the four robots form a unique network with one leader (i.e., the green-lighted robot with the red hat). Figure 5(b) shows the trajectories of the four robots while going to the *rendez vous* position. Successively, the network splits into two sub-networks of two robots each (Figure 5(c)). Both networks have their own leader (i.e., the green-lighted robot with the blue hat and the green-lighted robot with the red hat). The two teams reach opposite sides of the arena to explore different areas. Figure 5(d) shows the tracked trajectories of the robots after the splitting of the network.

The graph in Figure 6 shows the time trace of the experiment. This graph can be easily obtained by the saved data of the logging system. In particular, in this figure, we show: the network configuration, i.e. the single leader network (Network AB), and the two sub-networks (Network A and Network B); the missions in progress; the robots assigned to each mission and their network; the activation of tasks. For all these information a line representing high (activation) or low (deactivation) state is depicted. In Table 1, the correspondences among missions, robot networks, robots and tasks are reported for better understanding the graph in Figure 6. It can be noticed that Mission 1, which involves all robots (i.e., Network AB), starts at sampling instant 7. This mission is accomplished at time 33, when each robot in the network terminates Task 1, that is, the *GoToGoal* task, in order to reach the *rendez vous* point at the centre of the arena. After Mission 1, the network splits into the two sub-networks A and B. As we note, network B starts Mission 2 before network A at the time instant 33. Obviously, the robots #2 and #4, which are members of network B, become busy when Mission 2 starts, performing, first, Task 1 and, then, at time 47, Task 3 (*WallFollowing*). At the instant 36, Mission 3 starts involving robots #1 and #3. Network A, first, performs Task 1 and, then, Task 2 (*ExploreEnvironment*). At completion of Mission 3, at time 68, the robots of network A become available and all the tasks they are involved in terminate. Finally, at time 74, Mission 2 is accomplished, robots #2 and #4 are released, and all tasks in the whole network are successfully terminated.

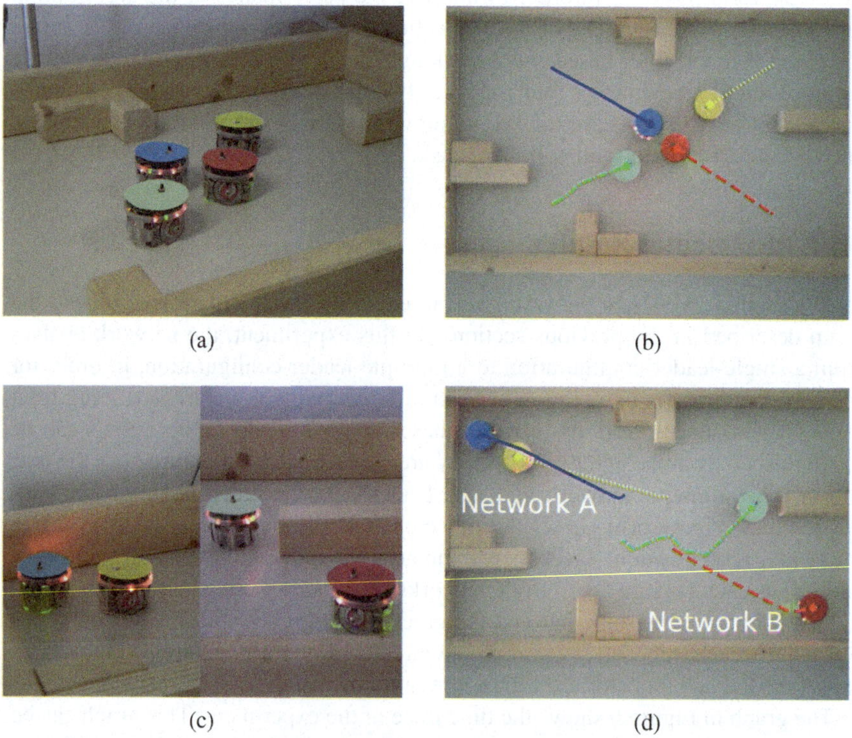

Fig. 5 (a)-(b) *Rendez vous* of the networked robots with a unique leader; (c)-(d) splitting of the network into two sub-networks with two different leaders to explore opposite sides of the arena

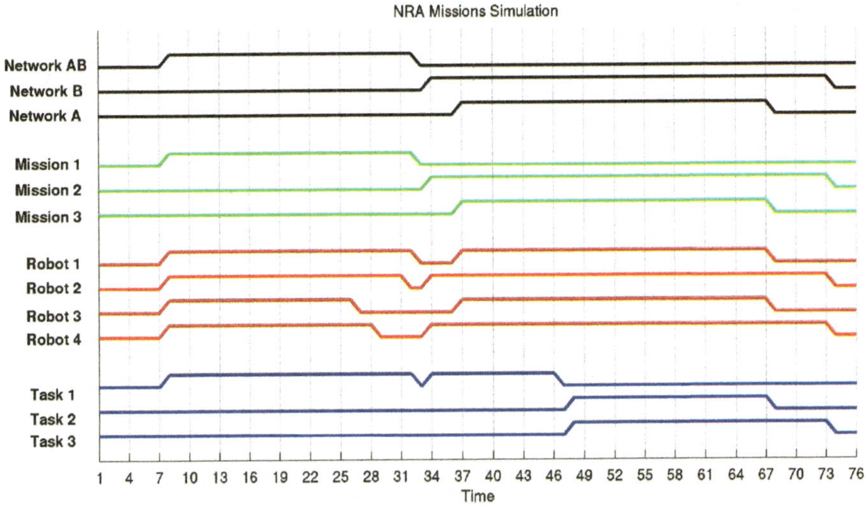

Fig. 6 Time trace graph for the experiment of Figure 5 .

Table 1 Correspondences among missions, robot networks, robots, and tasks

Mission ID	Network Name	Robot ID	Task ID
1	AB	1,2,3,4	1
1 2	B	2,4	1,3
3	A	1,3	1,2

4 Conclusions

We described an experimental set-up for multi-robot applications. The small sized, relatively simple robots E-Puck are utilized. A USB camera connected to a high speed PC is adopted for robot monitoring and global positioning. Image processing provides the positions of the robots to the control algorithms. The latter are based on discrete event framework, for the high level control, and a task and behaviour-based distributed control framework at the lower level.

We believe that this test bed is a useful experimental facility, which can be employed for testing multi-robot coordination and control algorithms in graduate and undergraduate courses as well as for research purposes.

References

1. Burgard, W., Moors, M., Stachniss, C., Schneider, F.: Coordinated multi-robot exploration. IEEE Transactions on Robotics 21(3) (2005)
2. Fierro, R., Das, A., Spletzer, J., Esposito, J., Kumar, V., Ostrowski, J.P., Pappas, G., Taylor, C.J., Hur, Y., Alur, R., Lee, I., Grudic, G., Ben Southall, B.: A Framework and Architecture for Multi-Robot Coordination. The International Journal of Robotics Research 21(10-11), 977–995 (2002)
3. Kumar, V., Rus, D., Singh, S.: Robot and sensor networks for first responders. IEEE Pervasive Computing 3(4), 24–33 (2004)
4. Vincent, R., Fox, D., Ko, J., Konolige, K., Limketkai, B., Morisset, B., Ortiz, C., Schulz, D., Stewart, B.: Distributed multirobot exploration, mapping, and task allocation. In: Annals of Mathematics and Artificial Intelligence. Springer, Netherlands (2009)
5. Pugh, J., Martinoli, A.: Multi-robot learning with particle swarm optimization. In: Proceedings of the Fifth International Joint Conference on Autonomous Agents and Multiagent Systems (2006)
6. Lerman, K., Chris Jones, C., Galstyan, A., Matarc, M.J.: Analysis of Dynamic Task Allocation in Multi-Robot Systems. International Journal of Robotics Research 25(3), 225–241 (2006)
7. Viguria, A., Maza, I., Ollero, A.: SET: An algorithm for distributed multirobot task allocation with dynamic negotiation based on task subsets. In: Proc. of 2007 IEEE International Conference on Robotics and Automation, Roma, Italy (2007)
8. Werner, F., Rckert, U., Tanoto, A., Welzel, J.: The Teleworkbench A Platform for Performing and Comparing Experiments in Robot Navigation. In: IEEE International Conference on Robotics and Automation (ICRA), Anchorage, AK, USA (2010)
9. Mondada, F., Franzi, E., Guignard, A.: The Development of Khepera. Experiments with the Mini-Robot Khepera, 7–14 (1999)
10. Mondada, F., Bonani, M., Raemy, X., Pugh, J., Cianci, C., Klaptocz, A., Magnenat, S., Zufferey, J.-C., Floreano, D., Martinoli, A.: The E-Puck, a Robot Designed for Education in Engineering. In: Proceedings of the 9th Conference on Autonomous Robot Systems and Competitions, vol. 1(1), pp. 59–65 (2009)
11. Brooks, R.A.: Intelligence without Representation. Artificial Intelligence 47, 139–159 (1991)
12. Payton, D.W., Keirsey, D., Kimble, D.M., Krozel, J., Rosenblatt, J.K.: Do whatever works: A robust approach to fault-tolerant autonomous control. Applied Intelligence 2(3), 225–250 (1992)
13. Tacconi, D., Lewis, F.: A new matrix model for discrete event systems: application to simulation. IEEE Control System Magazine 17(5), 62–71 (1997)
14. Di Paola, D., Gasparri, A., Naso, D., Ulivi, G., Lewis, F.L.: Decentralized Task Sequencing and Multiple Mission Control for Heterogeneous Robotic Networks. In: Proc. of IEEE International Conference on Robotics and Automation (2011)
15. Di Paola, D., Naso, D., Turchiano, B.: A Heuristic Approach to Task Assignment and Control for Robotic Networks. In: Proc. of the IEEE International Symposium on Industrial Electronics, ISIE (2010)

iRov: A Robot Platform for Active Vision Research and as Education Tool

Abdul Rahman Hafiz and Kazuyuki Murase

Abstract. This paper introduces an autonomous camera-equipped robot platform for active vision research and as an education tool. Due to recent progress in electronics and computing power, in control and agent technology, and in computer vision and machine learning, the realization of an autonomous robots platform capable of solving high-level deliberate tasks in natural environments can be achieved. We used iPhone 4 technologies with Lego NXT to build a mobile robot called the iRov. iRov is a desk-top size robot that can perform image processing onboard utilizing the A4 chip which is a System-on-a-Chip (SoC) in the iPhone 4. With the CPU and the GPU processors working in parallel, we demonstrate real-time filters and 3D object recognition. Using this platform, the processing speed was 10 times faster than using the CPU alone.

1 Introduction

There is no general platform for developing embedded systems such as autonomous camera-equipped robot systems. The existing platforms are mostly based on pre-specified models, which are difficult to obtain for different robot tasks. For designing and developing a general platform for autonomous camera-equipped robot systems we propose for the first time iRov robot, a platform which is based on iPhone 4 technology with Lego NXT. This robot platform is excellent for broad range of active vision related researches, behavior-based robots, image and signal processing. More than that, it can be used as an educational platform, due to its low price and ease of use.

In this paper we introduce the design of iRov robot, and show how this robot can perform image filters, change its perspective to the object, and recognize 3D

Abdul Rahman Hafiz · Kazuyuki Murase
University of Fukui, Graduate School of Engineering, Department of Human and Artificial Intelligence System, University of Fukui, Japan
e-mail: abdul@synapse.his.u-fukui.ac.jp,
murase@synapse.his.u-fukui.ac.jp

objects and that in real-time. The next section reviews some of the existing robot platforms that can be used in active vision and introduce the concepts of designing such robots. Section three introduce iRov robot. Section four introduces an experiment using iRov robot. Section five shows the experimental results. Section Six discusses the performance and the significance of the platform, and finally section seven concludes this paper.

2 Background

Due to the advancement in digital camera technology, cameras have been used as sensors in many robots platforms, but its has many limitations, such as the size of the frames, the frames rate, in addition to the limitation of the parallel computation capability. These are some examples that available currently.

2.1 Desk-top Size Robots

The E-puck [1] is a desk-top size robot and it is widely used for research purposes, especially in behavior-based robots. It is equipped with a camera with a resolution of 640 x 480 pixels. The full flow of information this camera generates cannot be processed by a simple processor like the dsPIC on the robot. Moreover the processor has 8k of RAM, not sufficient to even store one single image. To be able to acquire the camera information, the frame rate has to be reduced as well as the resolution. Typically we can acquire a 40 x 40 subsampled image at 4 frames per second. This size of image is enough to study and realize insect like vision such as optical flow, but it is not useful in studying high level visual functions that exist in mammalian vision systems.

2.2 Laptop Based Robots

Another type of robots is bigger in size and it is operated by a laptop computer. The base Pioneer 3-DX platform arrives fully assembled with motors with 19 cm wheels [2]. Adding a laptop equipped with camera this robot can be used as platform for robot vision research. The hardware of such robots does not have the flexibility to be reconfigured according to different vision tasks. Moreover, because of its large size it is quite impossible to operate in the desk-side environment.

2.3 Robot Design for Active Vision Research

To design a robot with active vision, that can realize mammalian like visual behaviors, and help in studying different high cognitive and vision function, the robot has to be equipped with high resolution camera that can operate in real-time and with high frame rate. In addition to that, if the image filters are performed by parallel microprocessor such as GPU, the robot would have the ability to perform in real-time, and change its attention and perspective of vision to interact with the users and the objects. Moreover, the robot hardware should have the flexibility to be reconfigured for every different experiments, such as interactive 3D object recognition or human-robot interaction.

3 iRov Mobile Robot

To reach our goal we decided to use iPhone 4 cameras and utilize the A4 chip for the vision research. Furthermore, iPhone 4 and new generation iPods are equipped with multi-touch display, speaker, external stereo microphones, accelerometer and gyroscope. which allow the robot to be aware of its body, and its head orientation relative to the environment, and be able to built correlations between different sensory input. Moreover, We built the robot body with Lego NXT parts. It made the robot flexible to be configured differently for different tasks, taking the advantage of the 3 connected servos.

In the design phase, we took into consideration to the robot's ability to change its perspective to the object and be able to manipulate it. To achieve that, iRov consists of two parts the head, and the body(Fig.1).

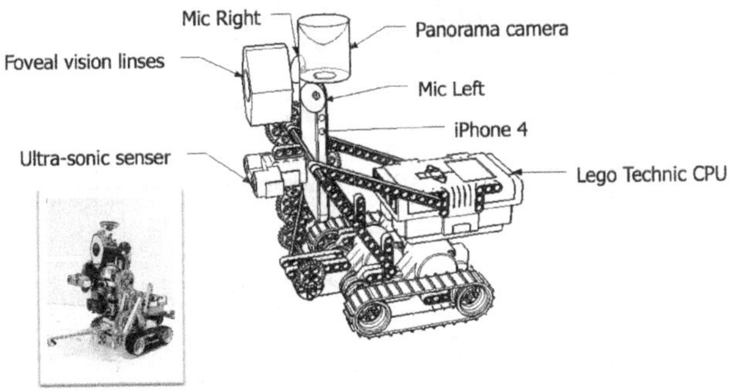

Fig. 1 Drawing of different parts of iRov robot

The head can rotate 180°so the foveal vision (see Sect. 3.2 for detail) can face up and down, to locate the object or the user in the center of its vision (fovea), while the body helps the robot to move to change the robot perspective and to manipulate the objects.

3.1 Characteristics of iRov

Because the iRov robot is small in size, it can perform in lab environment. Fig. 2 shows the setup consists of the user, the robot, and the stationary server, which establish the connection between the iPhone and the Lego NXT. This will be used for high computational load like object recognition.

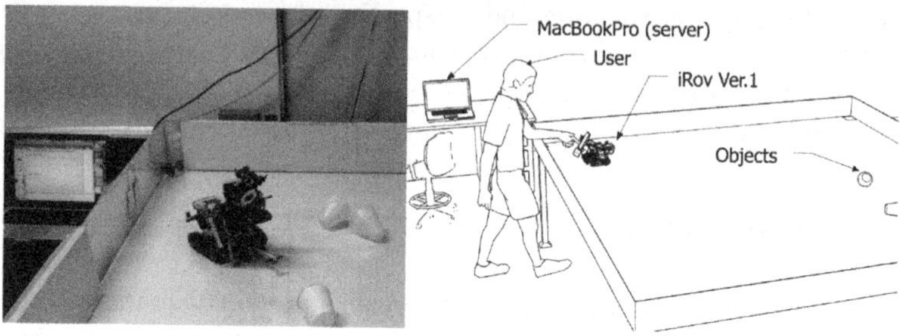

Fig. 2 Typical experiment setup. The user interacts with the robot by using the interactive touch display

Since the iPhone 4 cannot send commands to the Lego NXT part directly we used a server computer to establish this connection between the iPhone and Lego NXT parts as shown in Fig.3.

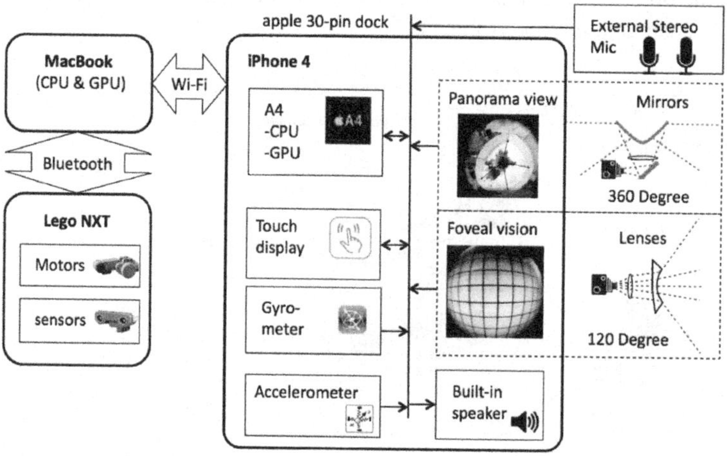

Fig. 3 The different parts of the robot and their connections

3.2 *Robot Cameras*

Mounted in the top of the robot is a panorama camera (the view through the camera is shown in Fig. 4). This camera helps the robot localize itself in the environment and avoid obstacles.

The range of the back-facing camera in iPhone 4 is 30°. In order to have wider view of the periphery, while maintaining the sharpness of the center, we used a special lens as shown in Fig. 5. It generates foveal representation that helps the robot to be aware of any movement in the periphery (120°). At the same time it allows to examine the detail of the object in the center (fovea).

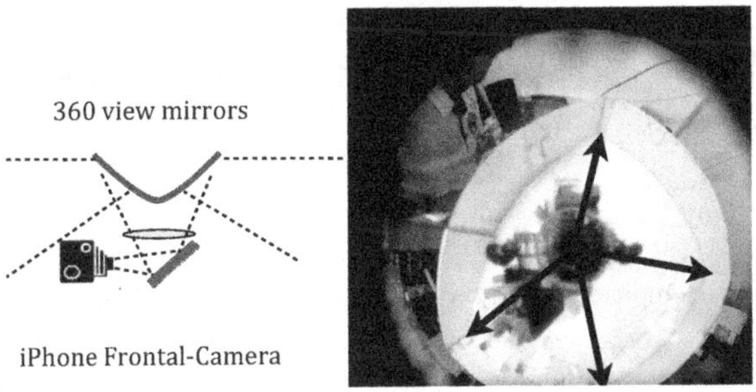

Fig. 4 Panorama view from iPhone 4 front facing camera

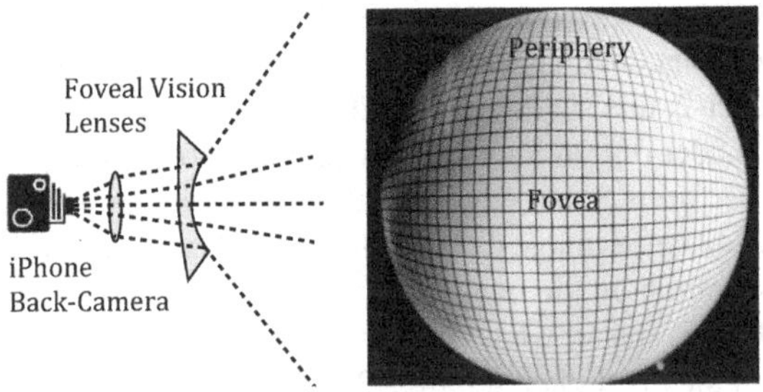

Fig. 5 Foveal view of the environment using special lenses (iPhone 4 back facing camera)

4 iRov Vision and 3D Object Recognition

By dividing the robot vision to foveal and periphery vision, we could implement an attention system similar to our previous work [3]. Since iRov has a high-speed image processing ability, iRov can recognize a simple 3D object and its orientation in real-time. To do that we implemented Hierarchical Chamfer Matching which is a parametric edge matching algorithm (HCMA) [4]. By utilizing the capability of parallel processing of the A4 chip we could achieve real-time image filtering.

Before applying HCMA we need to generate edges of the 3D model templates and the real edges of the object. First the model templates are programmatically generated (Fig. 6). Second the image in the robot fovea has to pass through filters such as Gaussian Smoothing and Edge Detection as Fig. 7 (b). Third Building distance image pyramid (Fig. 7 (c))[5]. Finally by applying HCMA[6], we can found the local minimum of the best match between the template and the detected edges (Fig. 7 (d)).

To perform these algorithms in real time we utilize the A4 Chip (SoC), which has a general purpose processor called the CPU, as well as a second processor called the GPU. With the two processors working in parallel, the device is capable of doing a lot more works at one time. But this doesn't happen automatically. The main program that runs in the CPU has to send small programs that runs in the GPU during runtime. These programs are called shaders, and perform the following tasks:

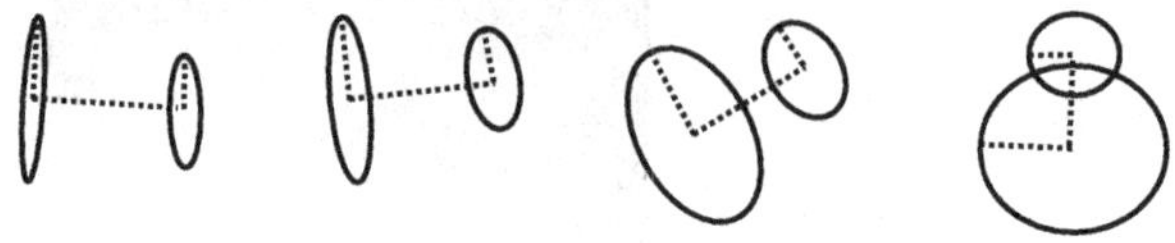

Fig. 6 From 0°to 90°template generation by considering camera perspective to the object

Shader-1 This shader smoothes the row buffer by gaussian blur filter and stores it in the buffer-1 at 20 fps.

Shader-2 This shader performs motion detection by: First comparing between the RGB of buffer-1 and buffer-2 for any change and adds the result to the alpha channel in the buffer-2. Second copy the RGB of buffer-1 to buffer-2, this shader runs at 20 fps too.

Shader-3 This shader runs at 5 fps to copy the result from the buffer-1 to buffer-3, then performs Sobel edge detection to the image in buffer-3. Finally it builds distance image pyramid for HCMA search algorithm which need four iteration on the image.

Shader-4 Finally this shader which runs at 5 fps generates the final result in buffer-4 from the pixel data in buffer-2 and buffer-3. Listing-1 shows the code of this shader as example.

Listing 1 Shader-4 source code

```
precision mediump float;
varying vec2 textureCoordinate;
uniform sampler2D MD;
uniform sampler2D ED;
void main()
{
  float r, func;
  vec4 fovea = texture2D(ED, textureCoordinate);
  vec4 periphery = texture2D(MD,textureCoordinate);
  r = sqrt(textureCoordinate.x * textureCoordinate.x
        + textureCoordinate.y * textureCoordinate.y);
  func = pow(2.7, -16.0 * pow(1.0 - r, 2.0));
  gl_FragColor = (r < 1.0) ?
                    func*periphery+(1.0-func)*fovea
                   :vec4(0.0, 0.0, 0.0, 1.0);
}
```

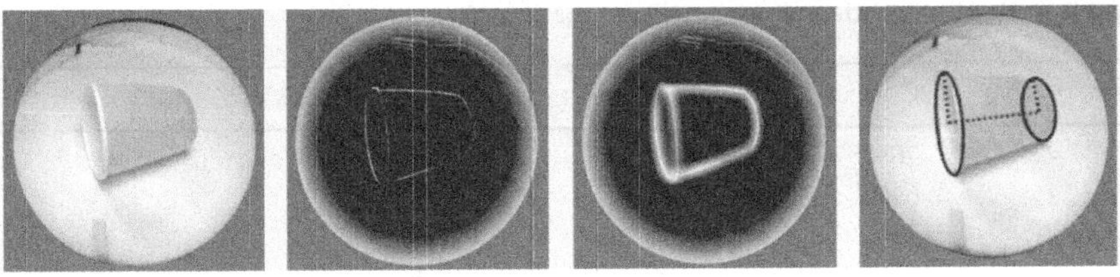

(a) (b) (c) (d)

Fig. 7 3D object recognition by applying Hierarchical Chamfer Matching (HCMA)

From this listing we can see how shader-4 generates foveal representation (Fig. 7 (b)) from the motion detection result (*MD*) and edge detection result (*ED*), by implementing equation 1, where r is the distance from the center of the view.

$$Output = ED * (1 - e^{-16(1-r)^2}) + MD * e^{-16(1-r)^2} \quad (1)$$

We can notice that the frame rate have been reduced to 5 fps(from 20 fps witch was necessarily for motion detection), nevertheless its not enough yet to be passed to the stationary server through wireless connection bandwidth. To solve this problem the image size have to be resized to 25% of the original size, as the final step that performed onboard. This result which contain the motion, and the edges information in the alpha channel(of the RGBA color representation) can be passed at 5 fps to the server, for attention and object recognition.

5 Experimental Results

The default value for the parameter τ , threshold for selection of start points, is set to 0.7. Other parameters using HCMA were determined by suggestions from [6]: The factor λ is set to 2; the intervals in x-coordinate and y-coordinate, u_x and u_y, are both set to 3, the radius r to define the neighborhood is 1; and the maximum level of distance pyramid L is set to 5, which means that image on the top of pyramid is 1/16 rescaled from the original image.

Table 1 is the result of the recognition rate for different perspective of the object from $0^\circ$to 90°.

Fig. 8 shows that by dividing the process to primary filters that perform in the robot and the secondary recognition stage that perform in the server we can achieve faster performance comparing with the work [6]. While the value of τ is increased, more candidate points are eliminated at the early stage.

6 Discussions

From Table 1, the 0°perspective of the object results in a low recognition rate. To overcome that the robot have to change it perspective to the object so that it can have multiple view points, by rotating its head vertically or its body horizontally.

Table 1 Result for different perspective of the object

perspective	Correct	Incorrect	Undetected
0°[a]	20%	9%	71%
30°	61%	11%	28%
60°	73%	15%	12%
90°	77%	4%	19%
Total	57.75%	9.75%	32.5%

[a] In this state, large part of the object edges that can be compared with the template model is occluded with the object itself.

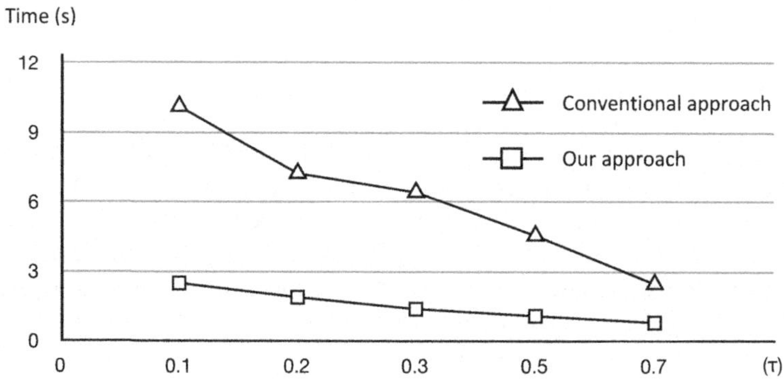

Fig. 8 The relationship between threshold and computational time in both our work and the work done in [6]

As we can notice, the performance in this platform rely on the high resolution cameras(640 x 480 pixels) that can be processed up to 60 fps. In addition to that, the accessibility to the hardware such as the CPU and GPU and other sensors like motion and rotation sensors, is much simpler than other platform thanks to iOS SDK[7].

The significance of this Robot platform fills in two main fields: First, as research tool for active vision robots, it takes the advantage of the mentioned performances and the flexibility of the robot to be reconfigured for different tasks (with respect to the hardware). Second, as education tool, because of the publicity of both the Lego NXT and iPhone in the education environment, it provides the excellent programming experience to the students through this robot platform, more than that the iOS Developer University Program allows the student to access to the platform free of charge[8].

7 Conclusion

iRov robot is a platform for developing autonomous camera-equipped robot. Which can help to produce flexible systems at high degrees of sophistication. This system helps developing the future robots. More than that, it helps in studying the human vision and cognition by synthetic approach. We demonstrated that by performing simple 3D object recognition, and by dividing the process between the CPU and the GPU on the iPhone 4 and the server computer, we achieved real-time performance.

We plan to release the hardware design and the programs that run in the iPhone, Lego NXT, and the server in our website [9], so that any one can take the advantage of this work in very short time. Now we are working in robot attention system and human robot interaction, which will demonstrate more of the iRov robot capabilities.

Acknowledgements. This work was supported by grants to KM from Japanese Society for Promotion of Sciences and from the University of Fukui.

References

1. Mondada, F., Bonani, M., Raemy, X., et al.: The e-puck, a Robot Designed for Education in Engineering. In: Proceedings of the 9th Conference on Autonomous Robot Systems and Competitions, vol. 1, pp. 59–65 (2009)
2. Filipescu, A., Susnea, I., Stancu, A.L., et al.: Path following, real-time, embedded fuzzy control of a mobile platform pioneer 3-DX. In: Proceedings of the 8th WSEAS International Conference on Systems Theory and Scientific Computation (ISTASC 2008), Rhodes (Rodos), Island, Greece, pp. 334–335 (2008)
3. Hafiz, A.R., Alnajjar, F., Murase, K.: A Novel Dynamic Edge Detection Inspired from Mammalian Retina toward Better Robot Vision. In: Proceedings of the 12th International Symposium on Robotics and Applications (ISORA 2010), World Automation Congress (WAC 2010), Kobe, Japan, pp. 1–6 (2010)
4. Thayananthan, A., Stenger, B., Torr, P.H.S., et al.: Shape context and chamfer matching in cluttered scenes. In: Proceedings of the IEEE Conference on Computer Vision and Pattern Recognition (CVPR), pp. 127–133 (2003)
5. Butt, M.A., Maragos, P.: Optimum design of chamfer distance transforms. IEEE Transactions on Image Processing 7, 1477–1484 (1998)
6. Zhang, Q., Xu, P., Li, W., et al.: Efficient edge matching using improved hierarchical chamfer matching. In: Circuits and Systems, ISCAS 2009, pp. 1645–1648 (2009)
7. Apple Inc., iOS Developer Center, `http://developer.apple.com/devcenter/ios/index.action` (Cited March 8, 2011)
8. Apple Inc., iOS Developer University Program, `http://developer.apple.com/programs/ios/university` (Cited March 8, 2011)
9. University of Fukui, Bio Science and Engineering Laboratory, `http://www.synapse.his.fukui-u.ac.jp/en/` (Cited March 8, 2011)

Autonomous Corridor Flight of a UAV Using a Low-Cost and Light-Weight RGB-D Camera

Sven Lange, Niko Sünderhauf, Peer Neubert, Sebastian Drews, and Peter Protzel

Abstract. We describe the first application of the novel Kinect RGB-D sensor on a fully autonomous quadrotor UAV. In contrast to the established RGB-D devices that are both expensive and comparably heavy, the Kinect is light-weight and especially low-cost. It provides dense color and depth information and can be readily applied to a variety of tasks in the robotics domain. We apply the Kinect on a UAV in an indoor corridor scenario. The sensor extracts a 3D point cloud of the environment that is further processed on-board to identify walls, obstacles, and the position and orientation of the UAV inside the corridor. Subsequent controllers for altitude, position, velocity, and heading enable the UAV to autonomously operate in this indoor environment.

1 Introduction

One of our research projects focuses on enabling micro aerial vehicles to autonomously operate in GPS-denied environments, especially in indoor scenarios. Autonomous flight in confined spaces is a challenging task for UAVs and calls for accurate motion control as well as accurate environmental perception and modelling. RGB-D sensors are relatively new sensor systems that typically provide an RGB color image along with distance information for each image pixel and thus combine the perceptual capabilities of RGB cameras with those of stereo camera or 3D laser measurement systems.

While sensor systems like the SwissRanger or PMD cameras have been successfully used in robotics and UAV applications, they are still very expensive (>6000 €). With the very recent release of the Kinect device – an accessory to the Microsoft Xbox video game platform – very cheap (≈ 150 €) RGB-D sensors are available to

Sven Lange · Niko Sünderhauf · Peer Neubert · Sebastian Drews · Peter Protzel
Department of Electrical Engineering and Information Technology, Chemnitz University of Technology, 09111 Chemnitz, Germany
e-mail: firstname.lastname@etit.tu-chemnitz.de

the robotics community. Although precise comparisons of performance and accuracy between the established RGB-D systems and the Kinect are not yet available, it is already foreseeable that the Kinect will be a valuable sensor for a variety of robotics applications.

Our paper explores the application of the Kinect on a quadrotor UAV to aid autonomous corridor flight. After a short introduction to the sensor's working principle, we present our UAV platform and its internal system and control architecture before real-world experiments and their results are described.

1.1 The Microsoft Kinect – A Valuable Sensor for Robotics Applications and Research

The Kinect RGB-D sensor was released by Microsoft in November 2010 as an accessory to its Xbox video game platform. Open source drivers are available from the OpenKinect project [9] or as part of the OpenNI framework [11] that can be interfaced using ROS [12].

In our work, we use the Kinect driver that is available as part of the ROS framework. This driver allows to request an RGB image, a depth image, a 3D point cloud, the raw IR image (all of resolution 640×480) and readings from the internal accelerometers of the device. Fig. 1(b) and Fig. 1(c) show a depth image along with the corresponding RGB image.

The device consists of two cameras and an infrared laser light source. The IR source projects a pattern of dark and bright spots onto the environment (see Fig. 1(a)). This pattern is received by one of the two cameras which is designed to be sensitive to IR light. Depth measurements can be obtained from the IR pattern by triangulation. According to the patent [2] held by PrimeSens Ltd., this is done by comparing the perceived IR pattern against a reference image and thereby determining the relative shift of groups of spots.

To compare the quality of the Kinect's depth measurements against other RGB-D devices such as the SwissRanger 4000 or PMD's CamCube, we determined the measurement repeatability. 3000 measurements were taken on a static planar object in distances of 2 and 4 meters. The standard deviations of these measurements were

(a) (b) (c)

Fig. 1 (a) The infrared light pattern projected by the Kinect sensor. (b) 640×480 depth image (c) RGB image of the same scene as captured by the Kinect.

Table 1 A basic comparison of three RGB-D systems. We determined the values for the repeatability of the Kinect's measurements by taking 3000 measurements in a distance of 2 and 4 meters respectively. All other values are taken from publicly available datasheets. Notice further that we reduced the weight of the Kinect to 200 g by removing the unnecessary housing.

Sensor	max. range	resolution	field of view in deg	repeatability in mm (1σ)	weight
Kinect	10 m	640×480	57.8×43.3	7.6 @ 2 m, 27.5 @ 4 m	440 g
SwissRanger 4000	8 m	176×144	43×34 / 69×56	4 / 6	470 g
PMD CamCube 3.0	7 m	200×200	40×40	3 @ 4 m	1438 g

determined to be 7.6 and 27.5 mm respectively, which is (especially at the larger distance) considerably larger than the values provided in the datasheets of the two established devices. However, considering the significantly lower price, we expect these values to be sufficient for most robotics applications. Table 1 compares the basic features of the Kinect, SwissRanger 4000 and PMD CamCube 3 RGB-D devices.

1.2 Related Work

RGB-D devices have been used by different groups of researchers for depth perception, SLAM and navigation on both ground based and aerial robots: Morris et al. [8] use a SwissRanger 4000 RGB-D camera for 3D indoor mapping planned for use on a quadrotor UAV. Henry et al. [5] describe an approach of using an RGB-D sensor to construct dense 3D models of indoor environments.

Autonomous navigation of UAVs in GPS-denied indoor environments using Hokuyo laser range finders instead of RGB-D devices has been demonstrated by [1] while [3] focused on porting SLAM algorithms that were previously developed for ground based robots to UAVs.

In previous publications [7] [6], we described our work with smaller autonomous UAVs (type "Hummingbird") whereas this paper contains updated material on the system architecture and controller structure on the new and larger "Pelican" quadrotor. To our knowledge, this paper is the first scientific publication that describes the application of the Kinect RGB-D device on an autonomous UAV.

2 Hardware and Software Architecture of Our UAV "Pelican"

The UAV we use in our project is a "Pelican" system (see Fig. 2(a)) that is manufactured by Ascending Technologies GmbH, Munich, Germany. This mid-size four-rotor UAV, or quadrocopter, measures 72 cm in diameter and can carry up to 500 g of payload for about 20 minutes.

The Pelican is propelled by four brushless DC motors and is equipped with a variety of sensors: Besides the usual accelerometers, gyros and a magnetic field sensor, a

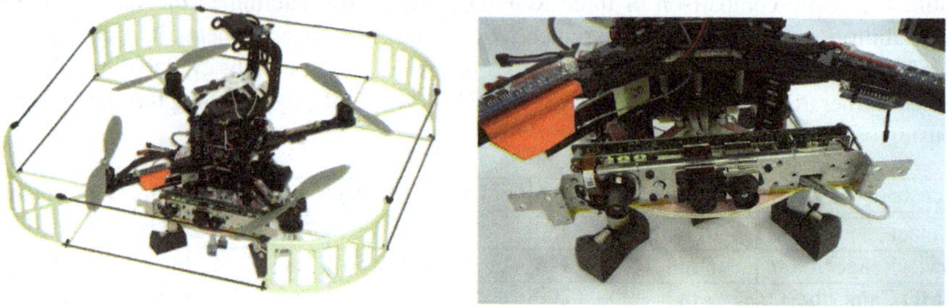

(a) (b)

Fig. 2 (a) Our modified *Pelican* quadrotor system. (b) The dismantled Kinect sensor mounted on the UAV.

pressure sensor and a GPS module provide input for AscTec's sophisticated sensor fusion algorithm and the control loop running at 1 kHz. Like on the smaller Hummingbird system, an "AscTec AutoPilot board" is responsible for data fusion and basic control of the UAV. More technical details on the AutoPilot Board, the Hummingbird and the controllers can be found in [4]. Especially outdoors where the control loop can make use of GPS signals to enter the GPS position hold mode, the Pelican is absolutely self-stable and requires no human pilot to operate. In most cases, the deviation from the commanded hover position is below 1 m. However, when GPS signals are not available, the UAV tends to begin drifting very quickly. Therefore, when the UAV should be operated in GPS-denied environments (for instance close to buildings or indoors) a position stabilization that is independent from GPS signals is required.

With respect to the Pelican's standard configuration, our system is using additional features offered by AscTec. A large propeller protection *(135 g)* is added to minimize the effects of contact with obstacles. An embedded PC system is used for onboard computing. It is equipped with a CoreExpress 1.6 GHz Intel Atom (Z530) processor board from Lippert Embedded Computers GmbH with 1 GB of RAM including a WiFi module *(100 g)*.

2.1 Additional Custom Made Hardware and Sensors

We extended the UAV's configuration using the available payload and equipped the quadrocopter with additional hardware. This includes an SRF10 sonar sensor *(10 g)*, a microcontroller board based on an ATmega644P *(25 g)*, an ADNS-3080 optical flow sensor board based on an ATmega644P as well *(25 g)*, and a Kinect RGB-D sensor *(200 g)*. Completely equipped the quadrocopter has an overall weight of 1675 g including LiPo batteries *(380 g)*, frame and cables. All components and their communication paths are shown in Fig. 3(a).

To ensure realtime execution, the controllers are implemented on a custom made microcontroller board. It is connected to the quadrocopter via USART for polling the quadrocopter's internal sensor readings and sending flight commands. The second

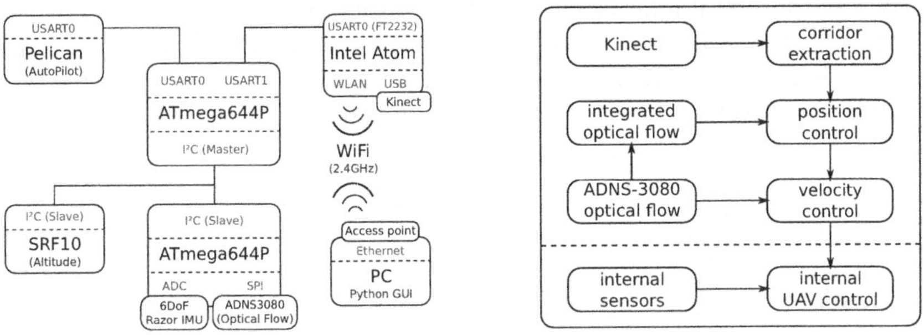

(a) (b)

Fig. 3 (a) Schematic diagram of the main components and their communication channels. (b) Cascaded controller structure and main sensors used. For reasons of clarity the altitude controller is not shown here. It operates parallel to the position and velocity controller and uses an SRF-10 sonar sensor to measure the altitude.

USART port of the microcontroller is connected to the Atom processor board for transmitting and receiving status messages and commands from a ground station as well as from the Atom processor board itself. Communication with the ground station is realized by the integrated WiFi module.

The sensors used in the controller architecture are connected to the microcontroller via I^2C bus, acting as slave devices. An SRF10 sonar sensor measures the current altitude over ground with high precision. Furthermore the Avago ADNS-3080 optical flow sensor combined with a small and light-weight camera lens determines the UAV's current velocity over ground. The sensor is not connected directly over the I^2C bus, but is interfaced via SPI to another ATmega644P which is acting as an I^2C slave device. Additionally, the optical flow sensor board is equipped with a high power LED to enhance the optical flow performance in less illuminated environments and a separate IMU sensor for high frequency gyro measurements.

2.2 *Controller Structure*

In order to achieve a stable flight behavior in GPS-denied areas, we used an altitude controller and a cascaded controller structure for position stabilization. Figure 3(b) shows the principal structure along with the main sensors used by the position and velocity controllers.

2.2.1 Altitude Controller

The altitude controller receives its input from an SRF10 sonar sensor. This off-the-shelf sensor commonly used in robotics projects provides accurate altitude information and offers considerable advantages because of its short minimum operation range of 3 cm. The altitude controller itself is implemented on the ATmega644P microcontroller board as a standard PID-controller. The control loop operates on a 25 Hz cycle and is able to stabilize the UAV's altitude with an accuracy of 3 cm.

The standard PID altitude controller is very sensitive to steps in its measurements or setpoint values, which are common when flying over obstacles. If they appear, an overshoot is the common reaction of the quadrocopter. For that reason we implemented a step detection combined with a ramp function for both cases.

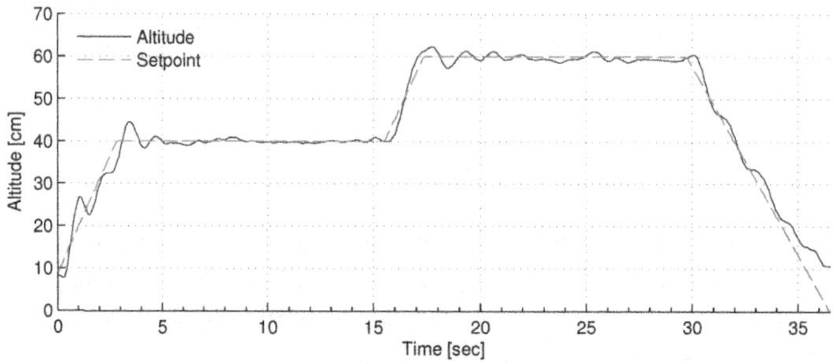

Fig. 4 Altitude plot showing a take-off, altitude-hold at 40 cm followed by an altitude-hold at 60 cm and a landing maneuver. The altitude setpoint is automatically increased and decreased in a ramp. Note that the sonar sensor is mounted about 10 cm above ground.

In Figure 4 we show the results of our altitude controller running on the Pelican UAV. The plot starts when controlled flight mode is activated and the human operator requested an altitude of 40 cm. Notice that the setpoint used by the altitude controller is increased using a ramp function to avoid overshoots caused by steps in the setpoint. After 16 seconds, the setpoint was increased to 60 cm and after an additional time of 13 seconds, landing was initiated. The implemented PID-controller is able to follow the setpoint very closely and stabilizes the UAV at the requested altitude with a maximum deviation of 3 cm.

2.2.2 Position and Velocity Controller

When no GPS signals are available, the only way to measure position and velocity with the standard platform is to integrate information from the onboard acceleration sensors and gyros. However, due to the noisy input signals large errors accumulate quickly, rendering this procedure useless for any velocity or even position control.

In our approach, an optical flow sensor facing the ground provides information on the current velocity and position of the UAV. The Avago ADNS-3080 we use is commonly found in optical mice and calculates its own movements based on optical flow information with high accuracy and a framerate of up to 6400 fps. After exchanging the optics and attaching an M12-mount lens with a focal length of 8 mm to the sensor, we are able to retrieve high quality position and velocity signals that accumulate only small errors during the flight. The sensor also provides a quality feedback which correlates with the number of texture features on the surface the sensor is facing. If the surface is not textured enough, the quality indicator drops and the velocity and shift signals become more noisy. Likewise the quality drops in cases of ill-illuminated textures, so we added a high power LED with directed light

(10° FOV) and a proper wavelenght *(623 nm)* matching the maximum responsivity of the optical flow sensor.

To calculate the velocity in metric units relative to the ground, the sensor resolution, FOV, height and changes in orientation are used. With respect to previous work [7] we increased the lens' focal length from 4.2 mm to 8 mm to obtain a higher spatial resolution and so improve velocity estimation. Because the sensor is now more sensitive to rotational changes, it is insufficient to use the quadrocopter's internal measurements for angle compensation which are available at a rate of about only 7 Hz. Therefore, we extended our optical flow sensor board by a simple 6 degrees of freedom IMU system to get gyro measurements at a much higher rate (> 50 Hz).

Both the velocity and position controller are implemented as standard PID and P controllers, respectively, and operate at a frequency of 25 Hz. In combination, they are able to stabilize the UAV's position and prevent the drift that quickly occurs when the GPS-based position hold mode is inactive.

3 Autonomous Corridor Flight Using the Kinect RGB-D Device

The Kinect driver of ROS provides a 640×480 3D point cloud that is downsampled (thinned) to approximately 3,000 points before it is processed further. This downsampling is done to increase the performance of subsequent steps in the algorithm. Details on our specialized and efficient downsampling algorithm are provided in section 3.1. After downsampling, large planar sections are found in the approximately 3,000 remaining points by applying a sample consensus (MLESAC) based parameter estimation algorithm. Fig. 5 visualizes the results. The Point Cloud Library [10] already provides convenient algorithms that extract the planes in their parameter form $ax + by + cz + d = 0$.

Given these parameters, the distance of the RGB-D device from each plane is calculated as $\Delta_i = |d_i| / \sqrt{a_i^2 + b_i^2 + c_i^2}$. Roll, pitch, and yaw angles are calculated from the plane parameters as well, e.g. the yaw angle ϕ_i is given by $\phi_i =$

Fig. 5 *(left)* RGB image of the corridor. *(mid)* Downsampled point cloud containing about 3,000 points. *(right)* Extracted planes of the walls. The planes have been classified into floor, ceiling, left, and right wall. The green arrow shows the intended motion direction computed from the position and orientation relative to the walls.

atan2$(c_i/d_i, a_i/d_i)$. According to these angles, the extracted planes are assigned to one of the following classes: *floor, ceiling, left, right, front*.

To keep the UAV aligned with the corridor axis and in the center of the corridor, motion commands $(dx, dy, d\phi)$ are calculated from the plane distances Δ_i and the yaw estimates ϕ_i.

dx controls the forward movement along the corridor. It is always set to 1 meter, as long as no obstacle or wall is detected in front of the UAV. The desired horizontal displacement (perpendicular to the corridor walls) is given by $dy = (w_{left} \cdot (a - \Delta_{left}) + w_{right} \cdot (\Delta_{right} - a))/(w_{left} + w_{right})$ where a is half of the corridor width. w_{left} and w_{right} are weight factors and are equal to the number of scan points in the left and right plane respectively. This way, the wall that is supported by more scan points has a stronger influence on the resulting motion command. The same weight factors are used to generate the desired rotation command $d\phi$ that keeps the UAV aligned with the corridor axis: $d\phi = -\sum w_i \cdot \phi_i / \sum w_i$

The motion commands are transformed into the body coordinate system of the UAV by using $d\phi$ and then sent to the position and heading controller on the ATmega644P microcontroller board for execution. Commands are sent at a rate of approximately 4 Hz. The cascaded position and velocity controllers we described in section 2.2.2 are able to follow the commanded trajectories and thus keep the UAV in the corridor center. Although the altitude above the floor level can be estimated from the plane parameters as well, altitude control relies solely on the SRF10 sonar sensor, as a high measurement frequency is crucial for stable altitude control.

A video that shows an example flight is available at our website `www.tu-chemnitz.de/etit/proaut/forschung/quadrocopter.html.en`. Fig. 6 shows the position estimates Δ_{left} and Δ_{right} inside the corridor while performing autonomous flight. According to these internal measurements, the maximum deviation from the corridor center was 35 cm. Mean velocity was $0.36\frac{m}{s}$. At the moment, no ground truth information is available for a more detailed and quantitative analysis. We are working towards integrating an external 3D position measurement system to be able to evaluate the system's performance in depth in future work.

3.1 Point Cloud Downsampling

Point cloud downsampling is an effective way to reduce the computational effort for subsequent processing steps. To achieve homogeneous coverage in the three dimensional space, downsampling has to be done in voxel space, contrary to downsampling in the image plane of the Kinects depth image. The PCL (Point Cloud Library) [10] already provides a mechanism for voxel downsampling. Table 2 shows the time consumption of the PCL algorithm for downsampling a point cloud of 307,200 points to about 3,000 points and compares it to the time consumption of subsequent steps in the overall algorithm. It can be seen that the downsampling takes approximately 25 times as long as the wall extraction and position estimation together and thus absolutely dominates the overall runtime. To reduce the effect of this bottleneck, we implemented a specialized downsampling algorithm for our application.

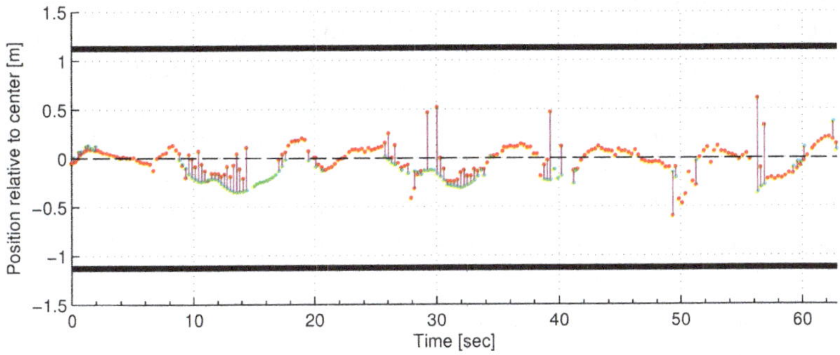

Fig. 6 Position estimates of the quadrotor within the corridor during autonomous flight, based on the Kinect measurements. The red points represent the calculated distances based on the left wall and the green points are based on the right wall. If two walls are visible, measurements are connected through a line. The corridor walls are shown as thick black horizontal lines. The maximum deviation from the corridor center was 35 cm.

While the PCL downsampling algorithm processes point clouds of various extents and data types, we use our prior knowledge about the range of the sensor, data type and data storage to speed-up the computations. Given the maximum range of the Kinect sensor and its field of view, we define a three dimensional grid of the desired resolution and size. While iterating the data of the input point cloud message, each grid cell contains just the Boolean value whether we have already processed a point inside the cell's corresponding space or not. Whenever an input point in an unoccupied grid cell is visited, it is stored in the resulting, downsampled point cloud. Further speed-up is achieved by directly processing the data in the ROS message to avoid the additional conversion from ROS message to PCL point cloud type.

By using our optimized downsampling routine, we gained an overall speed-up of factor 5.

Table 2 Time consumption comparison for point cloud downsampling methods and the other steps of the corridor extraction measured on the onboard Atom-based embedded system.

Step	Time in ms
Point cloud downsampling and conversion with PCL	492
Point cloud downsampling and conversion using our own method	89
Wall extraction, position estimation, trajectory generation	19

4 Conclusions and Future Work

Our first experiences with the novel Microsoft Kinect RGB-D device clearly showed that the sensor can be applied beneficially to mid-sized UAVs like the Pelican. We expect to see the Kinect and similar devices being used in a variety of robotics application in the near future, both in research and industry.

We demonstrated how the Kinect's depth data can be used to autonomously navigate a UAV in a corridor, without using any external sensor equipment like motion capture systems. The sourcecode for efficient point cloud downsampling and trajectory generation is available to the community as part of our ROS package at http://www.ros.org/wiki/tuc-ros-pkg. The next natural step in our work will be to extend the UAV's capabilities by enabling it to use the 3D data to avoid static and moving obstacles, and plan trajectories in less well defined indoor spaces, including large open areas as well as cluttered and confined spaces.

Acknowledgements. This work has been partially funded by the European Union with the European Social Fund (ESF) and by the state of Saxony.

References

1. Achtelik, M., Bachrach, A., He, R., Prentice, S., Roy, N.: Autonomous Navigation and Exploration of a Quadrotor Helicopter in GPS-denied Indoor Environments. In: First Symposium on Indoor Flight Issues (2009)
2. Freedman, et al: US2010/0118123A1 (2010)
3. Grzonka, S., Grisetti, G., Burgard, W.: Towards a Navigation System for Autonomous Indoor Flying. In: Proc. of IEEE International Conference on Robotics and Automation, ICRA 2009, Kobe, Japan (2009)
4. Gurdan, D., Stumpf, J., Achtelik, M., Doth, K.-M., Hirzinger, G., Rus, D.: Energy-efficient Autonomous Four-rotor Flying Robot Controlled at 1 kHz. In: Proc. of IEEE International Conference on Robotics and Automation, ICRA (2007)
5. Henry, P., Krainin, M., Herbst, E., Ren, X., Fox, D.: RGB-D Mapping: Using Depth Cameras for Dense 3D Modeling of Indoor Environments. In: Proc. of International Symposium on Experimental Robotics, ISER 2010 (2010)
6. Lange, S., Protzel, P.: Active Stereo Vision for Autonomous Multirotor UAVs in Indoor Environments. In: Proc. of the 11th Conference Towards Autonomous Robotic Systems, TAROS 2010, Plymouth, Devon, United Kingdom (2010)
7. Lange, S., Sünderhauf, N., Protzel, P.: A Vision Based Onboard Approach for Landing and Position Control of an Autonomous Multirotor UAV in GPS-Denied Environments. In: Proc. of the International Conference on Advanced Robotics, ICAR (2009)
8. Morris, W., Dryanovski, I., Xiao, J.: 3D Indoor Mapping for Micro-UAVs Using Hybrid Range Finders and Multi-Volume Occupancy Grids. In: Proc. of Workshop on RGB-D: Advanced Reasoning with Depth Cameras in Conjunction with RSS 2010 (2010)
9. OpenKinect (2010), http://www.openkinect.org/
10. The Point Cloud Library (2010), http://www.pointclouds.org/
11. OpenNI (2011), http://www.openni.org/
12. ROS (2011), http://www.ros.org/

Segmentation of Scenes of Mobile Objects and Demonstrable Backgrounds

Frederic Maire, Timothy Morris, and Andry Rakotonirainy

Abstract. In this paper we present a real-time foreground–background segmentation algorithm that exploits the following observation (very often satisfied by a static camera positioned high in its environment). If a blob moves on a pixel p that had not changed its colour significantly for a few frames, then p was probably part of the background when its colour was static. With this information we are able to update differentially pixels believed to be background. This work is relevant to autonomous minirobots, as they often navigate in buildings where smart surveillance cameras could communicate wirelessly with them. A by-product of the proposed system is a mask of the image regions which are demonstrably background. Statistically significant tests show that the proposed method has a better precision and recall rates than the state of the art foreground/background segmentation algorithm of the OpenCV computer vision library.

1 Introduction

The segmentation of moving objects in a fixed camera scene is still a developing area of research because of the many conflicting goals of background model maintenance [11]. Distinguishing background and foreground is a fundamental task of many computer vision applications such as the analysis of video streams of road traffic [1, 6], the distributed control of mobile robots in an environment equipped

Frederic Maire
FaST QUT and NICTA QRL, Brisbane QLD 4001
e-mail: f.maire@qut.edu.au

Timothy Morris
FaST QUT, Brisbane QLD 4001
e-mail: t5.morris@student.qut.edu.au

Andry Rakotonirainy
CARRS-Q QUT, Brisbane QLD 4001
e-mail: r.andry@qut.edu.au

with static cameras [9], or the monitoring of people in public places [3]. The challenges for building an accurate background model include dealing with dynamic lighting and background motion [11] such as swaying trees or water waves. Segmentation in videos is predominately achieved by comparing the current frame to some learned background model [4]. If pixels do not fit well the statistical model, they are classified as foreground. The accuracy of the background model directly affects any subsequent image processing step.

The most popular foreground-background segmentation algorithms rely on the adaptation of a statistical model. The statistical model is usually a mixture of Gaussians or a collection of bins. The statistical model must try to meet two conflicting objectives; on one hand it must adapt rapidly to react to sudden changes in lighting conditions (sun hiding behind a cloud for example), on the other hand it must have enough inertia in order not to forget how the background looks like behind a slow moving foreground object. The time scale of the adaptation of the statistical model critically depends on its learning rate. The experiments presented in Section 4 demonstrate that the state of the art foreground–background segmentation algorithm of the OpenCV computer vision library struggles to simultaneously segment slow moving and fast moving objects. The method that we introduce in Section 3 aims at addressing this problem.

2 Previous Work

Simple background segmentation methods like [12] use a single Gaussian whose parameters are updated recursively. More sophisticated methods accommodate backgrounds exhibiting multi-modal characteristics with a mixture of Gaussians [10]. These algorithms are capable of learning a statistical model for dynamic backgrounds like waves on water or swaying tree branches [1]. A number of variations of this method have been proposed. For example, in [13] the number of Gaussians per pixel can adaptively change, and in [7] a learning procedure that improves the segmentation accuracy and model convergence rate is proposed. Other improvements include the modeling of each background pixel with a set of code words [5], and the utilization of a histogram of features per pixel [8]. In [2] the neighbourhood of a pixel is modeled using local binary pattern histograms.

Figures 6, 7 and 8 were obtained with the OpenCV library implementation of the state-of-the-art background segmentation algorithm introduced in [8]. The top-left images of these figures show how moving objects smear the image associated to the statistical background model. Figure 8 shows how slow moving or stopping vehicles get integrated into the background model. In particular, slow moving elongated homogeneous blobs like the bus do no always get their interior properly segmented because the statistical model gets habituated too quickly to the interior colour of the blob. The method that we introduce in Section 3 addresses these problems.

3 Proposed Method

Our approach to create a more robust model exploits the following property of the bird eye view of most environments; if a pixel p of the image has not changed significantly from time $t-\Delta$ up to time $t-1$, and if at time t the pixel p is covered by a blob that we can trace in the short-term video memory (that is, we have observed the blob moving in the last few frames), then pixel p at time $t-\Delta$ was with high probability a background pixel. Indeed, the most likely explanation of the evolution of the colour of pixel p is that at time $t-\Delta$ pixel p corresponded to a patch of the ground and that a mobile object ran over it at time t.

Knowing that a pixel is likely to be background allows us to differentiate the way the pixel model is updated. The more confident we are that a pixel is background, the larger its learning rate should be. The pseudo-code below outlines our algorithm. Our statistical model consists of a matrix of individual normalized colour histograms for each pixel.

Steps 3 and 4 can be replaced by the computation of the likelihood of the observed pixel values, and thresholding these probabilities. However, we found that the computationally simpler method of retrieving the most likely image from the statistical model works suficiently well. It is indeed easy to keep track of which bin of a histogram is the most populated during Steps 15 and 16.

Algorithm 1. Proposed method for foreground-background segmentation

1: **while** a new frame is available **do**
2:     grab next frame F_t time-stamped t
3:     retrieve most likely image I_M from the statistical scene model
4:     threshold the difference $\|F_t - I_M\|$ into a binary image I_B
5:     apply a morphological close operator to I_B
6:     replace the blobs of I_B by their convex hulls
7:     attempt to merge neighboring convexified blobs
8:     **for all** blob B of F_t **do**
9:         create a set S of feature points belonging to the blob
10:         track the feature points of S in F_{t-1}
11:         **if** any blob B' from F_{t-1} contains a feature point for S **then**
12:             Let B inherit some properties of B' {like the age of the blob}
13:         **end if**
14:     **end for**
15:     update statistical scene model
16:     update the provable background model
17: **end while**

A short-term video memory in the form of a circular buffer of frames enables a straightforward blob tracking. For each frame a number of attributes are computed and some recorded; the age of each blob (that is, the number of frames since its first detection), a binary mask of where motion was detected, a list of feature points (corner like points) detected in the moving blobs, and the contour of each blob.

When a new frame is grabbed, a number of processing steps are performed. The first step is the computation of a mask of foreground pixels I_B. This segmentation is achieved with a simple thresholded frame difference $\|F_t - I_M\|$ where F_t denotes the current frame, and I_M the most likely image according to the adaptive statistical model of the scene. A morphological close operation on I_B helps remove image noise. Next, in lines 6 and 7, we approximate the blobs with their convex hulls. The convex hull of a blob is a better approximation of the blob than its best fitting rotated rectangle or its best fitting ellipse. But the convex hull is nevertheless simpler than the original contour of the blob in terms of the number of points of the contour.

We try to merge neighbouring convex blobs B_i and B_j the following way; we consider the contour C_U of the convex union U of B_i and B_j, then for each pair of diametrically opposite vertices on C_U, we scan the line segment joining these two vertices. If all the scanned segments are likely foreground, then B_i and B_j are replaced by the convex blob U.

In order to track the blobs, we perform a search of a set of *good features to track* in the region of the new frame F_t restricted to I_B, and try to match these feature points in previous frames by calculating the optical flow for this sparse feature set using the iterative Lucas-Kanade method with pyramids (implemented in the OpenCV library). Good features are located by examining the minimum eigenvalue of each 2 by 2 gradient matrix, and features are tracked using a Newton-Raphson

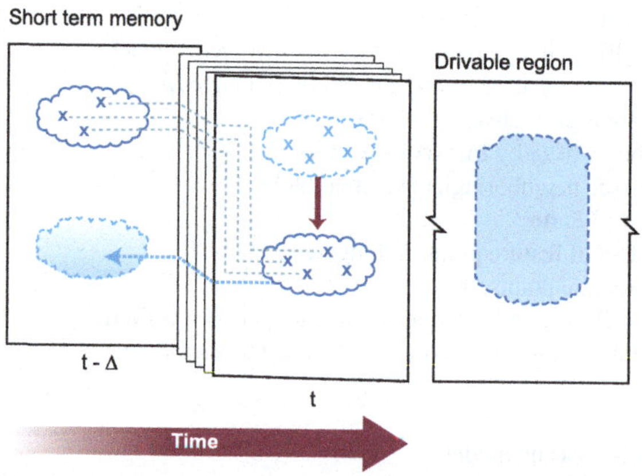

Fig. 1 The short-term video memory collects information about moving blobs. When a foreground blob is detected in frame F_t, features inside this blob are tracked backward temporally to determine whether the blob corresponds to a moving object. If the pixels in the corresponding footprint of the blob in frame $F_{t-\Delta}$ did not change significantly from time $t-\Delta$ to time t, then these pixels are assumed to be background. The traces of the moving blobs are accumulated into a *driveable* mask.

method of minimizing the difference between the two windows. Multi-resolution tracking allows for relatively large displacements between images. Each blob of the current frame either inherits the attributes (internal identification number and drawing colour) from the matched blob (if any) in the previous frame, or is classified as a new blob. To finish the main loop, we update the statistical model of the scene. The update also refines the driveable region model as illustrated in Figure 1.

4 Experimental Results

We have implemented using the OpenCV computer vision library the method described in the previous section. The OpenCV library provides optimized functions for many of the image processing tasks our method needs to complete (like the detection of feature points, the computation of the convex hulls and the extraction of the contour of a blob). In particular, finding distinctive points that can be tracked by the Lucas Kanade method does not require much programming. Lucas Kanade method is a widely used differential method for optical flow estimation that runs in real time.

We have tested our system on two videos. The system runs in real time (more than 25 frames per second on a laptop). Videos comparing our proposed method to the state-of-the-art background segmentation method available to anyone through the OpenCV library *CvFGDStatModel*. The results of these tests can be viewed on-line. Per image, our method was about 5 ms slower than *CvFGDStatModel*.

Video name	Video URL	Method used
Intersection	`http://www.youtube.com/watch?v=c5b52L00xUE`	Proposed method
Highway	`http://www.youtube.com/watch?v=jxyY2Rs11FQ`	Proposed method
Intersection	`http://www.youtube.com/watch?v=UeBr_7Kn2hU`	CvFGDStatModel
Highway	`http://www.youtube.com/watch?v=8X893aZGFy4`	CvFGDStatModel

Precision and recall are two widely used metrics for evaluating the correctness of pattern recognition algorithms. In the context of our application, a true positive is a detected blob that corresponds to a foreground object, a false positive is a detected blob which does not correspond to a foreground object. A false negative is a foreground object which has not been detected. The *precision* is the number of true positives divided by the sum of the number of true positives and the number of false positives. The *recall* is the number of true positives divided by the sum of the number of true positives and the number of false negatives. The performance of the two tested algorithms are summarized in the table below. For our experiments, we labeled by hand the moving objects in the two videos. The blobs corresponding to captions in the videos were ignored as they could be discarded easily with a mask in a preprocessing step.

Video name	Method used	Precision	Recall
Intersection	Proposed method	1.00	0.83
Highway	Proposed method	1.00	0.97
Intersection	CvFGDStatModel	0.80	0.68
Highway	CvFGDStatModel	0.97	0.71

Running statistical one-tailed *t-test* for paired samples from each video with respect to the two methods shows that the proposed method performs better with a statistical significance of a *t-value* level less than 0.01 for all cases.

Figures 2, 3, 4, 6, 7 and 8 are all divided in four subfigures. The top left subfigure shows the original frame, the top right subfigure is the most likely image according to the statistical model of the scene. The bottom left binary subfigure shows the intermediate segmentation (after pixel classification with the statiscal background model and morphological close). The bottom right subfigure shows the final segmentation.

Fig. 2 Frame 68 processed by the proposed method. The cyclist was just detected (bottom left subfigure), but this blob is not old enough to appear in the segmented image. The most likely image (top right subfigure) is clean.

Fig. 3 Frame 269 processed by the proposed method. The blob corresponding to the cyclist is old enough to appear in the segmented image.

Fig. 4 Frame 204 processed by the proposed method. One of the cars stopped at the traffic light has not been tagged in the segmented image. The colour of the car is too similar to the colour of the road. Although a blob corresponding to the front of this car can be seen on the bottom left image, it is too recent as no blob was detected for this car on the previous frame.

Fig. 5 Provable background image quantized with 64 gray level values.

Fig. 6 Frame 68 processed by the OpenCV library method. The cyclist is missing from the binary image. Slow moving objects smear the model image.

Fig. 7 Frame 269 processed by the OpenCV library method. The cyclist is still missing from the binary image. Moreover some cars are completely missed.

Fig. 8 Frame 204 processed by the OpenCV library method. When cars slow down at the traffic light, they become part of the background.

5 Conclusion

In this paper we have presented a new foreground-background segmentation method that exploits blob motion to learn a more robust statistical model of the environment. We have designed and implemented in C++ (using OpenCV) a prototype for a scene analysis system. The approach introduced in this paper is applicable to any environment that is intrinsically two dimensional. Fixed networked smart cameras which look down on the ground could assist autonomous mobile robots in their navigation task. A robust background segmentation algorithm like the one proposed here is highly desirable for these environments.

Acknowledgements. The first author would like to thank Ms Irina Gordienko for her invaluable help. The algorithms for the experiments were implemented and evaluated using open-source software, including Linux-based operating systems. The authors would like to extend their thanks to all open-source developers for their efforts.

References

1. Cheung, S.-C.S., Kamath, C.: Robust techniques for background subtraction in urban traffic video. In: Panchanathan, S. (ed.) Society of Photo-Optical Instrumentation Engineers (SPIE) Conference Series, Presented at the Society of Photo-Optical Instrumentation Engineers (SPIE) Conference, vol. 5308, pp. 881–892 (January 2004)
2. Heikkilä, M., Pietikainen, M.: A texture-based method for modeling the background and detecting moving objects. IEEE Transactions on Pattern Analysis and Machine Intelligence 28(4), 657–662 (2006)
3. Huwer, S., Niemann, H.: Adaptive change detection for real-time surveillance applications. In: Proceedings. Third IEEE International Workshop on Visual Surveillance, pp. 37–46 (2000)
4. Kim, H., Sakamoto, R., Kitahara, I., Toriyama, T., Kogure, K.: Robust foreground extraction technique using background subtraction with multiple thresholds. In: Optical Engineering, vol. 46, SPIE (September 2007)
5. Kim, K., Chalidabhongse, T.H., Harwood, D., Davis, L.: Real-time foreground-background segmentation using codebook model. Real-Time Imaging 11(3), 167–256 (2005), Special Issue on Video Object Processing
6. Lai, A.-N., Yoon, H., Lee, G.: Robust background extraction scheme using histogram-wise for real-time tracking in urban traffic video. In: 8th IEEE International Conference on Computer and Information Technology, CIT 2008, July 2008, pp. 845–850 (2008)
7. Lee, D.-S.: Effective gaussian mixture learning for video background subtraction. IEEE Transactions on Pattern Analysis and Machine Intelligence 27(5), 827–832 (2005)
8. Li, L., Huang, W., Gu, I.Y.H., Tian, Q.: Foreground object detection from videos containing complex background. In: MULTIMEDIA 2003: Proceedings of the Eleventh ACM International Conference on Multimedia, pp. 2–10. ACM Press (2003)
9. Losada, C., Mazo, M., Palazuelos, S., Redondo, F.: Adaptive threshold for robust segmentation of mobile robots from visual information of their own movement. In: IEEE International Symposium on Intelligent Signal Processing, WISP 2009, August 2009, pp. 293–298 (2009)
10. Stauffer, C., Grimson, W.: Adaptive background mixture models for real-time tracking. In: Computer Vision and Pattern Recognition (CVPR), vol. 2, pp. 246–252 (1999)

11. Toyama, K., Krumm, J., Brumitt, B., Meyers, B.: Wallflower: principles and practice of background maintenance. In: The Proceedings of the Seventh IEEE International Conference on Computer Vision, vol. 1, pp. 255–261 (1999)
12. Wren, C.R., Azarbayejani, A., Darrell, T., Pentland, A.P.: Pfinder: real-time tracking of the human body. IEEE Transactions on Pattern Analysis and Machine Intelligence 19(7), 780–785 (1997)
13. Zivkovic, Z.: Improved adaptive gaussian mixture model for background subtraction, vol. 2, pp. 28–31(August 2004)

A Real-Time Event-Based Selective Attention System for Active Vision

Daniel Sonnleithner and Giacomo Indiveri

Abstract. In real world scenarios, guiding vision to focus on salient parts of the visual space is a computationally demanding tasks. Selective attention is a biologically inspired strategy to cope with this problem, that can be used in engineered systems with limited resources. In *active* vision systems however, the stringent real-time requirements limit the space of solutions that can be achieved with conventional machine vision techniques and systems. We propose a hybrid approach where we combine a custom neuromorphic VLSI saliency-map based attention system with a conventional machine vision system, to implement both fast contrast-based saccadic eye movements in parallel with conventional visual attention models that use high-resolution color input images. We describe the system and characterize its response properties with experiments using both basic control visual stimuli and natural scenes.

1 Introduction

Selective attention is the strategy used by a wide range of animals [3, 6, 19, 23] to cope with the problem of processing high amounts of sensory inputs in real-time. Rather than attempting to process everything in parallel at once, selective attention allows the system to process the most relevant parts of the sensory input sequentially [9, 22]. For example, in primates selective attention plays a major role in determining where to center the high-resolution central foveal region of the retina for visual processing [21], by biasing the planning and production of saccadic eye movement sequences [2, 13].

This is a highly effective strategy for optimizing the use of computing resources that is often also used in artificial sensory-motor systems. In particular this strategy

Daniel Sonnleithner · Giacomo Indiveri
Institute of Neuroinformatics, University of Zurich and ETH Zurich,
Winterthurerstrasse 190, CH-8057 Zurich, Switzerland
e-mail: {daniel.sonnleithner,giacomo}@ini.phys.ethz.ch

has been adopted by a large number of research projects within the field of robotics and machine vision (see Frintrop et al. 2010 [12] for a recent survey). However, as vision is computationally intensive, selective attention models have been applied mainly to *passive* vision systems (i.e., machine vision systems operating on static images). *Active* vision systems on the other hand have extremely stringent requirements, as they often need to carry out all of the sensory processing in real-time. The real-time requirements together with additional constraints on size and power consumption of the computing hardware are still limiting the application of selective attention models to active-vision systems and mobile robotics.

To overcome this problem, we developed an active vision framework based on a dedicated hardware solution that can carry out the planning and production of camera movements in real-time, interfaced to a conventional machine vision system. The conventional machine vision system is composed of a standard color camera interfaced to a workstation for executing machine vision algorithms, while the custom hardware past is composed of hybrid analog/digital Very Large Scale Integration (VLSI) chips that implement real-time models of sensory processing systems and neuromorphic models of spiking neurons and cortical neural networks [16]. In particular the neuromorphic multi-chip system presented in this paper comprises a Selective Attention Chip (SAC [1]) inspired by saliency-based models of attention [18] and a Dynamic Vision Sensor (DVS [20]) inspired by the fast transient pathway of mammalian retinas. The DVS is a low-resolution vision sensor that responds to temporal contrast changes in the sensor's field of view in real-time and is not sensitive to color. Both, conventional color imager and custom DVS are mounted on a motorized pan-tilt-unit which orients them towards the most salient stimuli, as computed by the SAC.

While the low-resolution custom vision system responds in real-time to moving stimuli (such as objects entering the sensor's field of view) and can be used to produce fast reactive motor outputs, the conventional high-resolution machine vision system can be used to carry out higher level processing tasks (such as object recognition) on the images being analyzed, in between saccadic camera movements.

The framework proposed is inspired by the mammalian visual system that uses a high-resolution color "device" (the retina's fovea) in parallel with a lower-resolution "device" (the retina's periphery) that responds mainly to moving or transient stimuli and is less sensitive to color and shape. Computation of a saliency map using mainly changes in contrast of a moving scene is supported by recent findings that demonstrate that motion and temporal change are strong predictors of human saccades [17].

In the next section we describe the active vision setup. In Section 3 we present experimental results that demonstrate the real-time capabilities of the selective attention system, and in Section 4 we discuss the results and present concluding remarks.

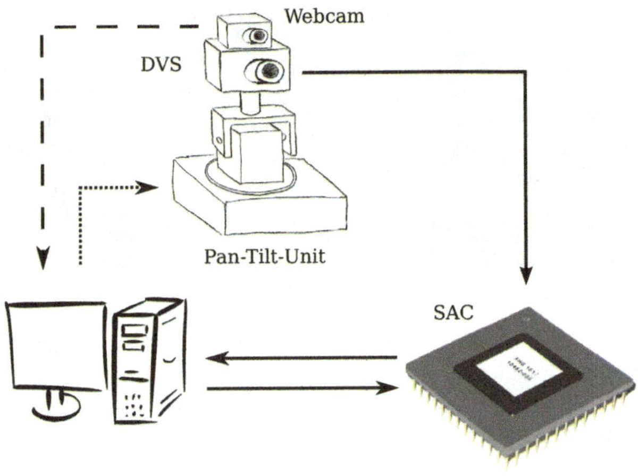

Fig. 1 Experimental setup diagram: both a high-resolution camera and a DVS chip are mounted on a pan-tilt-unit, controlled by a workstation. The camera is directly connected to the workstation, while the DVS sends its outputs to the SAC. The SAC processes the DVS data, computes the location of the most salient input, and transmits this information to the workstation. The workstation is then used to drive the pan-tilt-unit so that the most salient location is centered in the DVS field of view. Solid lines represent AER connections, the dashed represents vision signals from the standard camera, and the dotted line represents motor control signals

2 The Active Vision Setup

The active vision setup, with standard machine vision components interfaced to custom neuromorphic devices is depicted in Fig. 1. The pan-tilt-unit orients both vision sensors toward salient stimuli. It can operate at speeds of more than $300^\circ/\mathrm{s}$ with a resolution of about 0.05°, and is controlled by the workstation via a serial interface. The high-resolution camera (a Logitec C200 web cam) is interfaced to the workstation via a standard USB connection and provides a 640×480 pixels video stream at 30 Hz. The DVS is the 128×128 pixel sensor described in Lichtsteiner et al. 2008 [20]. This sensor responds to temporal changes in the logarithm of local image intensity, thus encoding relative temporal changes in contrast, rather than absolute illumination (as in the conventional camera).

Thanks to the logarithmic compression, the DVS is able to detect contrast changes as low as 20 % with a dynamic range spanning over 5 decades. Each pixel in the DVS performs this computation independently (local gain control), allowing the DVS to optimally respond to scenes with non-homogeneous illumination (e.g., outdoors or in environments with uncontrolled illumination). An important feature of the DVS, which makes it radically different from the sensors used in conventional machine vision approaches is the way it transmits output signals: signals are not scanned out on a frame-by-frame basis. Rather, the address of a pixel is transmitted on a shared digital bus, as soon as that pixel senses a difference in contrast.

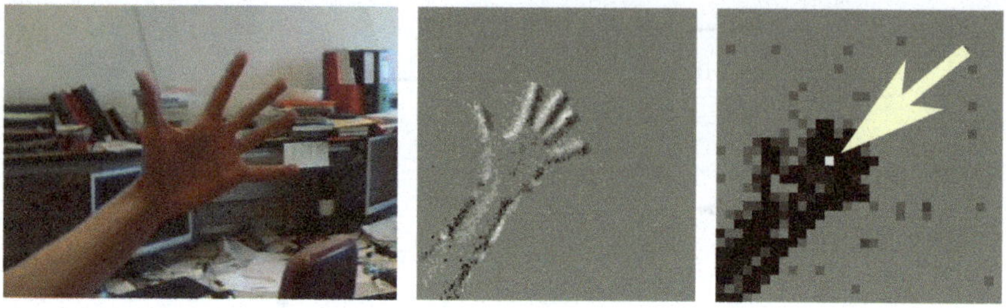

(a) (b) (c)

Fig. 2 Output example of standard vision sensor, DVS and SAC. (a) Image acquired from the high-resolution color camera. (b) Same scene recorded with the dynamic vision sensor. Resolution is 128 × 128; white dots represents increase, black dot decrease in contrast respectively. (c) Same scene as it is seen from the SAC at a resolution of 32 × 32 pixel. Black dots represent the input given also to the SAC. The white pixel (indicated with the arrow) is the output of the SAC and represents the location with the current highest saliency.

This "event" is written on the bus as it happens, in a completely asynchronous fashion. Each pixel address is written on the bus in real time, and potential conflicts (cases in which multiple pixels attempt to access the shared bus at the same time) are managed by an on-chip arbiter. This asynchronous communication protocol is based on the address-event Representation (AER) [4, 7]. As the DVS only transmits data when pixels sense sufficient contrast changes, redundancy in the data is strongly reduced (e.g., no data is transmitted and no bandwidth is used when there is no change in the visual scene). This produces a sparse image coding and optimizes the use of the communication channel, as well as the post-processing and storage effort. This, combined with the real-time asynchronous output nature of the DVS ensures precise timing information and low latency [20] yet requires a much lower bandwidth than used by frame-based image sensors of equivalent time resolution [8].

In general, AER systems convert analog signals into streams of stereotyped non-clocked digital pulses (spikes) and encode them using pulse-frequency modulation (spike rates). When a spiking element on an AER VLSI device generates an event, its address is encoded and instantaneously put on a digital bus. In this way time represents itself, and analog signals are encoded by the inter-spike intervals between the addresses of their sending nodes. By converting analog signals into this digital representation, we can take advantage of the high-speed digital communication tools and exploit the flexibility offered by digital systems. The Selective Attention Chip presented here, and the overall system use this AER scheme for both *communicating* and *processing* events that travel across the system's computational stages. Indeed, the SAC is using the same representation to receive the DVS signals, process them, and transmit the outcome of the selective attention processing.

The events produced by the DVS are transmitted to the SAC for computing in real-time the position of the salient target(s). The events generated by the SAC are then transmitted to a workstation for further processing that results in driving the

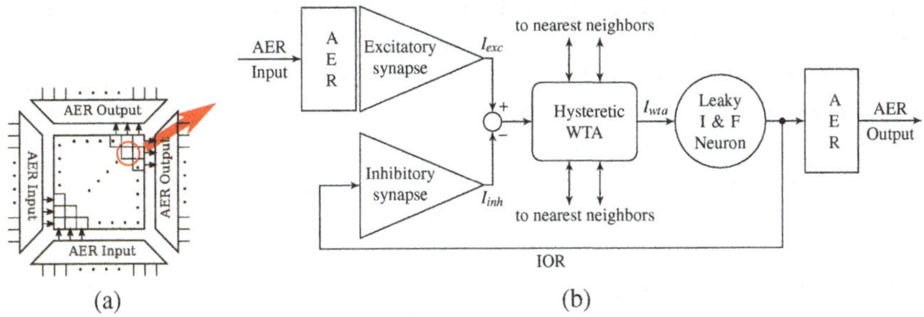

Fig. 3 Selective Attention Chip (SAC) diagram. (a) The SAC consists an array of 32×32 pixels providing its computational resources and communicates with other hardware via AER receiver-transmitter circuits. (b) Block diagram of an SAC pixel. Each pixel receives AER spikes from the input bus and competes for saliency by means of a hysteretic winner-take-all network connected to its neighbors via lateral connections. The winning pixel sends its address to the output AER bus and self-inhibits via a local inhibitory synapse. All blocks are implemented with hybrid analog/digital circuits described in Bartolozzi et al. 2009 [1].

pan-tilt-unit. Figure 2 shows the outputs of both vision sensors, as well as the output of the SAC, in response to the same input stimulus.

All of the asynchronous address-event traffic is managed by custom Field Programmable Gate Array (FPGA) boards [11] and a look-up table based "mapper" that assigns destination addresses to each source address [10]. In this way events produced by different pixels on the DVS can be mapped to one or more pixels on the SAC e.g. to implement log-polar or retinotopic mappings. Similarly events produced by other AER sensors, such as the silicon cochlea [5] can be used to create more complex saliency maps. Events produced from algorithms executed on workstations can also be used to shape or modulate the saliency map, for example to model the effect of top-down influences on the selective attention competitive process (see Section 3).

2.1 The Selective Attention Chip

The saliency map is constructed by the input circuits of the SAC, which integrate the incoming address-events and carry out further processing on them. The chip has been described in detail in Bartolozzi et al. 2009 [1]. It comprises an array of 32×32 pixels with AER digital circuits as well as analog neuromorphic circuits that implement silicon synapses, neurons, and additional signal processing stages. Figure 3(b) shows the block diagram of an SAC pixel: each pixel in the array receives input sequences of spikes which encode the saliency of the corresponding pixel in the visual scene; an input excitatory synapse integrates the spikes into an excitatory current I_{exc} which is then fed into a hysteretic Winner-Take-All (WTA) circuit [14]. The hysteretic WTA network compares the input currents of all pixels and activates only the pixel receiving the largest input current, while suppressing the output of all other pixels. The winning pixel will then produce a constant output current I_{wta}, which is independent of the input, and source it to the pixel's leaky Integrate and Fire (I&F) neuron. This circuit, fully characterized in Indiveri et al. 2006 [15], produces voltage pulses (spikes) at a rate which is proportional to it's input current. Each time a spike is emitted from a neuron, the address of the spiking pixel is encoded on a digital bus, instantaneously. The output AER circuits manage the asynchronous transmission of address-events to the other components of the selective attention system. In parallel, the spikes of the I&F neuron are sent to the pixel's inhibitory synapse which produces a negative current I_{inh}. This implements a negative feedback loop in which the current integrated from the output spikes I_{inh} is subtracted from the external input current I_{exc}. The net input current to the winning pixel therefore decreases until a different pixel wins the competition for saliency. This self-inhibition implements a known mechanism in selective attention models named *inhibition of return* (IOR). It allows the network to shift from the currently attended stimulus to a different one, selecting sequentially the most salient regions of the input space in order of decreasing salience, reproducing the attentional scan path [18].

2.2 Mapping Events to the Selective Attention Chip

The custom FPGA boards and the mapper device developed to manage the AER traffic [10] allow us to define arbitrary connectivity patterns for implementing different mappings from one or more AER sensors to the SAC. As the mapping tables are stored in the main memory of a PC motherboard, we have access to large amounts of fast memory (2 GiB in this system) and can program a wide variety of mappings.

As the DVS and the SAC have different resolutions, the use of the address-event mapper is extremely useful: the DVS uses a 15 bit address space to encode the position of its 128×128 pixels as well as the polarity of the pixel's sensed contrast change resulting either to an increase or decrease of contrast (on- or off- event). In contrast, the SAC's 32×32 pixel array uses only 10 bits to encode its pixel addresses.

The mapping used throughout this paper was linear, i.e. both the x- and the y-value of the DVS output addresses are divided by 4 and mapped topographically to the SAC, irrespective of the event polarity (each SAC pixel has a receptive field corresponding to a 4×4 pixel area on the vision sensor). However, as the mapper can be reconfigured easily, this infrastructure is also useful to explore alternative mappings (e.g. retinotopic) from the dynamic vision sensor to the SAC, or to fuse inputs coming from different AER sensors (e.g. multiple vision sensors, or vision and auditory sensors).

2.3 Controlling the Pan-Tilt-Unit

In addition to being monitored by the workstation (e.g. to evaluate the system performance), the SAC's output address-events are used to control the pan-tilt-unit movements: the workstation processes the SAC output address-events to orient both vision sensors, moving them toward the salient regions in the visual space. As the control algorithms are executed on the workstation, many different strategies can be flexibly explored.

In the current experiments we adopted a control strategy defined as follows: the visual field is subdivided into five main regions (top, left, bottom, right and center); if an event is generated by pixels belonging to the center, the system does not move the actuator; if the event belongs to one of the other regions, a motion vector is calculated for both pan and tilt as

$$\Delta\alpha_i = \frac{e_j}{31} \cdot \beta_i - \frac{\beta_i}{2}, \quad i \in \{\text{pan, tilt}\},\ j \in \{\text{x,y}\}, \tag{1}$$

where e represents the event's x or y address, $\Delta\alpha_i$ denotes the changes of angle that have to be applied to the pan-tilt-unit, and β represents the angle of view of the DVS. Note that the highest possible address value is 31.

An alternative control algorithm that we adopted calculates $\Delta\alpha$ for every event produced (irrespective of the region it belongs to) and performs an appropriate thresholding, rather than first checking the region of origin and then performing the calculation for events coming only from border regions.

3 Experiments

To characterize the selective attention system, we conducted a set of basic control experiments. In a first set of experiments we measured the response of the SAC to different stimulus conditions, without activating the pan-tilt-unit motors (see also Sonnleithner et al. 2011(b) [25]). In a second set of experiments, we activated the control loop and used the events produced by the SAC to orient the vision sensors (see also Sonnleithner et al. 2011 [24]).

3.1 Covert Attention Experiments

To examine the SAC's response to different visual inputs, we stimulated the DVS by presenting different patterns on an LCD screen, and analyzed the SAC output address-events. The DVS was stimulated by three blinking black rectangles on a white background of the LCD screen. We used blinking frequencies ranging from 5 to 30 Hz. The size of the rectangles was chosen such that in most of the cases only one pixel in the SAC was stimulated.

Due to the "real-world" conditions used in this experiment, namely the refresh rates of the LCD screen, the mapping of the 128×128 DVS pixels to the 32×32 SAC pixels, the variability in the illumination conditions, and the mismatch and inhomogeneous properties of both DVS and SAC VLSI circuits, the spike-trains received by each SAC pixel do not have a regular 5 to 30 Hz frequency. Rather, they are inhomogeneous, with periods of bursting activity interleaved by periods of noisy low frequency. The inter-burst frequencies are proportional to the visual stimuli blinking frequencies.

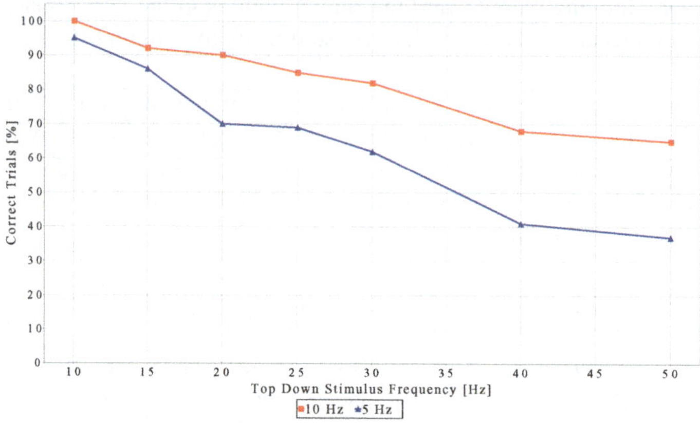

Fig. 4 Percentage of correct trials for different distractor frequencies. The X-axis represents the top-down stimulus frequencies. In a correct trial, the SAC reports the location of the distractor rather than the top-down stimulus location. The Y-axis shows the percentage of correct trials.

This experimental setup was chosen as a compromise between "natural" scene stimuli (that would be used in typical operating conditions), and well controlled stimuli (e.g., produced by function generators or computers), in order to determine the system's settings, for optimal operation in natural conditions, while having good control of the stimulus properties.

In these control experiments the selective attention system is expected to detect the rectangle that blinks with the highest frequency (i.e. the salient target) and ignore the two distractors blinking with a lower common baseline frequency. As done in psycho-physics experiments, we set parameters in our experiments at threshold, so that the system would not select the right target 100% of the times, and measured the equivalent of psychometric curves on the artificial system, by gradually increasing the difference between baseline stimulus frequencies and target stimulus frequencies. We ran two sets of experiments with different baseline frequencies: one with 5 Hz, and the other with 10 Hz (see Fig. 5). Furthermore we repeated the experiments with an additional input generated synthetically on the workstation, as a sequence of extra address-events merged to the stream of address-events coming from the sensor, to apply the concept of top-down attention to the system.

3.1.1 Experiment Description

Each experiment comprises three 5 s lasting runs. Before the beginning of each experiment run, the system was reset to an initial state: the weights of the input excitatory synapses were set to zero, the WTA circuit bias current was turned off and the leak of the output neurons was set to max. At the onset of each run these parameters were reset to their default values.

To account for mismatch effects from both DVS and SAC circuits, we chose the locations of the three black rectangles randomly for each experiment, but kept them fixed for each of the experiment's runs. During the three runs, the target was permuted among the three locations. We swept the target frequency from the baseline value (either 5 or 10 Hz) up to 30 Hz. Higher target frequencies could not be used, due to interference with the monitor or system refresh rate. For each target frequency chosen, we repeated multiple trials of the experiments and calculated the percentage of correct choices made by the SAC. To estimate how reliable the selection of the correct target is, we repeated the same set of experiments, using the same randomly picked locations, multiple times (see error-bars in Fig. 5).

As a next step, we set appropriate weights to the inhibitory synapses to activate the inhibition-of-return mechanism in the winning WTA cell. This feature should allow the system to scan through all salient regions (i.e., the three blinking rectangles), but ideally the location of the strongest stimulus should be chosen more often than the distractors.

Finally, we were interested to test if the concept of "top-down attention" is applicable to our system and to see how it would influence the performance of the detection of the salient target. We simulated top-down influence by using a computer-generated stimulus that provides an additional input to the location of the target rectangle, and measured its effect on the selection process. The stimulus was chosen such that it would not always win the competition process against the distractors, if presented in isolation (without the visual target). Therefore we generated an artificial 15 Hz Poisson spike train that stimulated an area of 3×3 SAC pixels centered at the location of the visual target and applied it in parallel to the visual "bottom-up" stimulus.

To calibrate the top-down stimulus in a way that it would not alter the bottom-up selection process if presented alone (i.e., to find the appropriate top-down stimulus frequency), we stimulated the SAC with the top-down Poisson spike train while displaying a visual stimulus corresponding to single rectangle blinking either at 5 Hz or at 10 Hz at a different spatial position, and evaluated the competition process. Then we counted the number of times the bottom-up visual stimulus was selected and related it to the total number of trials. The results of these control experiments are shown in Fig. 4. Since there is a significant drop at 20 Hz top-down stimulus frequency, we chose maximum frequency of 15 Hz for the top-down spike-train.

3.1.2 Results

For each experiment run, we recorded both the input events mapped to the SAC and the SAC output events. For each target-distractor frequency pair we counted the runs where the selective attention system chose the target stimulus correctly and related it to the total conducted runs. The percentage correct results are summarized in Fig. 5. As expected, when all three stimulus rectangles are blinking at the same (baseline) frequency the system picks one location at random (33% correct trials). This happens for both sets of experiments, with different baseline frequencies. As the difference between the target and the distractor frequencies increases, the percentage of correct runs increases. The drops in performance at 20 Hz for the 5 Hz baseline frequency correspond to an absolute target frequency of 25 Hz. Therefore

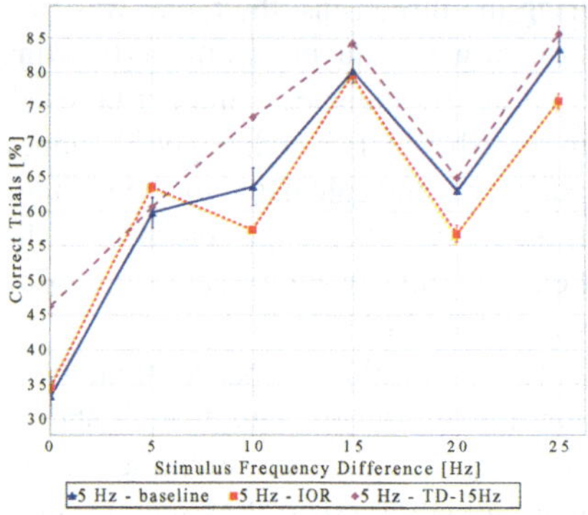

(a) 5 Hz distractor frequency

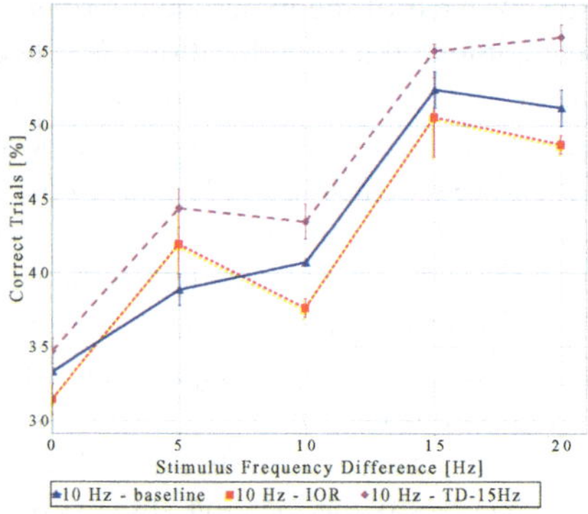

(b) 10 Hz distractor frequency

Fig. 5 Percentage of correct trials for different baseline distractor and target stimulus frequencies. The X-axis represents the *difference* between the distractor baseline frequency and the target blinking frequency. The Y-axis represents the percentage of correct trials. The dotted lines report the results of experiments with the IOR mechanism activated. Dashed lines show the results obtained with the additional top-down input. Error bars represent the standard deviation. There is a drop in performance at 20 Hz for the 5 Hz distractor frequency experiments. As this corresponds to an absolute stimulus target frequency of 25 Hz, the drop in performance is most likely due to artifacts due to interference with by the power line or the screen's refresh rate.

it is most likely due to artifacts induced by interference with the power line or the screen's refresh rate.

When activating the SAC's IOR mechanism, the system's performance is less regular. This is expected since this mechanism introduces additional dynamics into the selection process.

As expected, the top-down stimulus can positively bias the selection process: the system's performance in choosing the correct rectangle increases for both baseline frequencies (see dashed lines in Fig. 5).

3.2 Overt Attention Experiments

In this section we describe experiments in which the active vision system orients the camera and the DVS toward salient regions. Specifically, we oriented the dynamic vision sensor toward a standard LCD screen and presented visual stimuli provided by a Java program that we developed for this purpose. The stimuli consisted of two blinking blobs on two fixed locations A and B (see Fig. 6). We chose stimuli locations A and B such that they lay both in the DVS field of view, and such that both axes of the pan-tilt-unit had to move (pan: about 12°, tilt: about 8.5°) in order to shift the DVS to center location B in its field of view, from location A.

At the beginning of the experiment, a blob blinking at a frequency of 10 Hz was presented at location A, and the DVS was centered on A. After 5 s, a blinking blob

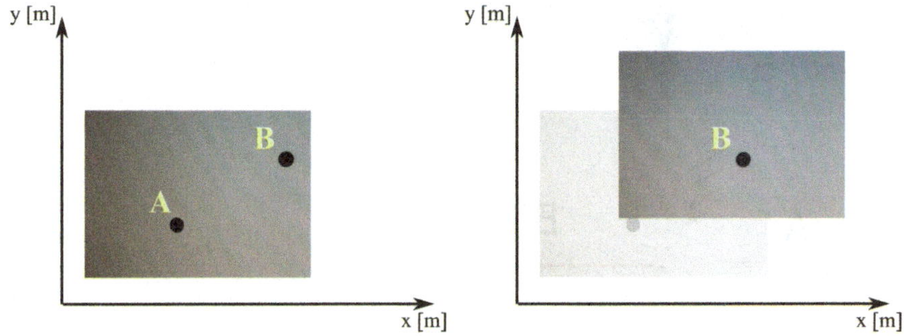

(a) Start position (b) End position

Fig. 6 Overt attention control experiment: (a) while the system is focusing on the bottom left dot A, the top right dot B appears and starts to blink. The system selects the new input B as the winner and eventually it makes a saccadic camera movement to centers the new target in its field of view (b). The system uses the DVS to calculate the field of view center, and the stimuli A in (a) and B in (b) are not in the center of the color vision sensor images because it is not perfectly aligned with the DVS.

of 20 Hz appeared at location B. At the same time, the blob at location A stopped blinking. After 5 s the blinking location was switched back, then blinking at a frequency of 30 Hz. The experiment ended after another 5 s. The increased frequencies made sure that the newer stimuli were always more salient than the preceding ones.

Both the stimulus data sent to the SAC and the output data produced by the SAC were recorded. Fig. 7 shows an example of raw address-event data: The plot's horizontal axis shows the experiment's time in seconds. Each dot in the figure represents the occurrence of an event. To represent the two dimensional structure of the chip, the pixels' x- and y-coordinates were collapsed on the y axis ($pos = x + 32y$).

During this control experiment the SAC's IOR feature was not enabled.

Measurement

The raw address-event data was analyzed to measure the active vision system's reaction times. To get a better visual representation of the data, the addresses that represented the blinking blobs were highlighted by colors (see Fig. 7). During the first phase (highlighted in blue), the system fixated the blinking blob at location A. At about 183.7 s the second blob at location B began to blink. In the raster plot, this phase is colored in pink. After a short time the system reacted on this new input and the pan-tilt-unit began to move. This phase can be easily identified by the high activity throughout all DVS addresses around 184 s. The arrows in Fig. 7(a) point to the clusters of spikes generated by the blob moving from B to B'. Finally, in the third phase of the experiment the system has centered the location B (colored in red, indicated with letter B').

On average, with the biologically plausible time-constants and settings used in these experiments, the system takes 128 ms ($\sigma = 25.3$ms) to shift from one location

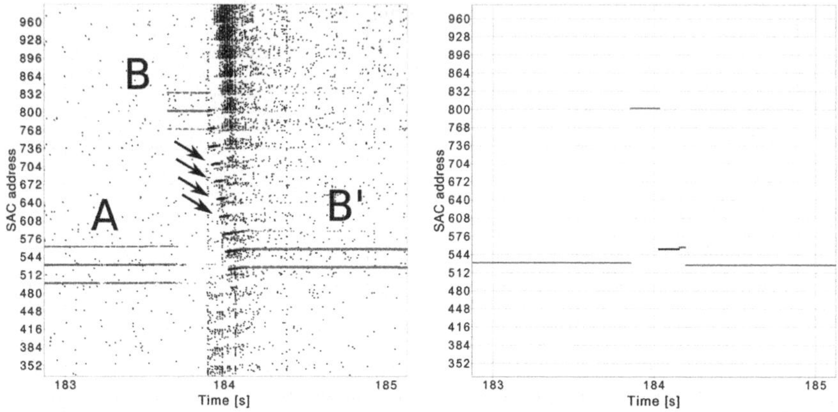

(a) SAC input spikes (b) SAC output spikes

Fig. 7 Raster plots of spikes representing the SAC input (a) and output (b). Each dot in the plot corresponds to an address-event. To represent the two dimensional structure of the chip, the pixels' X- and Y-coordinates were collapsed on the Y axis ($pos = X + 32Y$). Arrows indicate the clusters of spikes generated from the blob at location B during the camera movement.

to the next. As observed in the raster plot of Fig. 7(b), and as expected by the WTA operation of the SAC, there is only one winner at a time. After the winner is chosen, the system takes 28 ms ($\sigma = 1.4$ms) to start a new saccadic camera movement (latency measured from the first output spike produced by the SAC). We used the significant increase in overall activity of the DVS to define the time of saccade onset. With the beginning of the onset of a saccadic camera movement we measure the final figure of merit: the time required by the pan-tilt-unit to center the new salient region in the DVS field of view. We define the end of such period by using the spikes produced by the SAC at the new location. For this time period the system requires 324 ms ($\sigma = 18.2$ms).

The overall time used by the active vision system to select a new target and move the sensors to center it in its field of view can be obtained by summing up the time of these three different phases. This results in less than 500 ms. Both SAC and motion latencies can be easily decreased and tuned to the experiment/system requirements. In this experiment we purposely biased the SAC to have biologically plausible response properties, which result in these relatively high latencies.

4 Conclusions

We presented an active vision system that combines the strengths and advantages of both classical machine vision approaches and custom neuromorphic VLSI technology. In this work we described the overall system and focused mainly on basic control experiments to demonstrate the non-conventional aspects of the active vision system (namely it's ability to select salient targets and orient the sensor towards them in real time). We carried out additional experiments in less controlled and more cluttered environments, comprising for example blinking LEDs as targets in an office environment, with people walking in the background as distractors, and verified the same qualitative response properties.

The neuromorphic part of the architecture exploits the features of both the DVS and the SAC to create a biologically plausible selective attention system, similar to what has been previously proposed [1]. The overall framework developed here however allows the user to experiment with different models and different approaches: the programmable mapper used allows users to easily change look-up/mapping tables, so that events produced by the vision sensor can be mapped with different one-to-one, many-to-one, and/or one-to-many mapping schemes (e.g., to explore the effect of retinotopic mappings). In addition, this allows multiple AER sensors and devices to contribute to the creation of the saliency map on the SAC input synapses, raising the possibility to easily explore sensory-fusion strategies in the context of active (motorized) selective attention setups. The main strength of the framework proposed here lies in the ability to interface the classical machine vision methods to the neuromorphic components of the system. On both bottom-up, saliency-based selective attention algorithms as well as high-level or object-based models can be run in the machine vision system, and their output, once converted into AER, can be fused with the address-events being transmitted by the real-time sensors and processing chips. In this way complex software models that use high-resolution color vision sensors can modulate the saliency map on the SAC, influence or bias the selective attention competition taking place in the SAC, and ultimately determine the sequence of saccadic "eye" movements, where the "eye" in our case is the sensory system composed of a slow (frame-based) high-resolution color sensor, and a fast (asynchronous) low-resolution contrast transient sensor.

Acknowledgements. This work was supported by the Swiss National Science Foundation Grant #121713: "Neuromorphic Attention" (nAttention).

The SAC was designed by Chiara Bartolozzi. The DVS was gratefully provided by Tobi Delbrück and Raphael Berner. We thank Emre Neftci, Sadique Sheik and Fabio Stefanini for developing part of the AER software framework, and Daniel Fasnacht for developing the AER mapper.

We are grateful to Prof. Joaquin Sitte for comments and feedback on the manuscript.

References

1. Bartolozzi, C., Indiveri, G.: Selective attention in multi-chip address-event systems. Sensors 9(7), 5076–5098 (2009), `http://www.mdpi.com/1424-8220/9/6/5076`, doi:10.3390/s90705076
2. Behrmann, M., Haimson, C.: The cognitive neuroscience of visual attention. Current Opinion in Neurobiology 9, 158–163 (1999)
3. Bernays, E.: Selective attention and host-plant specialization. Entomologia Experimentalis et Applicata 80(1), 125–131 (1996)

4. Boahen, K.A.: Point-to-point connectivity between neuromorphic chips using address-events. IEEE Transactions on Circuits and Systems II 47(5), 416–434 (2000)
5. Chan, V., Liu, S.C., van Schaik, A.: AER EAR: A matched silicon cochlea pair with address event representation interface. IEEE Transactions on Circuits and Systems I 54(1), 48–59 (2007), Special Issue on Sensors
6. Culham, J., Brandt, S., Cavanagh, P., Kanwisher, N., Dale, A., Tootell, R.: Cortical fMRI activation produced by attentive tracking of moving targets. J. Neurophysiol. 81, 388–393 (1999)
7. Deiss, S., Douglas, R., Whatley, A.: A pulse-coded communications infrastructure for neuromorphic systems. In: Maass, W., Bishop, C. (eds.) Pulsed Neural Networks, ch. 6, pp. 157–178. MIT Press (1998)
8. Delbrück, T.: Frame-free dynamic digital vision. In: Hotate, K., et al. (eds.) Proc. of the Intl. Symp. on Secure-Life Electronics, University of Tokyo, vol. 1, pp. 21–26 (2008)
9. Desimone, R., Duncan, J.: Neural mechanisms of selective visual attention. Annu. Rev. Neurosci. 18, 193–222 (1995)
10. Fasnacht, D., Indiveri, G.: A PCI based high-fanout AER mapper with 2 GiB RAM look-up table, 0.8μs latency and 66 mhz output event-rate. In: Conference on Information Sciences and Systems, CISS 2011, Johns Hopkins University, pp. 1–6 (2011), http://ncs.ethz.ch/pubs/pdf/Fasnacht_Indiveri11.pdf, doi:10.1109/CISS.2011.5766102
11. Fasnacht, D., Whatley, A., Indiveri, G.: A serial communication infrastructure for multi-chip address event system. In: International Symposium on Circuits and Systems, ISCAS 2008, pp. 648–651. IEEE (2008), http://ncs.ethz.ch/pubs/pdf/Fasnacht_etal08.pdf, doi: http://dx.doi.org/10.1109/ISCAS.2008.4541501
12. Frintrop, S., Rome, E., Christensen, H.: Computational visual attention systems and their cognitive foundation: A survey. ACM Transactions on Applied Perception 7(1), 1–46 (2010)
13. Hoffman, J., Subramaniam, B.: The role of visual attention in saccadic eye movements. Perception and Psychophysics 57(6), 787–795 (1995)
14. Indiveri, G.: A current-mode hysteretic winner-take-all network, with excitatory and inhibitory coupling. Analog Integrated Circuits and Signal Processing 28(3), 279–291 (2001), http://ncs.ethz.ch/pubs/pdf/Indiveri01.pdf
15. Indiveri, G., Chicca, E., Douglas, R.: A VLSI array of low-power spiking neurons and bistable synapses with spike–timing dependent plasticity. IEEE Transactions on Neural Networks 17(1), 211–221 (2006), http://ncs.ethz.ch/pubs/pdf/Indiveri_etal06.pdf, doi:10.1109/TNN.2005.860850
16. Indiveri, G., Linares-Barranco, B., Hamilton, T., van Schaik, A., Etienne-Cummings, R., Delbruck, T., Liu, S.C., Dudek, P., Häfliger, P., Renaud, S., Schemmel, J., Cauwenberghs, G., Arthur, J., Hynna, K., Folowossele, F., Saighi, S., Serrano-Gotarredona, T., Wijekoon, J., Wang, Y., Boahen, K.: Neuromorphic silicon neuron circuits. Frontiers in Neuroscience 5, 1–23 (2011), http://www.frontiersin.org/Neuromorphic_Engineering/10.3389/fnins.2011.00073/abstract, doi:10.3389/fnins.2011.00073
17. Itti, L.: Quantifying the contribution of low-level saliency to human eye movements in dynamic scenes. Visual Cognition 12(6), 1093–1123 (2005)
18. Itti, L., Koch, C.: Computational modeling of visual attention. Nature Reviews Neuroscience 2(3), 194–203 (2001)

19. Kastner, S., De Weerd, P., Desimone, R., Ungerleider, L.: Mechanisms of directed attention in the human extrastriate cortex as revealed by functional MRI. Science 282(2), 108–111 (1998)
20. Lichtsteiner, P., Posch, C., Delbruck, T.: An 128x128 120dB 15μs-latency temporal contrast vision sensor. IEEE J. Solid State Circuits 43(2), 566–576 (2008)
21. Miller, M.J., Bockisch, C.: Where are the things we see? Nature 386(10), 550–551 (1997)
22. Mozer, M., Sitton, M.: Computational modeling of spatial attention. In: Pashler, H. (ed.) Attention, pp. 341–395. Psychology Press, East Sussex (1998)
23. Pollack, G.: Selective attention in an insect auditory neuron. Jour. Neurosci. 8, 2635–2639 (1988)
24. Sonnleithner, D., Indiveri, G.: Active vision driven by a neuromorphic selective attention system. Proc. of International Symposium on Autonomous Minirobots for Research and Edutainment, AMiRE 2011, 1–10 (2011), `http://ncs.ethz.ch/pubs/pdf/Sonnleithner_Indiveri11b.pdf`
25. Sonnleithner, D., Indiveri, G.: A neuromorphic saliency-map based active vision system. In: Conference on Information Sciences and Systems, CISS, Johns Hopkins University, pp. 1–6 (2011), `http://ncs.ethz.ch/pubs/pdf/Sonnleithner_Indiveri11.pdf`, doi:10.1109/CISS.2011.5766145

The ARUM Experimentation Platform: An Open Tool to Evaluate Mobile Systems Applications

Marc-Olivier Killijian, Matthieu Roy, and Gaetan Severac*

Abstract. This paper present the ARUM robotic platform. Inspired by the needs of realism in mobile networks simulation, this platform is composed of small mobiles robots using real, but attenuated, Wi-Fi communication interfaces. To reproduce at a laboratory scale mobile systems, robots are moving in an 100 square meters area, tracked by a precise positioning system. In this document we present the rational of such simulation solution, provide its complete description, and show how it can be used for evaluation by briefly explaining how to implement specific algorithms on the computers embedded by the robots. This work is an application of multi-robotics to research, presenting solutions to important problems of multi-robotics.

1 Objectives

In this paper, we present the ARUM robotic platform[1] targeted at evaluating performance, resilience and robustness of mobile systems. To obtain an efficient evaluation platform, three specific criteria were considered: **Control conditions** (real time monitoring, repeatability, flexibility, scalability), **Effective implementation** (easiness of configuration, devices autonomy, portability, low cost, miniaturization), and **Realistic environment** (network scale, traffic load, node mobility, positioning, radio broadcast behaviour). To our knowledge this platform is the only one to date to integrate all these features in a single environment.

Marc-Olivier Killijian · Matthieu Roy
LAAS-CNRS, 7 avenue du colonel Roche FR-31077 Toulouse, France
e-mail: roy@laas.fr

Gaetan Severac
ONERA (DCSD), The French Aerospace Lab FR-31055 Toulouse, France,
Universite de Toulouse
e-mail: gaetan.severac@onera.fr

* First and Corresponding author.

[1] ARUM stands for "an Approach for the Resilience of Ubiquitous Mobile Systems".

Fig. 1 A picture of the the ARUM Platform

Indeed, current evaluation strategies for distributed and mobile systems can be split in five categories:

- **Simulators.** Simulators are cheap and fast to set up, with almost no limitation in the number of nodes. Due to their scalability and simplicity, they are well suited for initial testing. Furthermore, they may speed up development of theoretical researches because since they allow a perfect monitoring and repeatability [21, 22]. Nevertheless, simulation is based on models of the running environment, and thus cannot reflect the real complexity of natural environments, particularly for radio communication and mobility pattern[7, 5, 3] .
- **Emulators.** Emulators are built to physically reproduce connections events using real wired network hardware[16, 19]. They provide features interesting for protocol implementation but they still use simulation to reproduce wireless communication behaviour and mobility[4].
- **Testbeds.** The ARUM platform we present in this paper can be classified in this category. Testbeds are closer to reality thanks to the use of real hardware. They exist since years now, from the historical MIT RoofNet[1], to the more recent MoteLab[2] service. Ideal to finalize and validate applications before real-life experimentations, testbeds provide much more realistic results than emulators or simulators. But they are also expensive, time consuming and limited by the physical resources/hardware used[14]. Because of those limitations, only a few of them implement real mobility. To the best of our knowledge, two platforms using mobile robots have been recently developed: MINT[17] and Mobile Emulab[10]. Original solutions that emulate mobility can also be found in the literature, like MOBNET [6] which varies the transmission power levels of fixed access points. It is interesting to notice that most of these platforms have to deal with large variations in the communication noise level because of environment perturbations.

[2] Harvard Sensor Network Testbed - http://motelab.eecs.harvard.edu

Such difficulties can be problematic during applications development phases, but they are representative of conditions encountered in the real life.

- **Hybrid simulators.** use both simulated networks and real devices, taking advantages and disadvantages of each[18, 23]. They are particularly suited to study, at a low cost, the interconnection of some real devices to a huge network, the latter being simulated.
- **Real live experiments.** This is, obviously, the more realistic kind of experimentation, but they present inherent technical problems which can bring more technical difficulties than scientific benefits[11, 15]. They are absolutely necessary for commercial applications, because it is impossible to truly simulate real environment yet. Yet, they are very expensive, error-prone, and they do not provide repeatability of experiments, due to the wide variability of real environments. As such, such platforms are not used in the context of research and education.

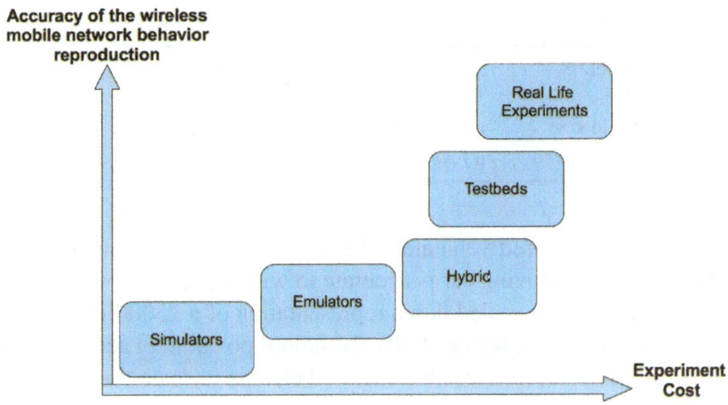

Fig. 2 Accuracy of evaluation solutions for mobile systems depending on their respective costs

Among all these technologies, there is no good or wrong solutions, the best choice depends on a specific needs and available resources, as shown in Figure 2.

In our case, both for scientific and for demonstration reasons, we decided to implement a testbed, the ARUM platform. Indeed our primary goal was to complement simulation and allow realistic evaluation of mobile systems, at a laboratory scale. It finally appeared to be a good platform for demonstration and education, since the platform can be used pedagogically to present various aspects and problems raised by mobile systems. The work describe here is an application of mini-robots to research, in a field different from robotics. However the solutions tested and implemented here can be applied to several important problems in multi-robotic field (e.g. positioning, mobility, communication...).

2 Design

To complete the goals presented in previous section, we designed an experimental evaluation platform composed of mobile devices. We dispose of a room of approximately $100m^2$ to emulate systems of different sizes, hence we decided to scale every parameter of the system to fit within our physical constrains. Technically speaking, each mobile device is built with : a programmable *mobile hardware* able to carry the device itself, a *lightweight processing unit* equipped with one or several *wireless network interfaces* and a *positioning device*. Hardware modelling required a reduction or increase of scale to be able to conduct experiments within the laboratory. To obtain a realistic environment, all services have been modified according to the same scale factor.

Table 1 Scale needs

Device	Real Accuracy	Scaled Accuracy
Wireless	range: 100m	range: 2m
GPS	5m	10cm
Node size	a few meters	a few decimeters
Node speed	a few m/s	less than 1m/s

In our case, we considered vehicular ad-hoc network experiments [13]. A typical GPS embedded in a moving car is accurate to within 5-20m. So, for our $100m^2$ indoor environment to be a scaled down representation of a $250000m^2$ outdoor environment (a scale reduction factor of 50), the indoor positioning accuracy needs to be $10-40$cm. Table 1 summarizes the required change in scale for all peripherals of a node.

We understand here that to meet those requirements some parts of the development were much more important. The focus was put on the reduced Wi-Fi interfaces, the precise positioning and the node mobility.The different parts of the platform will be detailed in the following section.

3 Technical Solutions

3.1 Mobility

To reproduce mobile systems conditions, the devices used in the platform must be mobile. But when conducting experiments, a human operator cannot be behind each device, so mobility has to be automated. This is why we considered the use of simple small robots in order to carry around the platform devices. The task of these robots is to implement the mobility of the nodes following a movement scenario.

A node, represented in figure 3, is implemented in the system using a laptop computer that is carried by a simple robotic platform, that includes all hardware devices, the software under testing and the software in charge of controlling robots

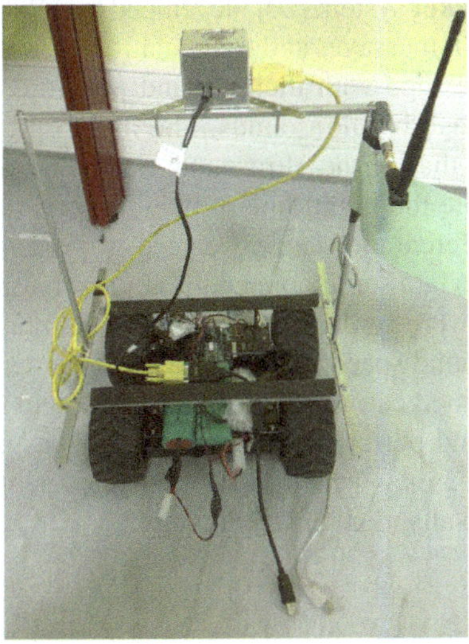

Fig. 3 A Mobile Node Picture, without the embedded computer

movements. Notice that software under testing and control software are totally independent, there are running on the same computer for practical reasons only.

For the mobile platform we use Lynxmotion 4WD rover. We selected it instead of other smaller robot (e.g. Lego Mindstorm) because this rover is able to carry a payload of 2 Kg during a few hours, running at a maximum speed of 1m/s. It is also relatively cheap (cf. table 2) and easy to build. We equipped it with infra-red proximity sensors to avoid collision, a top deck to support the laptop, a positioning system and a modified Wi-Fi interface.

The motion control software, running on the carried laptop, communicates speeds orders (linear speed and angular speed) to the robot. The mobility patterns are drawn by an operator for each mobile robot, using a dedicated software, that sends it to the mobile nodes control software. This enables flexibility – each node has its own mobility pattern – and repeatability – a pattern can be saved and replayed.

3.2 *Localization*

Positioning is a critical point of the platform. Firstly, we need to reproduce the kind of information produced by actual market solutions such as GPS, pondered by our scale factor. Secondly, we need a precise and real-time position of the mobile node to allow an accurate motion control of the robot. Our specifications required a precision within the centimetre and a minimum refresh of 2 Hz. Several technologies are currently available for indoor location [9], mostly based either on scene analysis (e.g. using motion capture systems) or on triangulation (of RF and ultrasound [20] or wireless communication interfaces [8]). During the building of the platform, we tried four different solutions.

We first tested the **Cricket** system [20], developed by MIT. Cricket is based on simultaneous ultrasound/RF messages and triangulation. Beacons fixed on the celling send periodically RF message with their ID and position, and, in the time, they send an equivalent message by ultrasounds. The flight time of the RF message is insignificant compare to that of the ultrasound. So the receiver, embedded on the mobile robot, can estimate the flight time of the ultrasound messages and and calculate, with at least 3 different messages received, its position. This position is then send to the mobile node via a serial connection. In theory, this system is very efficient, but in practice we were confronted to important limitations due to ultrasound disturbances. The ultrasound speed in the air can change depending of the temperature, so the results obtained can vary in the same way, and the ultrasound are also very sensitive to noise and perturbations. Neon lights was perturbing the systems and robots vibration, when they were moving, generated a lot of disturbances in the results. Finally we had to abandon this technology.

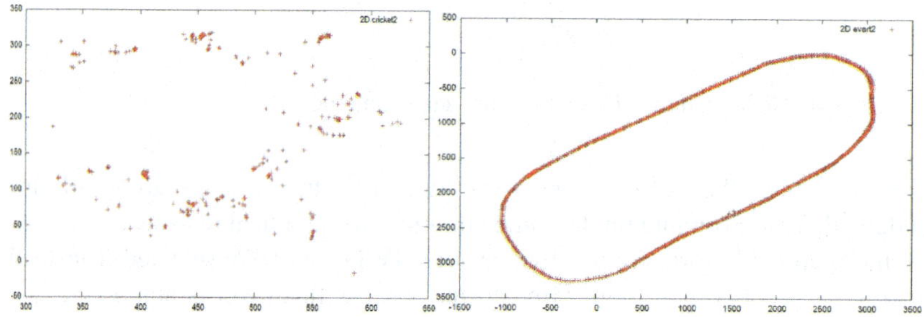

Fig. 4 Comparative results of Ultrasounds (left) and Infra-red (right) positioning systems. A robot is tracked by the two different systems while following the same circuit drew on the floor

To reach our desired level of accuracy for indoor positioning, we then used a dedicated motion capture technology that tracks objects based on real-time analysis of images captured by fixed infra-red cameras. **The Cortex system**[3] is able to localize objects at the millimetre scale. This technology uses a set of infra-red cameras, placed around the room, that track infra-red-visible tags. All cameras are connected to a server that computes, based on all cameras images, the position of every tag in the system. We equipped our small robots with such tags to get their positioning information. The figure 4 shows compared results of the ultrasounds and infra-red systems. Although the precision attained was more than enough for our needs, the system has some drawbacks: the whole system is very expensive (in the order of 100kEuros), calibration is a tedious task, and infra-red signals cannot cross obstacles such as humans.

The localization system currently used is the **Hagisonic StarGazer technology**[4]. It is also based on infra-red camera but they are small and embedded on-board on the

[3] Cortex Motion Capture - http://www.motionanalysis.com

[4] Hagisonic - http://www.hagisonic.com/

mobile robots. They locate themselves by tracking statically placed infra-red-visible tags. With Hagisonic, a camera needs to see only one single tag to be able to calculate its position and the precision is about a few millimetres, with a frequency of 10 Hz. So this technology, more affordable, was plenty satisfying our requirements.

An **Ultra-Wide-Band-based localization system** (UWB), by Ubisense[5], has also been deployed and used for the experiments. Localization is performed by 4 sensors, placed in the room at each corner, that listen for signals sent by small tags that emit impulses in a wide spectrum. Such impulses can traverse human bodies and small obstacles, so the whole system is robust to external perturbation, but, from our preliminary measurements, attainable precision is about 10cm. The next step will be to couple this technology with the Hagisonic camera system, resulting in a localization system with better properties: it will be relatively cheap, robust to external perturbations such as obstacles, and will have most of the time a precision about the order of a centimetre.

To keep our experimental platform positioning system generic, despite the numerous different technologies used, we developed a **position server**, accessible via the supervision wireless network of the experimentation room. Two kinds of clients can communicate with it, using standard XML messages. A client can be a *position provider* (Cortex, Hagisonic, UWB...) and send to the server the position of one or several mobiles or the client can be a *position consumer* (supervision application, motion control software...) and ask to the server the position of one or several mobiles. Using this strategy, it is possible to change the technology of one system, provider or consumer of position, and the modification will remain transparent to all the other devices.

3.3 Scenario Drawing Interface

To have adequate experimental conditions, the mobile nodes of the platform need to follow and repeat defined mobility scenario. But first, an operator has to define the mobility scenario. We developed a graphical user interface to draw, configure, visualize and manage mobility patterns. Now the interface is a complete program composed of 7 different tabs, figure 5 shows a screen-shot. The *Point* tab where users can define passage points on a map. The *Route* tab to edit robots routes using the passage points previously created, the trajectory will be calculate from those routes, it is also possible to specify time constraints and add some pauses in the route execution. The *Simulation* tab permits to see an overview of the robots movements, you can select the different routes you want to visualise. The *Rovers* tab is used to configure the TCP/IP connexion attributes to communicate with the robots. You can select the route each robot will execute in the *Association* tab. The *Upload* tab is designed to send the routes to the corresponding robots and start, stop or pause their executions. Finally the *Remote Control* tab allows to manually control a robot, with a virtual joystick, and give the possibility to dynamically define a route from its movements. The interface is coded in Java and can be run on any computer

[5] Ubisense - http://www.ubisense.net/en/products/precise-real-time-location.html

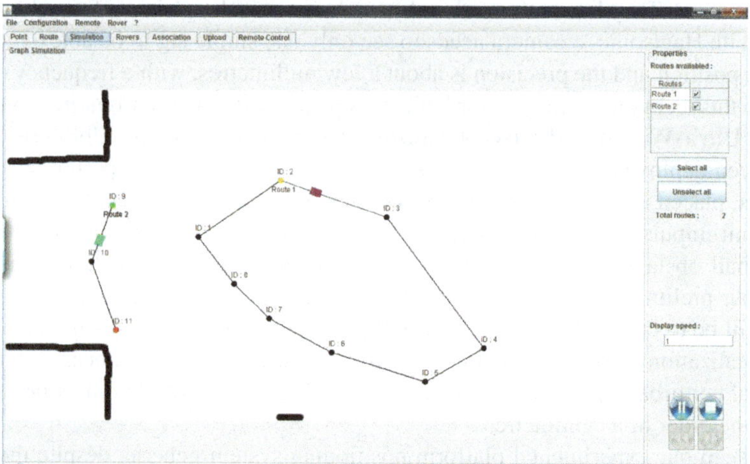

Fig. 5 Screen-shot of the mobility scenario drawing interface

connected to the supervision network. The mobility scenarios and movement orders are send to the mobile node via "Java Remote Method Invocation" (Java RMI).

3.4 Motion Control and Trajectory Computing

As you can see on figure 6, the mobility scenarios defined with the GUI are sent to the robot via Java Remote Method Invocation" (Java RMI). A robot control program, coded in Java, receives a scenario description or the movements orders from the scenario drawing interface. This program is running on the embedded computer of the mobile node. The mobility scenarios are then converted into commands and sent to the robot motion control environment. This environment is composed of GenoM[6] modules in charge of computing the final trajectory and controlling the robot speed to follow it. Proximity infra-red captors are continuously polled to stop the robot if an obstacle is detected.

We chose the GenoM environment, developed at the LAAS-CNRS laboratory, because it is an open source solution, already functional and still maintained by the robotic community. GenoM is a tool to design real-time software architectures. It is more specifically dedicated to complex on-board systems, such as autonomous mobile robots or satellites. It allows to encapsulate the operational functions on independent modules that manage their execution. The functions are dynamically started, interrupted or (re)parametrized upon asynchronous requests sent to the modules. A final reply that qualifies how the service has been executed is associated to every request. The modules are automatically produced by GenoM using a common generic model of module and a synthetic description of the considered module. At the end a set of modules composes an open, communicant and controllable system.

[6] GenoM - https://softs.laas.fr/openrobots/wiki/genom

Such environment can be very powerful and it used in our laboratory to control complex robots. In this platform, where the environment is controlled, the mobility is very simply implemented and we only used two GenoM modules. The module "Pilo", to compute the final trajectory (using Euler spirals) and the module "Loco" which controls the robot speed to follow the trajectory. The "Loco" module uses the position system to get in real time the position of the mobile node and send required linear and angular speed orders to the motors control card of the robot, through a serial connexion.

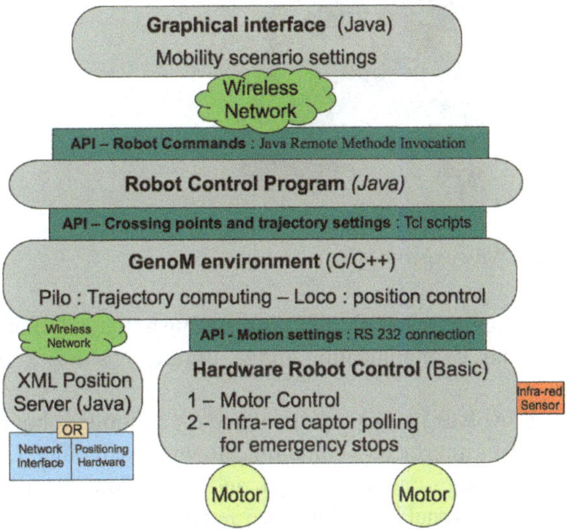

Fig. 6 Mobile Node architecture Overview

3.5 Reduced Wireless Communication

The communication range of the participants (mobile nodes and infrastructure access-points) has to be scaled according to the experiment being conducted. For our first experimentation, the scale factor had to be 50 (cf. Table 1) but, ideally, the communication range should be variable. Some Wi-Fi network interface drivers propose an API for reducing their transmission power. But the implementation of this feature is often rather limited, or ineffective, at a single room scale. A satisfying solution consists in using signal attenuators[7] placed between the Wi-Fi network interfaces and their antennas. The necessary capacity of the attenuators depends on many parameters such as the power of the Wi-Fi interfaces and the efficiency of the antennas, but also on the speed of the robot movements, the room environment, etc. As it is impossible to predict or calculate the Wi-Fi radio wave propagation we

[7] An attenuator is an electronic device that reduces the amplitude or power of a signal without appreciably distorting its waveform.

Fig. 7 The attenuation WiFi experiments

conducted empirical experimentation[12] to establish the relationship between signal attenuation and communication range, figure 7 show a picture of an experimentation.

This experiment involves two laptops mounted on a mobile robot platforms and using an external Wi-Fi interface to communicate with each other. One of the two nodes is static and the other one moves back and forth. Equivalent attenuators are attached between each external Wi-Fi interface and its antenna. The mobile platform moves along a line, stops every 20cm for 5min and performs a measurement at every stop, figure 7. For each measurement, the moving laptop joins the ad-hoc network created by the fixed one, measures the communication throughput and then leaves the ad-hoc network. The time for joining the network is logged, as is the measured throughput. A complete experiment is composed of 100 repetitions of a return trip along the 5m line. This data is logged and statistically analysed off-line, leading to figure such as the one presented on figure 8. To validate that those attenuated results correspond to real Wi-Fi propagation behaviour, we reproduce the same experiment, outdoor, without attenuation and we obtain exactly the same kind of graphic shape, distances range excepted of course. Thanks to those multiple results we can now use the adequate attenuation depending on the specification of the experiments and we can certify that this will be representative to real Wi-Fi connexion conditions.

3.6 Supervision Network

For the communications between the collaborative algorithms tested on the platform, the attenuated Wi-Fi interfaces previously presented are used. So the internal

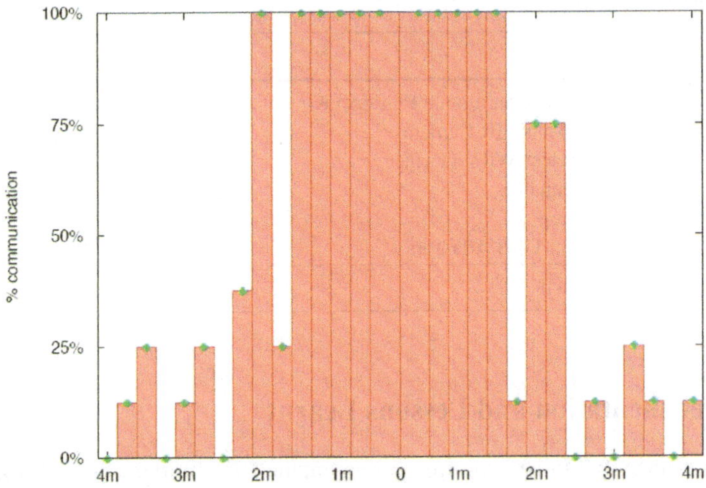

Fig. 8 The attenuation WiFi experiments

wireless card of the embedded computers is available for monitoring. Connected to a LAN access point, it provides direct access to each computer without disrupting the current experiment. This system is used to send monitoring information to the robot (e.g. position, commands, ...) and to retrieve data from the mobile nodes in real time, allowing a dynamically overview and analysis of the tested algorithms.

3.7 Implementation and Price

It is interesting to consider that all the different parts of a mobiles nodes (localization, trajectory planning, robot control, communication, ...) are connected thought clearly defined and documented interface, so it is easy and fast to change one of this part to make the platform evolve, without re-designing everything (cf. figure 6). For example we envisage buying a new localization system and changing the Lynxmotion robot for a Roomba[8] development mobile platform. Anyone interested in reproducing our evaluation platform in a laboratory can reuse some parts of interest and modify others. Full documentation and sources of the platform are available at: http://projects.laas.fr/ARUM/. As an indication, Table 2 sums the actual price of the different parts of a mobile node.

[8] Roomba Devel - http://www.irobot.com/images/consumer/hacker/roombascispecmanual

Table 2 Platform devices Costs

Device	Price ($)
Linxmotion mobile Platform Kit	1 000
Hagisonic IR Camera	1000
Wireless WiFi interface	50
Attenuators	70
Laptop	1200
Serial-USB Adaptors	30
Total	3350

4 Experimentation and Lessons Learnt

To evaluate our ARUM platform, we experiment the **Distributed Black-Box application**, or DBB for short. This work was conducted in the course of the European project HIDENETS[9]. The application developed provides a virtual device, whose semantics is similar to avionics black-boxes, that tracks cars history in a way that can be replayed in the event of a car accident. It ensures information is securely stored using replication mechanisms, by means of exchanging positions between cars. This architecture is a partial implementation of the HIDENETS architecture and has been detailed in the project deliverable [2].

The ARUM platform was used to emulate the network of communicating cars. Through this work, the global performance of the evaluation platform was validated. The modularity and repeatability of the mobility patterns was used to test and improve the DBB algorithms in controlled situations. The use of real, power reduced Wi-Fi interfaces allowed realistic results; we monitored during the experiment wireless signal variations similar to real wireless network behaviour in difficult conditions (maximum range limit, noise perturbation, obstacle, ...).

A very precise positioning system was used both by the tested cooperative algorithms and the robot motion control software, without disturbances. With hindsight we have to admit that, even if we get positive results, we had to deal with a lot of contingencies. The total labour cost of the platform development was more consequent than expected. Some parts of the development would have been impossible to reduce - attenuated Wi-Fi scaling, positioning systems tests - but if we had to rebuild all from scratch we would probably choose a mobile robotic platform that already has motion control implemented - such as the Roomba Devel platform.

However, now that the platform is finished and validated, it can be used as a tool "out of the box"; anyone can come to the laboratory to implement algorithms on the mobile nodes. As showed in Section 3, it is possible to interface any code with the different parts of the mobile node - communication, positioning, monitoring, ... - and to easily program a mobility pattern. All the parts of the platform are segmented by

[9] HIghly DEpendable ip-based NETworks and Services - http://www.hidenets.aau.dk/

software interfaces, defined in the documentation[10], so it can quickly be handled and adapted by anybody interested. Additionally the sources of each software module can be downloaded to build a similar platform in another laboratory.

Even if it is not the primary function of this platform, we noticed that the versatility and the easiness of use of this platform makes it an interesting educational tool. All the different parts of it, presented in Section 3, can be used, studied and replaced by students. The localization, the mobility scenario computing, the motion control and trajectory calculation or the reduces wireless communication, could support interesting university work.

5 Conclusion

This article started by pointing out the difficulties of evaluation of application for mobile devices systems. It presented the difficulties encountered to emulate a realistic mobile network environment at a laboratory scale. Those observations motivated the development of a testbed platform designed to evaluate distributed applications. Thanks to use of mini-robots, this platform, named ARUM, appears to provide an interesting compromise between resources consumption (in terms of manpower) and accuracy of results, appropriate to complement simulation. The whole architecture is described part by part to ease reuse by researchers or in an educational context, while reducing the waste of time and money in development and tests.

References

1. Aguayo, D., Bicket, J., Biswas, S., De Couto, D.: Mit roofnet: Construction of a production quality ad-hoc network. In: MobiCom Poster (2003)
2. Arlat, J., Kaâniche, M. (eds.): Hidenets. revised reference model. LAAS-CNRS, Contract Report nr. 07456 (September 2007)
3. Hamida, E.B., Chelius, G., Gorce, J.-M.: Impact of the Physical Layer Modeling on the Accuracy and Scalability of Wireless Network Simulation. Simulation 85, 574–588 (2009)
4. Beuran, R., Nguyen, L.T., Miyachi, T., Nakata, J.: Qomb: A wireless network emulation testbed. In: IEEE GLOBECOM 2009 (2009)
5. Cavin, D., Sasson, Y., Schiper, A.: On the accuracy of manet simulators. In: POMC 2002: Proc. of the second ACM International Workshop on Principles of Mobile Computing, pp. 38–43 (2002)
6. Lan, K.C., Perera, E., Petander, H.: Mobnet: the design and implementation of a network mobility testbed for nemo protocol. In: 14th IEEE Workshop on Local and Metropolitan Area Netw. (2005)
7. Chin, K.-W., Judge, J., Williams, A., Kermode, R.: Implementation experience with manet routing protocols. ACM SIGCOMM Computer (2002)
8. Correal, N.S., Kyperountas, S., Shi, Q., Welborn, M.: An uwb relative location system. In: Proc. of IEEE Conference on Ultra Wideband Systems and Technologies (November 2003)

[10] ARUM platform - http://projects.laas.fr/ARUM/

9. Hightower, J., Borriello, G.: A survey and taxonomy of location systems for ubiquitous computing. IEEE Computer (2001)
10. Johnson, D., Stack, T., Fish, R., Montrallo Flickinger, D., Stoller, L., Ricci, R., Lepreau, J.: Mobile emulab: a robotic wireless sensor network testbed. In: IEEE INFOCOM (2006)
11. Kiess, W., Mauve, M.: A survey on real-world implementations of mobile ad-hoc networks. Ad Hoc Networks (5), 324–339 (2005)
12. Killijian, M.O., Powell, D., Roy, M., Séverac, G.: Experimental evaluation of ubiquitous systems: Why and how to reduce wifi communication range. In: DEBS 2008 (2008)
13. Killijian, M.O., Roy, M., Séverac, G., Zanon, C.: Data backup for mobile nodes: a cooperative middleware and experimentation platform. In: DSN 2009 WADS (2009)
14. Kropff, M., Krop, T., Hollick, M., Mogre, P.S., Steinmetz, R.: A survey on real world and emulation testbeds for mobile ad hoc networks. In: Proceedings of 2nd IEEE International Conference on Testbeds and Research Infrastructures for the Development of Networks and Communities, TRIDENTCOM 2006 (2006)
15. Langendoen, K., Baggio, A., Visser, O.: Murphy loves potatoes: experiences from a pilot sensor network deployment in precision agriculture. In: International Parallel and Distributed Processing Symposium, p. 155 (2006)
16. Matthes, M., Biehl, H., Lauer, M., Drobnik, O.: Massive: An emulation environment for mobile ad-hoc networks. In: Proceedings of the Second Annual Conference on Wireless On-demand Network Systems and Services, WONS 2005 (2005)
17. Mitchell, C., Munishwar, V.P., Singh, S., Wang, X.: Testbed design and localization in mint-2: A miniaturized robotic platform for wireless protocol development and emulation. In: Proc. First International Conference on Communication Systems and Networks, COMSNETS (2009)
18. Osterlind, F., Dunkels, A., Voigt, T.: Sensornet checkpointing: Enabling repeatability in testbeds and realism in simulations. Wireless Sensor Networks, 343–357 (2008)
19. Pužar, M., Plagemann, T.: Neman: A network emulator for mobile ad-hoc networks. Technical Report 321, Dept. of Informatics, University of Oslo (2005) ISBN 82-7368-274-9
20. Smith, A., Balakrishnan, H., Goraczko, M., Priyantha, N.B.: Tracking moving devices with the cricket location system. In: 2nd International Conference on Mobile Systems, Applications and Services (Mobisys 2004), Boston, MA (June 2004)
21. Stojmenovic, I.: Simulations in wireless sensor and ad hoc networks: matching and advancing models, metrics, and solutions. IEEE Communications Magazine 46(12), 102–107 (2008)
22. Boleng, J., Camp, T., Davies, V.: A survey of mobility models for ad hoc network research. Wireless Comm. and Mobile Computing 2, 483–502 (2002)
23. Zhou, J., Ji, Z., Bagrodia, R.: Twine: A hybrid emulation testbed for wireless networks and applications. In: IEEE INFOCOM 2006 (2006)

Embodied Social Networking with Gesture-enabled Tangible Active Objects

Eckard Riedenklau, Dimitri Petker, Thomas Hermann, and Helge Ritter*

Abstract. In this paper we present a novel approach for Tangible User Interfaces (TUIs) which incorporates small mobile platforms to actuate Tangible User Interface Objects (TUIOs). We propose an application that combines gestural interaction and our actuated Tangibles, Tangible Active Objects (TAOs), for social networking. In our approach TUIOs represent messages with which the user can trigger actions with through gestural input using these objects. We conducted a case study and present the results. We demonstrate interaction with a working prototype of our embodied social networking client.

1 Introduction

In the communication age digital exchange of information and keeping in touch with each other is getting more and more important. The growing information society relies on social networks that emerge for diverse kinds of communities and interest groups. Additionally gesture-enabled devices such as smart phones and smart pads allow the users to stay connected wherever they go. At work or at home, however, the connectedness may disturb the daily workflow. Embedding the interaction within the everyday environment may help making the user experience unobtrusive and ubiquitous [1]. In this paper we present a prototype of a system that embodies social networking in actuated gesture-enabled tangible objects.

Tangible Interaction is a subfield of Human-Computer Interaction (HCI). Researchers in this field search for new ways of interaction with digital information

Eckard Riedenklau · Dimitri Petker · Thomas Hermann · Helge Ritter
Ambient Intelligence / Neuroinformatics Group, CITEC, Bielefeld University, Universitätsstraße 21-23, 33615 Bielefeld, Germany
e-mail: `{eriedenk,dpetker,thermann,helge}@techfak.uni-bielefeld.de`

* This work has been supported by the Cluster of Excellence 277 Cognitive Interaction Technology funded in the framework of the German Excellence Initiative.

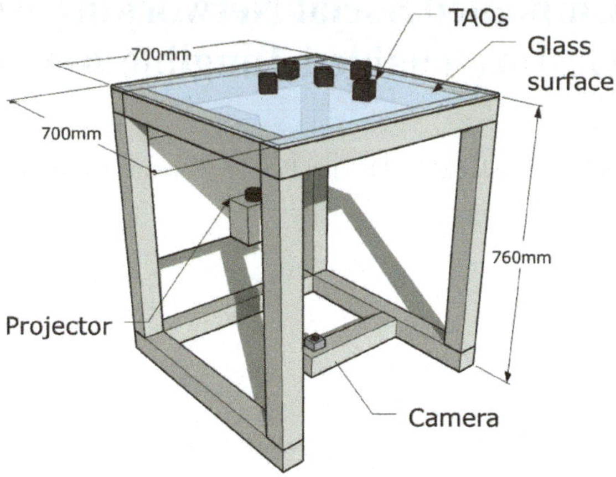

Fig. 1 The principal setup design. Our tDesk is equipped with a glass surface overlaid with a projection foil on which the projector mounted behind the table can project. A Firewire camera underneath the table allows visual tracking of the TAOs.

and functionality, keeping aloof from the traditional terminal consisting of display, keyboard, and mouse. This can be achieved by embodying those data with physical, graspable objects which users can interact naturally with using their everyday manipulation skills [2, 3].

Most TUIs use rigid, motionless objects, which the user can manipulate. The system itself is unable to move those objects. Therefore researchers built actuated objects. Pangaro et al. [4] created the Actuated Workbench, a system which incorporates a grid of individually controllable electro magnets that enables ferromagnetic objects to be moved across a tabletop surface. Weiss et al. [5] elaborated on this technology to create a versatile set of widgets for interactive tabletops. Rosenfeld et al. [6] created actuated Tangible Objects differently, integrating small mobile robotic platforms into their objects to enable the system to save and restore arrangements of the objects on the interactive surface.

Gesture-based interaction is another hot topic in HCI research. It has frequently been applied to consumer products such as web-browsers or smart phones. Drawing shapes on a touch screen with fingers or the mouse triggers commands, such as 'go back' or 'reload page'. As finding easily understandable gestures is not trivial, Wobbrock et al. already put alot of effort in collecting user defined gestures [7], evaluating different sets of gestures [8], and defined the guessability of such symbolic input [9]. RoboTable by Krzywinski et al. [10] enables the user to control mobile robots in a mixed-reality game scenario with motionless TUIOs. The authors claim that their system supports finger gestures or gestural input with passive TUIOs, but unfortunately they do not explain if and how gestural input is used in their approach.

2 Tangible Active Objects and Tangible Desk

Our system is based on the Tangible Active Objects (TAOs) [11] which are used on the tDesk (formerly known as Gesture Desk [12]). The tDesk is an interactive table equipped with a projector and a camera underneath a table-top glass surface equipped with projection foil, as shown in Fig. 1. We use it as a platform for interactive scenarios, such as multi-touch applications or TUIs. The TAOs act as a TUI. They contain small low-cost robotic platforms which allow actuation of these tangible objects. The TAOs' housing of these small robots are 3D printed cubes with an edge length of 5 cm ($\approx$ 2"). We attached visual markers underneath the TAOs for visual tracking, as depicted in Fig. 4. Like many small robotic mobile platforms, actuation is realized with a differential drive. An Arduino pro mini board[1] for rapid-prototyping of electronic systems controls this drive and XBee modules[2] allow wireless communication and remote control. Because of their modular design of the TAOs can be extended easily.

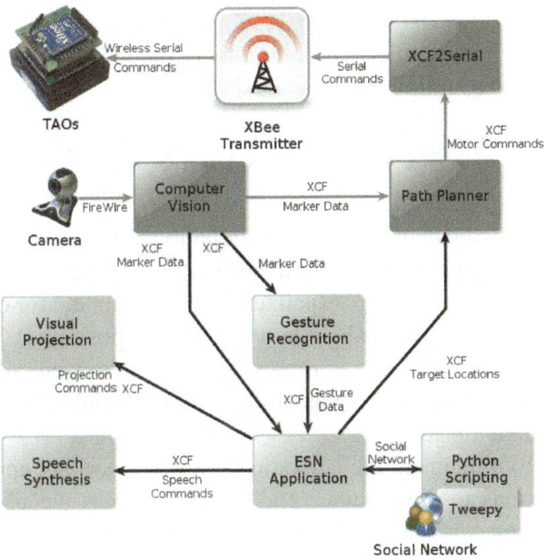

Fig. 2 This collaboration diagram describes the flow of information between the processes in our modular software architecture. The base modules used in this application are printed in dark gray, whereas the new modules are shown in light gray.

The software modules running on the host computer are organized in independent processes, communicating over the XML enabled Communication Framework (XCF) [13]. Fig. 2 depicts the software modules and their collaboration. A computer vision program analyzes the camera image from the Firewire camera mounted underneath the glass surface on which the TAOs interact to track markers attached underneath the TAOs.

A path planning module takes the marker information and target requests of application modules, computes trajectories and navigation commands and navigates the TAOs. Since the system has a complete overview of the scene, it is possible to compute attracting (target position) and repelling (other TAOs) force fields for each TAO and navigate it via gradient descent through this force field described by Latombe [14].

1 http://www.arduino.cc
2 http://www.digi.com/products/wireless/point-multipoint/xbee-series1-module.jsp

We adapted Latombe's approach of potential fields to harmonic potentials. Here the attractive potential is defined as following where λ is a factor for the strength of the field, j is the index of the navigating TAO and t is the current target position of that TAO.

$$E_{att}(\mathbf{x}_j) = \lambda \|\mathbf{x}_j - \mathbf{x}_t\|^2 \tag{1}$$

The repulsive forces are defined using a Gaussian, centered at the obstacles' positions, where again j is the index of the navigatin TAO and i the index of all other TAO which are obstacles to this TAO.

$$E_{rep}(\mathbf{x}_j) = -\mu \sum_{i \neq j} \exp\left(-\frac{1}{2\sigma^2}(\mathbf{x}_i - \mathbf{x}_j)^2\right) \tag{2}$$

The resulting forces of both potential fields are easily computed using the gradient defined as following.

$$\nabla E(\mathbf{x}_j) = 2\lambda(\mathbf{x}_j - \mathbf{x}_t) - \mu \sum_{i \neq j} \frac{(\mathbf{x}_j - \mathbf{x}_i)}{\sigma^2} \exp\left(-\frac{1}{2\sigma^2}(\mathbf{x}_i - \mathbf{x}_j)^2\right) \tag{3}$$

The vector that results from Equation 3 can then be used to compute the corresponding navigation commands that are relayed over the serial port to the wireless transmitter. Figure 3 visualizes the potential field for on single TAO. The TAO at position 0.33,0.33 is about to navigating to position 0.66,0.66. At position 0.25,0.75 is another TAO that is represented as an obstacle and corresponds in a hill in the three-dimensional visualization. This potential field is updated iteratively every new camera image.

To implement the embodied social networking application we added further extensions to our system. The system enables the user to physically interact with messages transmitted over a social network, such as Twitter[3] through gesture enabled TAOs. First of all, beside a speech synthesis module that can read interaction specific information to the user, we added back-projection capabilities to augment TAOs with visual information, such as messages, opened links, the different areas on the interactive surface and fields for textural input (via keyboard), etc. For this we replaced the previously used visual markers (derived from the reacTiVision markers [15], see Fig. 4(a)), to make the system more robust in combination with back-projection and independent from additional illumination that could be interfered by the projection. In a separate project we developed a self-luminescent visual marker based on infrared Light Emitting Diodes (LEDs). We arranged the LEDs on a 28×28 mm sized Printed Circuit Board (PCB) which exactly fits into the bottom of the TAOs' body. In the top left corner of the PCB seven LEDs define the position and orientation of the TAO, other six LEDs encode its ID as depicted in Fig. 4(c)). To track these new markers we used the corner detection algorithm, proposed by Chen He [16]. After finding the corners of the markers it is easy to determine which of the six ID encoding LEDs are illuminated, since their position relatively to the

[3] http://www.twitter.com

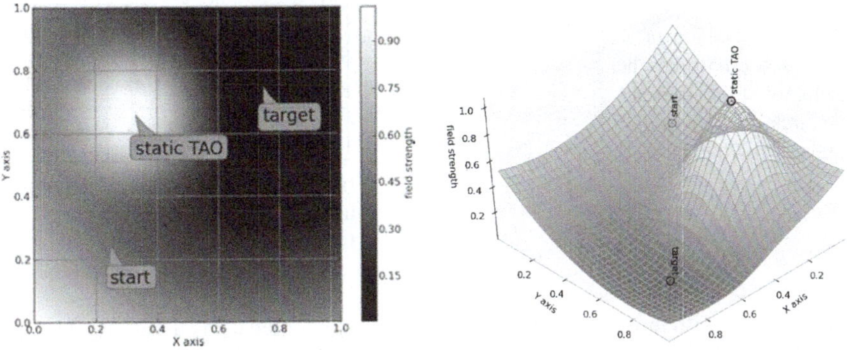

(a) 2D visualization of the potential field. The target position of the navigating TAO is the darkest area (attracting), where as obstacles correspond in brighter areas (repulsing).

(b) 3D visualization of the potential field. On it's way from the start to the target position the TAO has to slightly navigate around the other TAO to avid collisions.

Fig. 3 Two visualizations of potential field for navigation. There is one static TAO at the position $0.33, 0.33$ and another one navigating from position $0.25, 0.25$ to $0.75, 0.75$.

Fig. 4 Visual Marker: LED arrangement and example configurations.

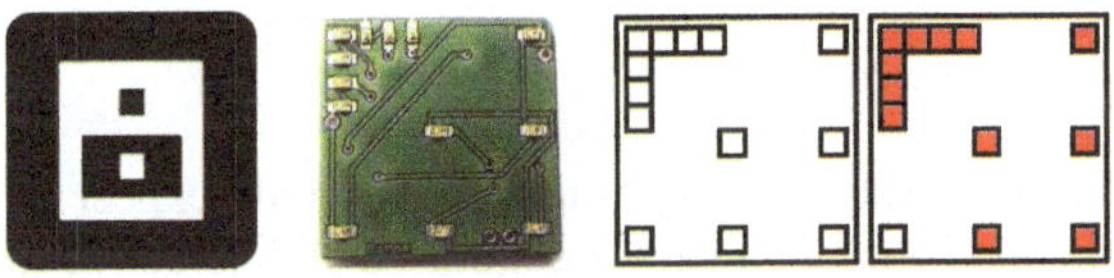

(a) An old marker

(b) The PCB

(c) Left: marker layout; right: Configuration for ID 1.

corner of the marker is fixed. This allows 2^6 possible ID configurations. The ability to change the ID of the markers on the fly through the software is a novel possibility in the field of TUIs. To make tracking and projection possible, we mounted an infrared filter in front of the camera to filter out the visual light from the projector.

For triggering actions during interacting with social network messages we chose gestural input. Moving the TAOs along a specific path, thereby executing a 'gesture' as known from mouse gestures enables a novel and easy understood means to trigger actions. For first experiments we used the gesture library LibStroke[4]. It provides very basic gesture recognition capabilities by segmenting gestures using a 3×3 grid of numbers as shown in Fig. 5(a). After performing the gesture shown in Fig. 5(b) LibStroke normalizes the trajectory according to the grid and outputs "1478963". The output in terms of numeric sequences allows further processing of the detected gestures through simple string comparison. Because LibStroke is designed for a single mouse cursor as input, we extended the library to cope with multiple input devices such as the TAOs.

[4] http://www. etla.net/libstroke

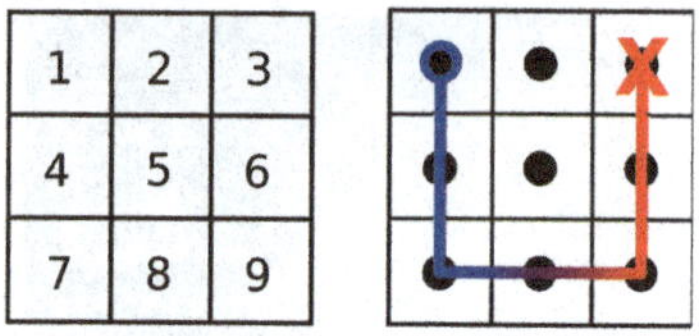

(a) Grid layout used by LibStroke (b) Example gesture

Fig. 5 Specification of Gestures in LibStroke; the example gesture (start at blue circle) results in the sequence "1478963"

Interaction with those messages can then be implemented using TAOs as a handle according to the container concept introduced by Ulmer et al. in the mediaBlocks system [17]. Thereby every TAO can 'contain' such a message. To interface with a social network, we utilized the Python library oauth-python-twitter2[5] which wraps the Twitter API. Furthermore we added two new modalities. A display module enables the system to visually present content, such as the messages, opened links, the different regions of the interactive surface and fields for textural input (with a keyboard) etc. In addition, a speech synthesis module reads information to the user.

3 Case Study

To investigate, which gestures users would expect to work with our embodied social networking client, we conducted an interactive case study. The subjects had to contemplate gestures according to a command selected randomly from a set of 11 commands. We asked the subjects to perform their gestures with one TAO, initially placed in the middle of the interaction area of the tDesk. In this study actuation is only used for automatically returning the TAO to the initial position after the subject finished the particular gesture to have the same initial situation for every trial. During the experiment we recorded the raw data of the trajectories and the output of the gesture recognizer. The gesture for each command was performed three times which results in a total of 33 trials per subject. During the experiment we also recorded the discussion between the subject and the experimenter for later analysis and transcription of gestures. After these trials we asked the subjects to fill out a small questionnaire to provide demographic information. Furthermore we asked our subjects if they already knew gesture-based interactions, e.g. mouse-gestures or finger-gestures and if they know and use social networks. We asked if the subjects could imagine to use such a system on their own desk and if they would accept standardized gestures or if they want an opportunity to define their own set of gestures.

[5] http://code.google.com/p/oauth-python-twitter2/

4 Results

We conducted the study described in Section 3 with 15 subjects, all from Europe. All of them got instructions in their native language (German or English). 20% of the subjects were female. The average age of the subjects was 32.3, the youngest was 23, the oldest was 61 years old. All subjects were right-handed. 9 subjects already knew touch- or mouse-gestures but only 3 of them were actually using them. Social networks were known by 13 subjects and 11 of these subjects used them.

Fig. 6 This plot depicts all gesture sequences, recognized by the gesture recognition module. All approx. 200 gesture occurrences (x-axis) are plotted against the commands (y-axis; in the same order as in Fig. 8). The darker a pixel is, the more often the combination of command and gesture occurred. It is normalized for better visibility.

Figure 13 depicts all approx. 200 gestures that were actually recognized by the gesture recognition module. On the x-axis all gestures made are plotted while the y-axis represents the 11 commands considered in our study. The plot is sparsely filled and there are only few spots where gestures occurred often. These spots are marked with their corresponding number sequence. There are so many gestures that do not very often occur because that the gesture recognition software was not able to recognize the more complex gestures correctly and produced only noisy, almost random sequences, as described later in the paper.

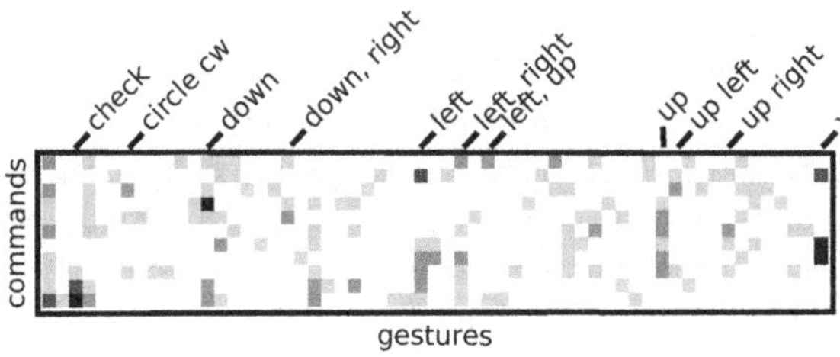

Fig. 7 Here all transcribed gesture occurrences (60) are plotted against the commands (in the same order as in Fig. 8).

Because the gesture recognition did not work well for complex gestures we transcribed the gestures from the collected data (recorded trajectories and audio). An overview over the transcribed data set collected in our experiments is visualized in Fig. 7. The plot is still sparsely filled (now because the subjects made these gestures seldom) and there are few frequent command-gesture combinations visible as darker

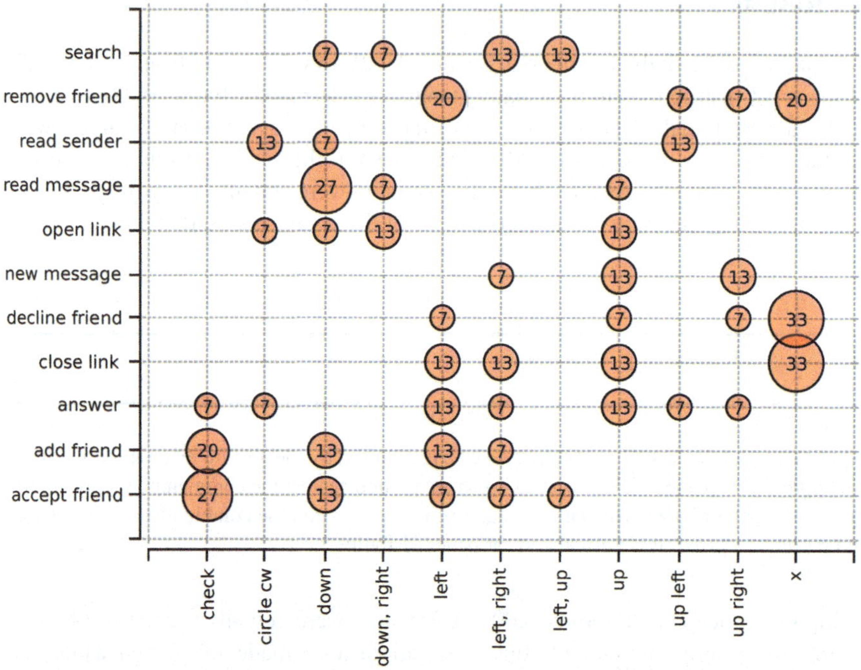

Fig. 8 This plot shows the (rounded) percentage of subjects that performed the (transcribed) gestures (x-axis) corresponding to the commands (y-axis). For a better overview we cropped away gestures with a score lower than 7%.

pixels. For better visibility we only consider the most frequently made gestures in the following.

Fig. 8 depicts the most frequently occurring gestures performed by the subjects together with the percentage of subjects that chose the particular gesture in combination with the corresponding command. For the semantically similar commands such as *accept friend* and *add friend* the 'check' gesture was chosen most frequently, which is quite natural. Also the gesture 'down' occurs frequently. Subjects preferring this gesture described it pulling something to themselves. For the *answer* command the preferred gestures are 'left' and 'up', both are metaphorically meant as sending something back (opposite of reading direction or away from oneself). For the commands *close link* and *decline friend* the 'x' gesture was chosen most often, which is again quite natural. Also for the command *remove friend* this gesture was preferred beside 'left'. For *new message* the most occurring gestures are 'up' and 'up right', where as the subjects came up with 'up' and 'down, right' most frequently for the command *open link*. The command *read message* got many correspondents in the gesture 'down'. Subjects stated that this symbolizes the process of reading a text line by line. The preferred gesture for the command *read sender* was 'circle cw' which means 'taking a closer look'. Some subjects stated that they located commands at corners or borders of the interactive surface and moved the TAO to one of these positions. This may result in different gestures if the starting point of the gesture is not located in the middle of the interaction area. One example is the *search* command. It found two winning gestures in 'left, right' and 'left, up'. Here referencing with the border of the interaction area was crucial for the subjects. For a better understanding, the example trajectories of gestures from our collected data are depicted in Fig. 9.

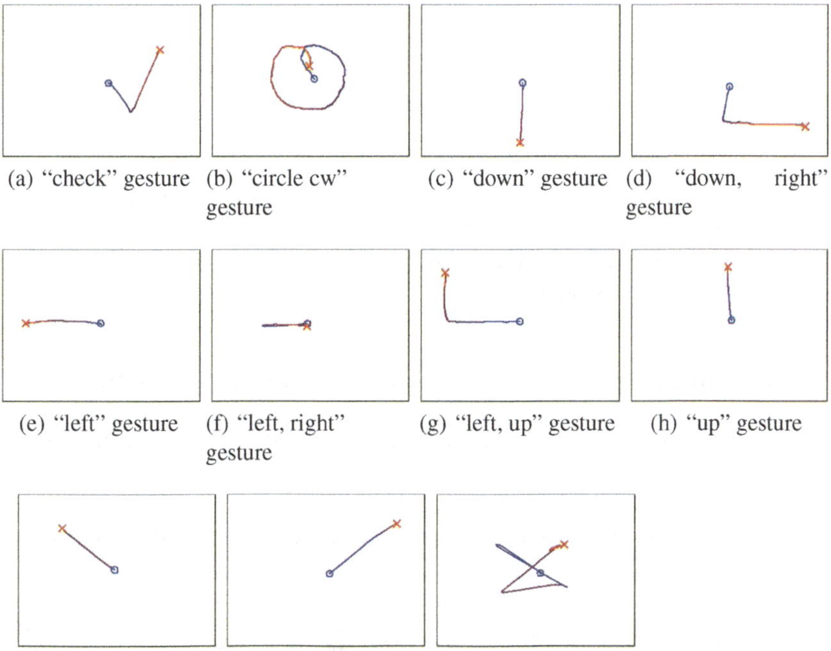

(a) "check" gesture (b) "circle cw" gesture (c) "down" gesture (d) "down, right" gesture (e) "left" gesture (f) "left, right" gesture (g) "left, up" gesture (h) "up" gesture (i) "up left" gesture (j) "up right" gesture (k) "X" gesture

Fig. 9 Visualizations of the winning gestures. A X symbol marks the start of the gesture movement, a circle symbol marks the end. The transition from start to end with respect to the elapsed time is represented by a gradient from blue to red.

The winning gestures are relatively basic up to medium complex. To illustrate the wide variety of different gestures (and ways of interpretation) made by the subjects we give an example of some complex gestures, which has been performed only once (see Fig. 10).

Another interesting result of our study is that some subjects tried to make gestures in ways we did not think of beforehand. For example the TAO was turned in place, which was not recognized, because the gesture recognition only works on trajectories of 2D positions. Furthermore subjects tried to lift the TAO once or repeatedly as a metaphor for clicking or wanted to shake or squeeze it. Obviously physical objects offer a much higher amount of flexibility for gestural commands.

(a) "mouth" gesture; used for the *read message* command

(b) "OK" gesture; used for the *accept friend* command

(c) "questionmark" gesture; used for the *read sender* command

Fig. 10 Some examples of gestures that were performed seldom. It shows the many different interpretation levels used to perform gestures going from symbolic or iconic metaphors to written letters and text.

From the questionnaire data we found that 4 subjects stated that they would use it, 3 could not imagine using it and 8 were unsure. To the question if self-defined or standardized gestures were preferred, 9 subjects stated that they would prefer self-defined gestures, 4 would prefer standardized ones and 2 were unsure.

5 Interaction Design and Implementation of the Interface

For our embodied social networking client we divided the table-top surface in four areas as depicted in Figure 11(a).

The actuation feature of the TAOs plays an important role in our application. Actuation is controlled by a finite state machine, which implements the state-graph depicted in Figure 12.

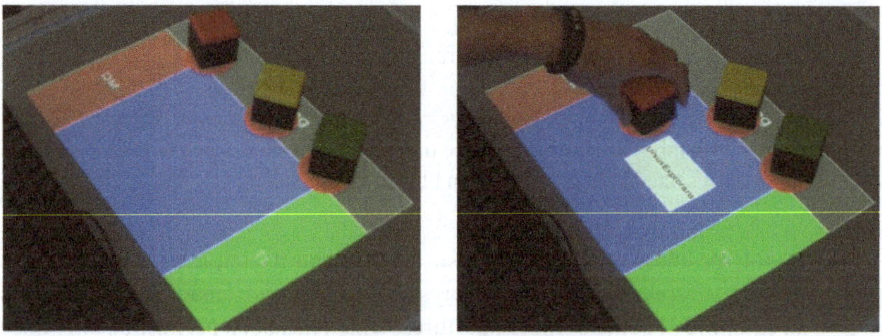

(a) Division of the tabletop surface: large interaction zone in the center, a message zone on each side for direct messages (left) and timeline events (right) and a waiting zone on top with three waiting TAOs.

(b) Picture of the final system. The user has just performed a gesture to view the name of a message's sender.

Fig. 11 Layout of the interactive area and the running system being used.

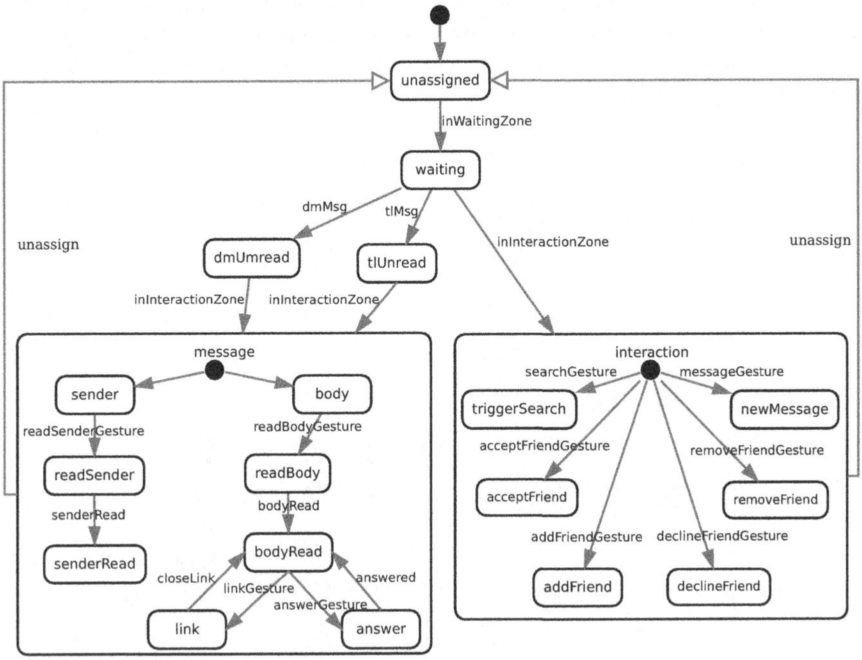

Fig. 12 State diagram: the implemented finite-state machine. Transition conditions contain message based events (such as *tlMsg*, indicating that a timeline message was assigned) and location based events (such as *inWaitingZone*, indicating that a TAO is in the waiting zone)

This results in the following behavior: Initially unassigned all TAOs are unassigned to any message and stay in the *waiting zone* until a new message is received from the social network. When a direct message (personal message from an other user) is received, it gets assigned to the leftmost TAO in the waiting zone. This TAO proceeds to the *direct message zone* area. When a timeline message (the timeline is a collected stream of postings from the user's friends) comes in it gets assigned to the rightmost TAO from the waiting zone. This TAO proceeds automatically to the *timeline message zone*. The TAOs in both message zones are ordered from bottom to top. The layout design of the interactive surface inherently maintains the chronological order of received messages as described later. When a TAO is taken out of the message zone other TAOs in the queue automatically rearrange downwards. The user can take a TAO embodying a message from the two message zones and put it into the *interaction zone* to interact with the embodied message: through different gestures, the user can instruct the system to present the message (visually and through speech), to view the message's author profile, to open (or close) a link included in the message, or to open an input field for answering.

Additionally the user can put an unassigned TAO from the waiting zone into the interaction zone. In this case no message is assigned to the TAO so there are other interaction opportunities: opening an input mask for writing for a new message, adding or removing a buddy from the user's contact list or for searching the social network's history for a specific topic. An example picture of a user interacting with the system is shown in Fig. 11(b). A video demonstration of this application is provided on our website.[6]

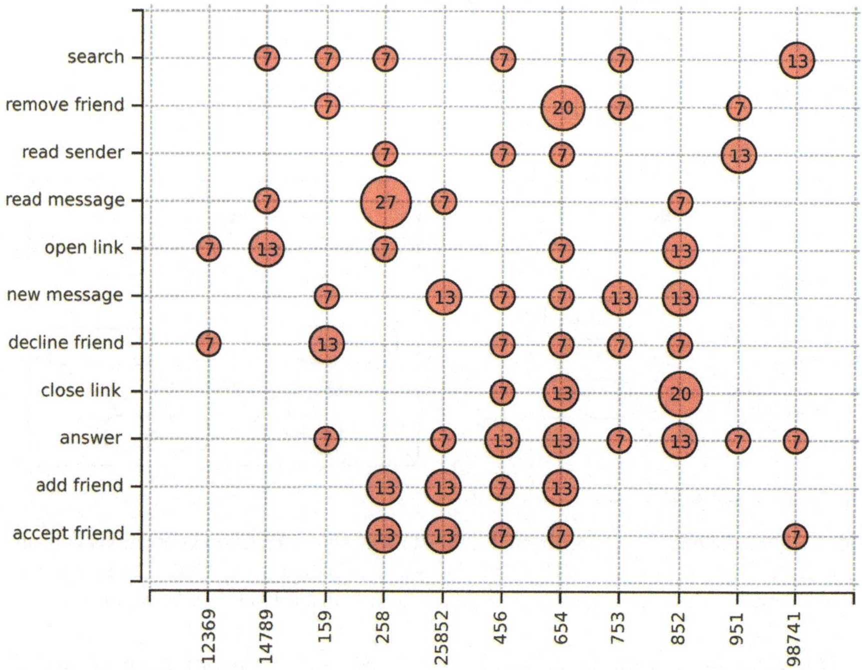

Fig. 13 Like Fig. 8, this plot shows the (rounded) percentage of subjects that performed the (actually recognized) gestures (x-axis) corresponding to the commands (y-axis). For a better overview we cropped away gestures with a score lower than 7%.

Because some of the transcribed gestures, described in Section 4, are not recognizable by the gesture recognition software, we decided to take a closer look at the recognized gestures until there is a recognition system that is able to recognize also the complex gestures. Figure 13 depicts the command/gesture correspondences we consider to be useful. For our current system implementation we chose the sequence 258 for the command *accept friend* and 25852 is used for *add friend*. For *answer* we allow the gestures 456. The *close link* command can be triggered with the gesture 852, where as the *decline friend* is triggerable with the sequence 159. We chose the sequence 753 for the command *new message* and 14789 for *open link*. It was obvious to chose 258 for the command *read message* and 951 for read sender. *Remove friend* can be triggered with the gesture sequence 654 and the sequence 98741 corresponds to the command *search*.

[6] http://www.techfak.uni-bielefeld.de/ags/ami/publications/RPHR2011-ESN/

Although this set of gestures differs for the transcribed gestures, there are still correspondences between these two sets where the winning gestures were quite simple. E.g. the command *read message* is the same gesture in both sets (258 = "down"). Also *remove friend* has the same gesture (654 = "left"). Since the sequence 753 corresponds to "up right", both gestures result in triggering the command *new message*, where as the sequence 951 is the same as "up left" which corresponds to the command *read sender*. The command *open link* has the gesture "down, right" that is the same as the sequence 14789. Because the sequence 98741 is the same as "left, up" the *search* command has also correspondences in both gesture sets.

6 Conclusion

The presented approach makes a novel contribution to the HCI research field. As our first prototype of the user interface concept we built a social networking client, which combines actuated TUIOs with gestural input. We conducted a study to investigate which gestures are suitable for interacting with a social network with TAOs. We already found tendencies for suitable gestures and furthermore got valuable feedback from our subjects for improvements of our system design and considerations for our assumptions on the interaction design. To our experience gestural input is a useful way to interact with TUIs. However this needs to be empirically verified in user studies. We also learned that users would like to use the richer interaction possibilities that physical objects offer for performing gestures such as lifting, rotating or shaking a TAO.

For complex gestures, such as 'x' the gesture recognition was not robust enough with LibStroke so that we plan to utilize another custom recognition framework, such as the Ordered Means Models developed in our research group [18]. This will enable the user to use e.g. objects' rotation in addition to the current translational gestures. For modalities that are not visually trackable, such as shaking (quickly) we need to create further extensions, such as additional integrated sensors. The modularity of our hardware and software makes such extensions easily applicable.

Acknowledgements. We thank Andreas Kipp, Matthias Schröder, and Dominik Weißgerber for helping creating the markers. We thank Christof Elbrechter for helping on the tracker and Dr. Dieta Kuchenbrandt for her advices on our study. We thank the German Research Foundation (DFG) and the Center of Excellence 277 Cognitive Interaction Technology (CITEC) who funded this work within the German Excellence Initiative.

References

1. Weiser, M.: The computer for the 21st century. Scientific American 272(3), 78–89 (1995)
2. Fitzmaurice, G.W., Ishii, H., Buxton, W.A.S.: Bricks: laying the foundations for graspable user interfaces. In: Proceedings of the SIGCHI Conference on Human Factors in Computing Systems, pp. 442–449. ACM Press/Addison-Wesley Publishing Co., New York (1995)

3. Ishii, H., Ullmer, B.: Tangible bits: towards seamless interfaces between people, bits and atoms. In: Proceedings of the SIGCHI Conference on Human Factors in Computing Systems, p. 241. ACM (1997)
4. Pangaro, G., Maynes-Aminzade, D., Ishii, H.: The actuated workbench: computer-controlled actuation in tabletop tangible interfaces. In: Proceedings of the 15th Annual ACM Symposium on User Interface Software and Technology, pp. 181–190 (2002)
5. Weiss, M., Schwarz, F., Jakubowski, S., Borchers, J.: Madgets: actuating widgets on interactive tabletops. In: Proceedings of the 23nd Annual ACM Symposium on User Interface Software and Technology, UIST 2010, pp. 293–302. ACM, New York (2010)
6. Rosenfeld, D., Zawadzki, M., Sudol, J., Perlin, K.: Physical Objects as Bidirectional User Interface Elements. IEEE Computer Graphics and Applications, 44–49 (2004)
7. Wobbrock, J., Morris, M., Wilson, A.: User-defined gestures for surface computing. In: Proceedings of the 27th International Conference on Human Factors in Computing Systems, pp. 1083–1092. ACM (2009)
8. Morris, M., Wobbrock, J., Wilson, A.: Understanding users' preferences for surface gestures. In: Proceedings of Graphics Interface 2010, Canadian Information Processing Society on Proceedings of Graphics Interface 2010, pp. 261–268 (2010)
9. Wobbrock, J., Aung, H., Rothrock, B., Myers, B.: Maximizing the guessability of symbolic input. In: CHI 2005 Extended Abstracts on Human Factors in Computing Systems, pp. 1869–1872. ACM (2005)
10. Krzywinski, A., Mi, H., Chen, W.: RoboTable: a tabletop framework for tangible interaction with robots in a mixed reality, pp. 107–114. ACM (2009), `http://portal.acm.org/citation.cfm?id=1690388.1690407`
11. Riedenklau, E.: TAOs - Tangible Active Objects for Table-top Interaction. Master's thesis, Faculty of Technology, Bielefeld University, Germany, supervized by Dr. Thomas Hermann, Prof. Helge Ritter, and Tobias Großhauser (June 2009)
12. Hermann, T., Henning, T., Ritter, H.: Gesture Desk – An Integrated Multi-modal Gestural Workplace for Sonification. In: Camurri, A., Volpe, G. (eds.) GW 2003. LNCS (LNAI), vol. 2915, pp. 369–379. Springer, Heidelberg (2004)
13. Fritsch, J., Wrede, S.: An Integration Framework for Developing Interactive Robots. In: Brugali, D. (ed.) Springer Tracts in Advanced Robotics, vol. 30 Springer, Berlin (2007)
14. Latombe, J.-C.: Robot motion planning, 3rd edn. The Kluwer International Series in Engineering and Computers Sci. Kluwer Acad. Publ., Boston (1993)
15. Bencina, R., Kaltenbrunner, M., Jorda, S.: Improved Topological Fiducial Tracking in the reacTIVision System. In: IEEE Computer Society Conference on Computer Vision and Pattern Recognition - Workshops, CVPR Workshops. IEEE (2005), `http://ieeexplore.ieee.org/lpdocs/epic03/wrapper.htm?rnumber=1565409`
16. Chen He, X., Yung, N.H.C.: Corner detector based on global and local curvature properties. Optical Engineering 47(5), 57008 (2008), `http://link.aip.org/link/OPEGAR/v47/i5/p057008/s1\&Agg=doi`
17. Ullmer, B., Ishii, H., Glas, D.: Mediablocks: physical containers, transports, and controls for online media. In: Proceedings of the 25th Annual Conference on Computer Graphics and Interactive Techniques, pp. 379–386 (1998)
18. Großekathöfer, U., Linger, T.: Neue Ansätze zum maschinellen Lernen von Alignments. Master's thesis, Faculty of Technology, Bielefeld University, Germany, supervized by Prof. Helge Ritter and Dr. Peter Meinicke (September 2005)

HECTOR, a New Hexapod Robot Platform with Increased Mobility - Control Approach, Design and Communication

Axel Schneider, Jan Paskarbeit, Mattias Schaeffersmann, and Josef Schmitz

Abstract. At the University of Bielefeld a new bio-inspired, hexapod robot system called HECTOR has been developed and is currently set up. To benefit from bio-inspired control approaches it is fundamental to identify the most important body aspects in biological examples and to transfer body features and control approaches as pairs to the technical system. According to this, the main functional characteristics of HECTOR as presented in this paper are the elasticity in the self-contained leg joint-drives with integrated sensory processing capabilities, actuated body joints and in addition a lean bus system for onboard communication.

1 Introduction

Within the field of autonomous minirobots for research and edutainment, bio-inspired robots fill an important gap by transferring biomechanical and neuro-ethological solutions for motion tasks from natural to artificial systems (constructive biomimetics). The synthesis of planned and goal-directed motions with evolutionary well-tested tools and methods from biological examples generates systems which can be different to those created by means of purely technological approaches. In this context, autonomous walking robots represent one sub-category. In particular, walking and climbing in uneven terrain is a challenging task for which several approaches exist. Many of these approaches are dominated by AI-methods. This, however, bears the risk of merely mimicking walking rather than understanding and using the underlying biological concepts.

Axel Schneider · Jan Paskarbeit · Mattias Schaeffersmann
University of Bielefeld, Mechatronics of Biomimetic Actuators, P.O. Box 100131, 33501 Bielefeld, Germany
e-mail: axel.schneider@uni-bielefeld.de

Josef Schmitz
University of Bielefeld, Biological Cybernetics, P.O. Box 100131, 33501 Bielefeld, Germany
e-mail: josef.schmitz@uni-bielefeld.de

In this paper we report on the design of the new hexapod robot HECTOR (**HE**xapod **C**ognitive au**T**onomously **O**perating **R**obot) which a) is bio-inspired in terms of biomechanical design and neurobiological control, b) serves as a testbed for cognitive (planning) approaches and c) allows a flexible integration of additional bio-inspired sensors, e.g. from the visual and tactile domain. The ultimate goal is to raise bio-inspired walking to a cognitive level while at the same time investigating and broadening the fundamentals of sensor-actor loops in the sense of even more stable gait generation in challenging walking situations. We also introduce a lean bus system for communication with those robot components related to walking (BioFlex Bus).

For the current robot design we decided in favour of a hexapod rather than a bipod setup. The reasons for this decision are manyfold. In two-legged walking a large effort has to be made to secure dynamical stability of the robot already during walking on flat terrain. Careful operation (falling has to be prevented at any costs) impedes the exploration of an otherwise large parameter space. In contrast, static stability in a six-legged machine can be guaranteed in most situations. The number of DoF (Degrees of Freedom) is large enough to allow redundant postural solutions in different movement situations. The number of closed kinematic chains that can be generated with arbitrary leg-pairs allows the robot to assume safe postures in different ways by generation of adequate ground forces. Furthermore, the redundancy of legs allows the front legs to be used for manipulation tasks while the remaining legs provide a safe foothold. The most important body features of the bio-inspired six-legged robot in this paper are designed following morphological details of the stick insect *Carausius morosus*.

Section 2 introduces the background of bio-inspired control of six-legged walking and exemplifies the relation to higher cognitive planning abilities. Section 3 introduces the mechanical setup of the robot, Sect. 4 explains details of the construction of the self-contained, elastic drives in the leg-joints and the arrangement of the servo-drives for the joints between body segments. Section 5 contains the arrangement of the communication system between the on-board computer and the drive- and sensor-components. The paper finishes with a discussion and an outlook on future work in Sect. 6.

2 From Bio-inspired Control of Walking to Planning

Walking seems to be a complex tasks which involves the control and coordination of many joints to propel a body forward in an efficient and organised manner. In most cases, the number of DoF in biological motor systems is higher than required to accomplish the motor task itself in just one way. This allows a wide variety of postural solutions including sub-configurations in the space of ground reaction forces and torque distributions. In contrast to this complexity, walking and even climbing in cluttered environments is easily accomplished by seemingly simple animals like insects. Without following the example of biological systems, motion control could be perceived as being a well suited problem for central control approaches. Full knowledge of all body details and the complete proprioceptive information could be used to pre-calculate movement trajectories of single legs or concatenate rhythmic motor primitives to achieve orchestrated leg movements. The result – of course – would be walking. The responsiveness to changes in the environment and the ability to negotiate even challenging substrates however would strongly depend on the imagination of the control engineer who might get lost in an ever-growing number of rules and exceptions.

Bio-inspired Control of Walking
Early successful solutions for the basic control of six-legged walking were based e.g. on the subsumption architecture as proposed by Brooks [5, 7] and used in the robot Genghis [6]. This architecture allows an incremental expansion of the control by additional behaviours organised hierarchically in layers. In this concept, the goals of higher level behaviours subsume those of lower level behaviours, giving the lower level behaviours a more reflex-like character. Biological research on different insect species has shown already in an early stage that legs have their own rhythmic behaviour which seemed to be weakly coupled to neighbouring legs (for stick insects see e.g. [24]). The first implementation of a corresponding distributed neural network controller in a six-legged robot by Beer and co-workers was based on Pearson's flexor burst-generator model [18, 17]. Beer and colleagues used one pacemaker neuron per leg with mutual inhibition to realise the inter-leg coupling of the sensor-driven rhythmic controllers (also formulated as neural networks) in ROBOT I (six legs, 2 DoF per leg) [3]. The concept of individual legs acting as autonomous agents which generate automatic transitions between swing phase and stance phase within one step-cycle based on local proprioceptive inputs like load and joint angles is also the basis for the WALKNET approach as formulated by Cruse and co-workers [11] (a review on the detailed biological foundations can be found in [15]). Early versions of WALKNET were tested on the walking machine MAX [19], later versions on the six-legged robot TARRY IIB [20]. Current and upcoming versions with cognitive expansions (e.g. reaCog see [13]) will be implemented on HECTOR. Besides the local organisation of swing and stance transitions, WALKNET mainly focusses on the coupling of legs by means of sparse information transfer between neighbours (legs) which is formulated in terms of coordination rules. Figure 1(a) depicts this decentralised control strategy with one separate control module per leg.

Coordination Rules
Biological research on the stick insect led to the formulation of six coordination rules [10]. The three most important rules for basic walking including their direction of influence are shown in Fig. 1(a) and described in Fig. 1(b).

Coordination rule 1 is directed from a leg (sender) to its anterior, ipsilateral neighbour leg (receiver) and ensures that this anterior neighbour does not lift off (transition to swing phase) as long as the sending leg is in swing. This is reached by enforced prolongation of the stance phase of the receiver by shifting its posterior extreme position (PEP) backwards for the duration of the swing phase of the sender. In current dynamics simulations of HECTOR, the PEP shift is equal to about 25%

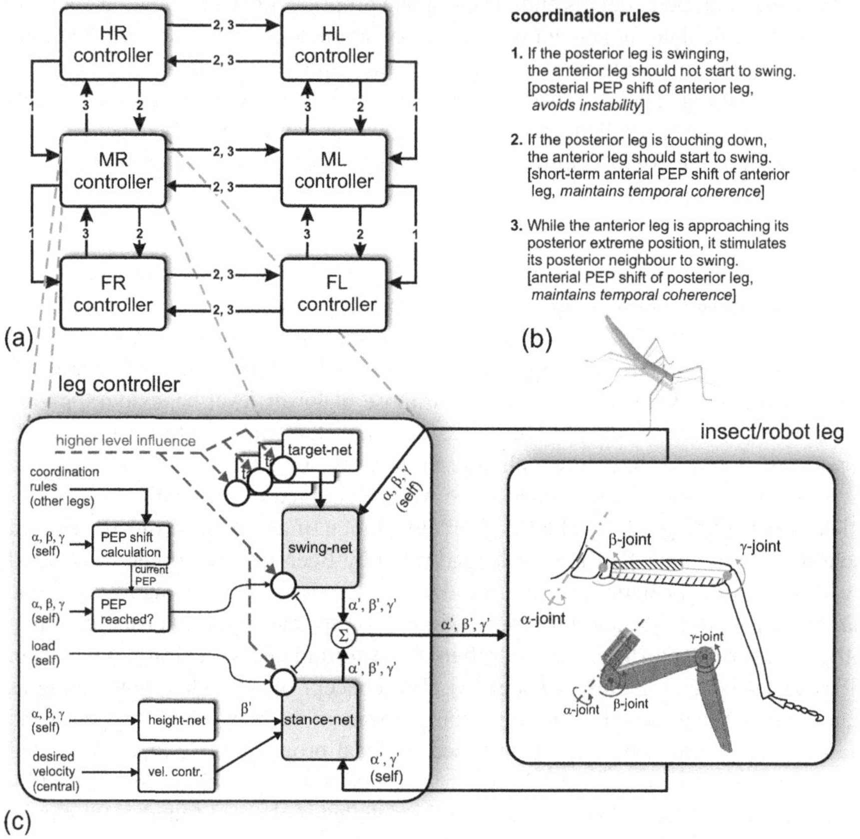

Fig. 1 (a) Insect-inspired walking assumes that each leg is an autonomous agent which exchanges sparse information with neighbouring legs to achieve coordinated movements (left hind, middle and front leg = HL, ML and FL; right side accordingly). (b) List of coordination rules as foundation for the sparse information exchange in (a). (c) Setup of a single leg controller. The core of the leg controller is a swing and a stance module which inhibit each other. Activation of swing or stance is influenced by local sensory information and coordination inputs from neighbouring legs. Modules can be activated by excitation of their activation units (bold circles).

of the normal stance amplitude. In literature, also values of up to 60% can be found [14]. This rule actively supports static stability of the walking agent.

Coordination rule 2 is also directed from a leg (sender) to its anterior, ipsilateral neighbour (receiver). The rule ensures that the anterior neighbour is facilitated to lift off as soon as the sending leg touches down. This is reached by a short time forward shift of the receiver's PEP which triggers lift off if the receiver has moved backwards already far enough in its stance phase. The duration of the short time forward shift is equal to 25% of stance duration of the sender [14]. In the dynamics simulation of HECTOR even a fixed value of ~0.2 s was sufficient. The amplitude of the PEP forward-shift was chosen to be 25% of the normal stance amplitude. Rule 2 also acts from a leg to its contralateral neighbour. The strength of the influence however is weaker than for the anterior, ipsilateral neighbour.

Coordination rule 3 is directed from a leg (sender) to its posterior, ipsilateral neighbour (receiver). Rule 3 enforces a swing movement of the posterior neighbour when the sender is about to reach its PEP during stance phase. A swing movement is considered to be fast and can thus be finished by the receiver while the sender moves the last bit towards its PEP. This effect is reached by a forward shift of the receiver's PEP. In the dynamic simulation of HECTOR a shift of about 12.5% of the normal stance amplitude was used. Rule 3 also acts between contralateral neighbours but again the influence is weaker than for the ipsilateral side.

Coordination rules 4, 5 and 6 do not have to be implemented necessarily for basic walking since they deal with exploitation of prior footholds (rule 4), distribution of load (rule 5) and introduction of correction steps to avoid stumbling (rule 6). They are not further described in this work. A detailed description can be found in [15].

Single Leg Controller

Figure 1(c) shows the details of a single leg controller. The two main modules are the swing and stance module. Both can be described and implemented as neural networks. Modules can be activated by excitation of their activation units (bold circles). As already described above, the first three coordination rules influence the position of the PEP of a single leg. The PEP shift is organised in the two upper boxes on the left side of the leg controller. According to the described influence of the coordination rules and the current leg position (represented by the three joint angles α, β and γ), the currently valid, shifted PEP is calculated (PEP shift calculation box). If the PEP is reached or the leg is even behind it (PEP reached box), the swing-net is activated via its activation unit. The swing-net activation unit inhibits the activation of the stance-net. The swing-net generates a swing movement of the leg. At the end of the swing movement the leg touches down and takes over its share of the body weight which leads to a load signal (left side of leg controller box). The load signal activates the stance-net which in turn inhibits the swing net in the now following stance phase. As part of the stance-net, the height-net basically controls the angle of the β-joint to regulate the height of the leg onsets. A velocity controller influences the retraction velocity of the α-joint during stance. In this way, alternating swing-stance-patterns are generated in a leg and inter-leg coordination is reached through the coordination rules.

Higher Level Control and Planning

Besides the above described decentral organisation of basic walking, also higher levels of control must be enabled to effect the low-level (reactive) layer of movement generation. In those controller extensions which are currently under development it is planned to mediate their access to the reactive layer by using the activation units as shown in Fig. 1(c). For instance the target-nets, that define the position a swing movement is aiming at, could be chosen to represent the actual position of the anterior neighbour leg or to follow a goal defined by a higher layer for example

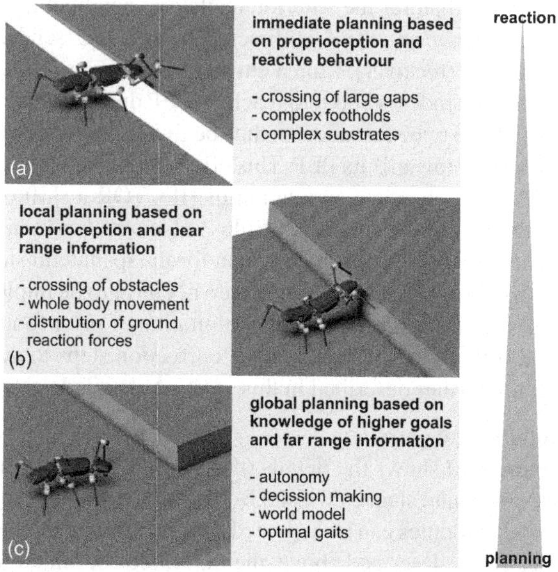

Fig. 2 (a) Based on reactive walking, simple planning based on proprioception allows crossing of large gaps with complex distributions of footholds with different substrates. (b) With growing complexity of the body actuation and sensorisation (near range), obstacle crossing by means of whole body actuation becomes possible. This is also related to the generation of suitable ground reaction forces. (c) Global planning and decision making (e.g. attack or avoid obstacle) requires a world model. Gait generation can then be regarded as optimisation problem also based on global parameters.

in climbing situations. Also the swing and stance generation can be intercepted to rigorously interrupt or softly influence basic walking.

Besides the fact that basic, reactive walking can be generated by the WALKNET framework described above, there are still important challenges that have to be mastered on the way to a universally deployable walking machine with cognitive abilities. Figure 2 gives an impression of some of these challenges.

Gap crossing is an important ability on the way towards autonomy for a six-legged robot like HECTOR. It has been shown that simple gap crossing can still be managed by WALKNET on the reactive level [4]. However, in natural environments it cannot be expected that gaps are that well structured and defined as it is shown in Fig. 2(a). In real world situations, a gap might occur which can only be traversed by using a certain order of probably even small footholds with uncertain substrate quality. Here, the question arises if the gap crossing ability has to be shifted to a planning instance which manages for example the position of the centre of gravity of the robot and which has to manipulate an internal model of the scene to solve the task incrementally. In order to tackle this task, HECTOR needs an increased freedom of movement for its legs and torque reserves to adopt even sprawled postures.

Obstacle crossing, as indicated in Fig. 2(b), is a task which is in parts related to gap crossing but which has a higher complexity since it introduces a foothold distribution across all three dimensions of the environment and – in the end – even inclined orientations of footholds. HECTOR will be equipped with actuated body-joints. This is the foundation for a rich repertoire of full body motions to adapt to the obstacle while crossing it. The high number of DoF allows the introduction

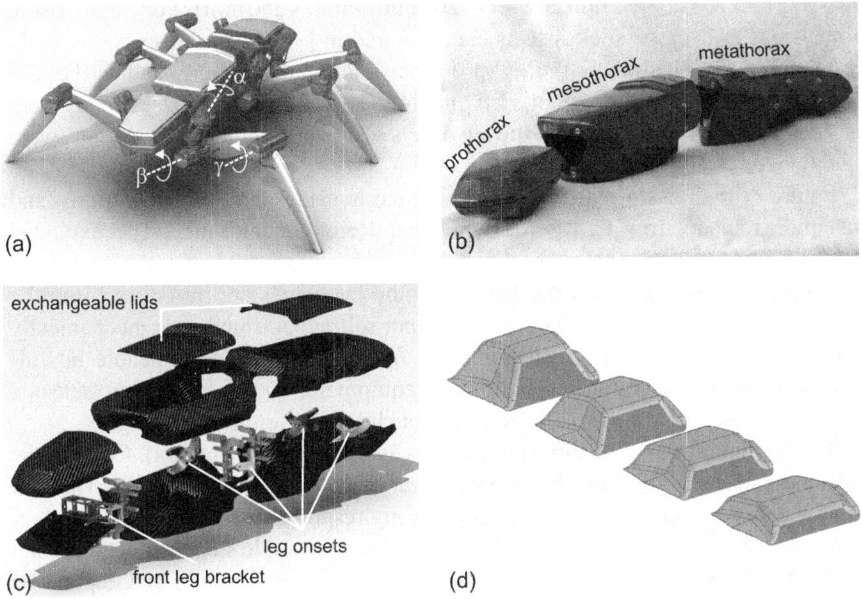

Fig. 3 (a) Rendered image of the robot with three axes of rotation for one leg. (b) Real image of the three housing parts for pro-, meso- and metathorax of the robot made from CFRP to achieve a lightweight construction. (c) Exploded view of the three housings showing the self-supporting structures, load transmission points of the leg onsets and exchangeable lids for the meso- and metathoracic segments. (d) Different mesothoracic lids which are reproducible with rapid-manufacturing methods (printing) for quick integration of additional sensors.

of additional goals like an optimisation of ground reaction forces etc. A higher, yet locally bounded, planning entity could take into account e.g. only immediately reachable footholds for movement planning. The internal model would have to be expanded from the simple body to additional representations of nearby objects (e.g. footholds).

Global planning, as shown in Fig. 2(c), is a task which requires a broader knowledge of the world (far range sensing), suitable world models and finally goals which have to be accomplished. In the aforementioned gap crossing paradigm, for instance, it might be a good idea to take a detour to circumvent the obstacle instead of directly attacking it. The decision might be guided by an overall optimality criterion like energy consumption, torque load minimisation etc.

3 Bio-mechanical Inspiration and General Robot-Layout

Important features of HECTOR were designed following morphological details of the stick insect *Carausius morosus*. A rendered image of HECTOR is depicted in Fig. 3(a). Body and leg lengths were scaled-up by a factor of approximately 20. This results in an overall length of the robot of ~950 mm and leg lengths without tarsi of ∼572 mm (coxa ∼32 mm, femur ∼260 mm, tibia ∼280 mm). For comparison, dimensions of average stick insects can be found in [9]. In stick insects, the body is divided into a head, three thorax (pro-, meso- and metathoracic) and further abdominal segments. For the robot, only the three leg-carrying thorax segments were copied. The relative distances of the leg onsets of front-, middle- and hind legs were maintained.

Figure 3(b) shows a real image of the three housing parts for pro-, meso- and metathorax made from CFRP (Carbon Fiber Reinforced Plastic). The exploded view in Fig. 3(c) shows that the housings for the meso- and metathorax have a self-supporting structure. At the leg onsets the leg forces are introduced into the CFRP-structure. The prothorax segment is not self-supporting but is mechanically connected to the bracket which holds the front legs. The exchangeable lids allow subsequent integration of additional equipment, such as different sensors. Figure 3(d) shows exchangeable mesothoracic lids in different stack heights.

The configuration of the three leg-joints is also depicted in Fig. 3(a). Close to the leg onsets at the body (hip), two joints are arranged close to each other. The first joint is the α- or subcoxal-joint which is mainly responsible for protraction and retraction of the leg during swing- and stance-phase of a single step. The second joint is the β- or coxa-trochanter-joint, mainly responsible for elevation and depression of the whole leg. The third and most distal joint is the γ- or femur-tibia-joint (knee), primarily responsible for the excursion of the leg. It can be seen that the rotational axes of the leg joints are not aligned with the axes of the cartesian coordinate system in the central body segment but are rather slanted and have the same spatial orientation as the joint axes in stick insects. Therefore, the effect of each axis α, β and γ during a full leg motion changes. For example, the α-joint movement also influences the body elevation.

Further details of the joint-axes' orientations are depicted in Fig. 4(a). It shows the pro- and mesothorax including the α-actuators of the front legs. The orientation of each α-axis can be characterized by fixed φ- and ψ-angles as indicated. To obtain the spatial orientation of the α-joint, a work plane (① in Fig. 4(a)) is assumed that is parallel to the x-z-plane in the body with an offset that is defined by the point of the leg onset. In the next step, this plane is rotated around the leg's onset point by the angle φ about the z-axis. The axis of the α-joint lies in the resulting plane, cuts the point of the leg onset and is tilted by the angle ψ relative to the vertical. The φ- and ψ-angles for the fore-, middle- and hind-legs as well as the maximal and minimal α- and β-joint-angles are listed in Table 1. For the front-legs the workspace is rather shifted to the front whereas the workspace for the hind legs is shifted rearwards. To assure the freedom of movement evoked by the listed α- and β-angles and all their combinations, while at the same time providing enough room for the internal components, the maximum outline of the robot had to be computed. This was done by a sequential process in which the α-joint was moved from front to rear accompanied by synchronous up and down movements of the β-joint. In this process, the femur segment cut through a virtual volume and deleted those voxels which were not part of the later housing. The results are shown in Fig. 4(d). The point of origin for the x- and z-axis corresponds to the lines ① and ② in Fig. 4(c),

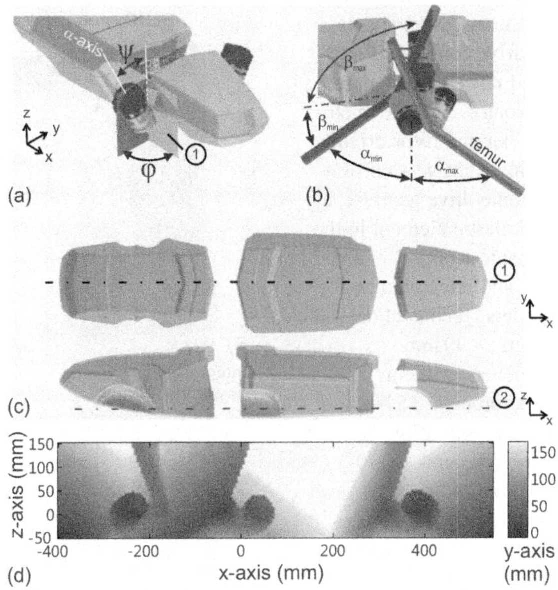

Fig. 4 (a) Pro- and mesothorax of the robot including α-actuators. The α-actuators' axis of rotation is first revolved around the z-axis by an angle φ relative to a plane ① that is parallel to the x-z-plane. Subsequently it is revolved by an angle ψ relative to the z-axis.
(b) Maximal positions of β-actuator and femur for the right middle leg.
(c) Top and side view of the robot housings.
(d) Maximal outline for the robot housings relative to the mirror plane ① depicted in (c). The point of origin for the z-axis is depicted in (c) as line ②.

respectively. The three black round areas in Fig. 4(d) indicate those volumes which the β-actuator penetrates during the above described process. The outer shape of the final robot housings (top and side view in Fig. 4(c)) is trimmed in such a way that it does not intersect with the virtual surface described above and at the same time leaves enough space for the internal components of the robot.

4 Drive Technology for Leg- and Body-Joints

Biological movement systems like stick insects are actuated by muscles which give them intrinsic, nonlinear compliance. In technical systems like robots, this compliance can be generated by pure control, by physically existing, elastic elements or by hybrid compliance. The new joint-drives (BioFlex Rotatory Drive), which are used in HECTOR, are designed to generate nonlinear compliance based on an integrated

Table 1 The table shows φ- and ψ-angles which define the angular orientation of the α-joint axis (see Fig. 4(a)) and maximal and minimal movement angles for α- and β-joints.

	φ	ψ	α		β	
			min	max	min	max
front legs	80	30	-50	100	-30	110
middle legs	90	30	-70	70	-30	110
hind legs	115	30	-90	50	-30	110

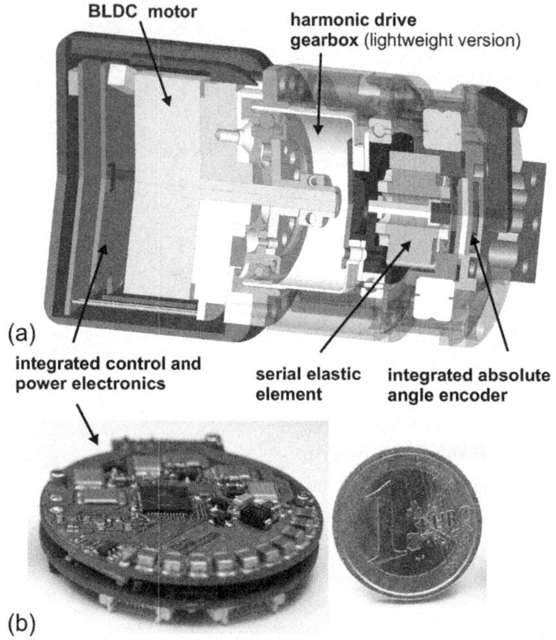

Fig. 5 (a) Section view of the elastic joint-drive used in the 18 leg joints. The drive consists of integrated power- and control-electronics, a BLDC motor with external rotor driving a lightweight version of a harmonic drive gearbox, a serial elastic element in the mechanical output of the drive and integrated angle encoders. Technical data:
- length: ∼90 mm
- diameter: ∼50 mm
- max torque: ∼15 Nm
- weight: ∼0.32 kg
- power/weight: ∼170-400 W/kg.

(b) Image of the power- and control-electronics integrated into the elastic joint drive as indicated in (a). One Euro coin to compare size.

serial-elastic element. A section view of the joint-drive is depicted in Fig. 5(a). Technical details are given in the caption. The setup of the joint-drive consists of a brushless DC motor (BLDC) with external rotor, a lightweight harmonic drive gearbox, a serial elastic element in the mechanical output and integrated encoders. To utilise the advantages of the BLDC motor, miniaturised compact electronic boards have been developed to fit into the back of the joint-actuator. Besides the power electronics, the board stack also contains the control electronics for processing of multiple sensory inputs and a microcontroller that is able to host local implementations of control approaches. The board stack is depicted in Fig. 5(b). The decision to incorporate a real, serial elastic element in the mechanical output is essential at least in four different ways. First, it protects the integrated gearbox from torque peaks which occur in insect-inspired walking at the end of the swing phase when the leg hits the ground. Second, impact forces which are exerted on an obstacle in a possible collision during the swing phase are attenuated. Third, a real elastic element reacts in real time as opposed to elasticity due to pure control which suffers from a finite control response time. Fourth, a real elastic element can store energy. The last point plays a dominant role in dynamic walking but a subordinate role in statically stable walking as it is the case for HECTOR. An additional important point for the adoption of compliance in bio-inspired joint-drives is that bio-inspired control approaches like

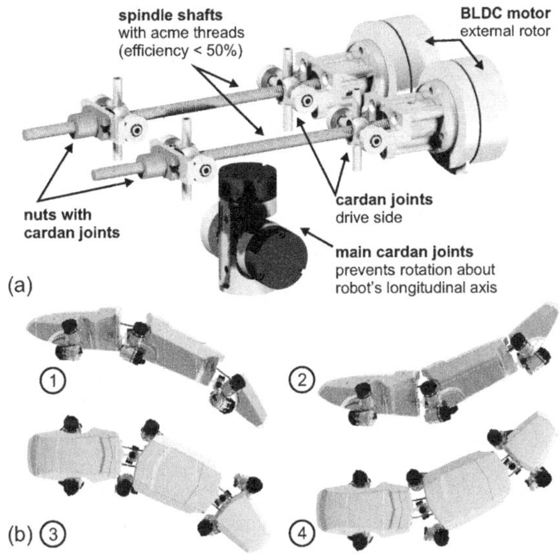

Fig. 6 (a) 2 DoF spindle drive which allows panning and tilting of two adjacent body segments. (b) Maximum downward ① and upward ② tilting by using both spindle drives between pro-/meso- and meso-/metathorax. The maximum right ③ and left ④ panning is also achieved by using both spindle drives but this time with opposite rotational directions.

Local Positive Velocity Feedback (LPVF, [21]) are based on compliant joints. These control approaches can be directly implemented in the on-board control electronics.

Stick insects in their natural habitats use the ability to pan and tilt their body segments to increase manoeuvrability. To introduce this ability also in the robot, a new 2 DoF spindle drive has been designed as shown in Fig. 6(a). Each actuator consists of a main cardan joint that prevents rotations about the robot's longitudinal axis and two spindle drives. The spindle drives are composed of shafts with acme threading and have an efficiency slightly below 50 % to be self-impeding (to retain joint-angles of the body without actuating the motors). Figure 6(b) depicts robot body postures for maximum downward and upward tilting as well as maximum left and right panning.

5 Communication Setup on the Robot

The drive systems introduced in Sect. 4 also have to integrate a bus system to exchange control and sensor information with other parts of the robot. The schematic depiction of the communication setup is shown in Fig. 7(a). A host computer (PC/104) is situated in the mesothorax for high-level control of all actuators (leg and body joint-drives). The communication between PC/104 (host) and the drives and other clients is realized with a newly developed bus system, the BioFlex Bus. Its hardware consists of a miniaturized bus master board which is connected to the host via high-speed USB. On the client side, the BioFlex Bus master possesses two galvanically decoupled channels which allow half duplex, single access communication via a differential RS-485 connection with 2 Mbit/s each. Since the bus system has to operate in electromagnetically noisy environments close to BLDC motors,

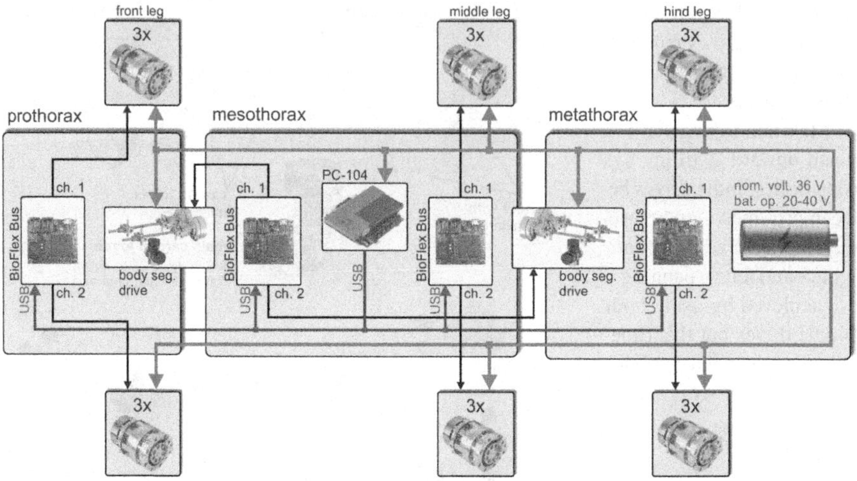

Fig. 7 (a) Schematic depiction of the robot from above showing the pro-, meso- and metathorax. Each thorax segment has two legs with three joint drives. The joint drives communicate with a bus master (BioFlex Bus) in each thorax segment. Each bus master has two channels (2 Mbit/s each) to connect to a maximum of 250 clients which are polled by the bus master to allow real-time operation. The bus masters are connected to the host computer (PC/104) in the mesothorax via high-speed USB. One additional bus master in the mesothorax is dedicated to the two spindle drive setups in the body joints.

bus systems like I^2C are not an option. The BioFlex Bus system was developed for several reasons. First, small size communication components had to be chosen on the client boards, especially in the self-contained joint-drives. Second, high-speed bus operation with short data packets (4 bytes payload, 4 bytes overhead) as well as medium-speed bus operation with long data packets (up to 153 bytes payload, 7 bytes overhead) was desired. Third, the BioFlex Bus uses an asynchronous master-controlled single access bus-sharing scheme that allows simple implementation as well as generation of real-time sensorimotor functions. Also, the communication of clients connected to the same channel is possible without packet repetition by the bus master.

The packet design and the configuration of the physical and network layer of the BioFlex Bus is visualised in Fig. 8. The BioFlex Bus master is connected to the host computer via USB and is treated as a virtual serial port on the host. Data segmentation happens in 8-bit chunks and is supplemented by start-, stop- and MPCM-bit (MPCM = Multi Processor Communication Mode, sometimes called Multi-Drop Mode). This elementary frame is shown in the top row of Fig. 8. The two middle rows show a short and a long packet with a 4 byte and an up to 153 byte payload container. A flag byte in the preamble (bottom row) contains the information whether the current packet is short or long and if the receiver is allowed to respond (permission for bus access).

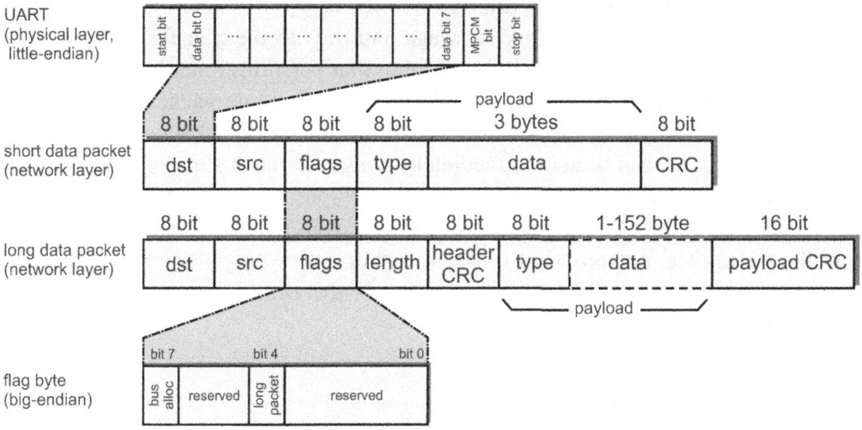

Fig. 8 Configuration of the physical and network layer of the BioFlex Bus. The BioFlex Bus master is connected to a PC via USB and is treated as a virtual serial port. This enforces segmentation of data into 8-bit chunks which make an 11-bit UART packet (shown in the top row). The two middle rows show the configuration of a short (4 byte payload) and long packet (up to 153 bytes payload). A flag-byte (bottom row) in the preamble informs the client of the packet length (short or long) and whether an answer is expected (bus allocation).

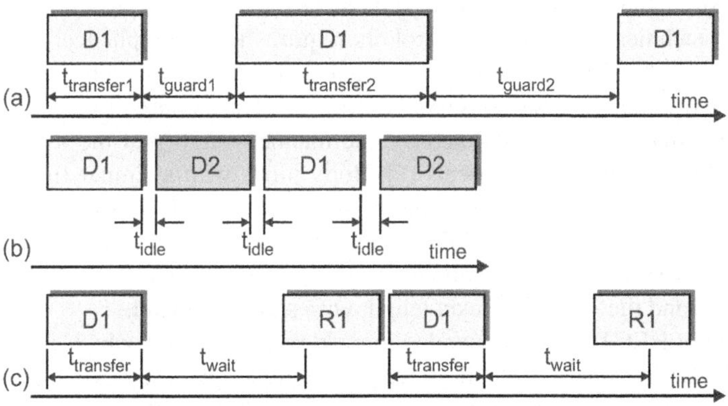

Fig. 9 Three examples for communication between BioFlex Bus master and clients (ch.1 and ch.2). (a) Communication between bus master and client 1. The time t_{guard} is needed by the client to perform CRC calculation and is proportional to data length. (b) If the bus master sends packets to clients (D1 = destination 1), the maximum communication speed is restricted by the time t_{idle} which is used by the bus master for other tasks than client-communications. (c) If the bus allocation bit is set in the flag-byte (see Fig. 8) the bus master waits for a response from the client which has to begin within a time t_{wait} after the end of transmission.

In Fig. 9, three examples for communication via the BioFlex Bus – between bus master and clients – are shown. Figure 9(a) shows the communication between bus master and client 1. After transmission of a packet, the client needs some time to process the received message. Therefore, the next possible transmission can be initialized by the bus master only after a safe time t_{guard}. The time interval t_{guard} is linearly related to the length of the message and thus to the transfer time t_{transfer}. The linear relation is due to the CRC-algorithm that performs linearly with respect to packet length. Figure 9(b) depicts the transfer of multiple packets to different destinations. After sending a packet to destination 1, the next packet with a different destination (e.g. D2) can be sent immediately. However, the bus master needs some time for other tasks than client communication which is reflected by an idle time t_{idle}. Finally, Fig. 9(c) shows a situation in which the bus master waits for a response from the client. The response has to begin within a time t_{wait} after the end of the bus master's transmission. If no response occurs the next queued packet will be transmitted.

6 Discussion and Future Work

In this paper, the new hexapod robot HECTOR was introduced. This robot adopts distinctive features of the biological example, the stick insect *Carausius morosus*. The main functional characteristics are the elasticity in the self-contained leg joint-drives and the introduction of actuated body joints. Most of the current walking machines like Scorpion [23] and Scarabeus [2] use inelastic joints. The introduction of compliant joints allows passive adaptation to the substrate. Moreover, many biological control strategies rely on elastic leg-joints. Therefore, the application of these strategies on technical joint-drive controllers requires joint-compliance. An example for this type of biological control strategy is LPVF [21, 22] which is based on the phenomenon of reflex reversals observed during the stance phase of walking (see e.g. [1]). The actuated body joints increase the manoeuvrability of the robot to a level comparable to that of stick insects [8]. Body joints with a similar functionality with 1 DoF can be found for example in Whegs II [16]. This paper also introduced the BioFlex Bus which is used for communication between components of the robot. Its main features are the small component size which is important for hardware integration and the variable packet length with small overhead.

With the new robot HECTOR, it is intended to tackle walking tasks which contain complex foothold positions and situations which require torque reserves e.g. for outstretched postures in gap crossing, steep walking or stair climbing. Besides walking based on low-level, reactive mechanisms and coordination rules as introduced in Walknet [12], the control system of HECTOR will be expanded by planning capabilities to reach predefined goals, to decide whether obstacles will be tackled or avoided and to invent strategies to cope with coordination problems on the level of leg coordination in difficult situations.

Acknowledgements. This work has been supported within the Excellence Initiative of the German Research Foundation (DFG Center of Excellence "Cognitive Interaction Technology", EXC277), by the Federal Ministry of Education and Research (BMBF) within the BIONA programme (ELAN-project to A.S.) and by an EU-FP7 grant (ICT-2009.2.1, No. 270182 to A.S. and J.S.).

References

1. Bartling, C., Schmitz, J.: Reactions to disturbances of a walking leg during stance. J. Exp. Biol. 203, 1211–1223 (2000)
2. Bartsch, S., Planthaber, S.: Scarabaeus: A Walking Robot Applicable to Sample Return Missions. In: Gottscheber, A., Enderle, S., Obdrzalek, D. (eds.) EUROBOT 2008. Communications in Computer and Information Science, vol. 33, pp. 128–133. Springer, Heidelberg (2009)
3. Beer, R.D., Chiel, H.J., Quinn, R.D., Espenschied, K.S., Larsson, P.: A distributed neural network architecture for hexapod robot locomotion. Neural Computation 4, 356–365 (1992)
4. Bläsing, B., Cruse, H.: Mechanisms of stick insect locomotion in a gap crossing paradigm. Journal of Comparative Physiology 190(3), 173–183 (2004)
5. Brooks, R.A.: A robust layered control system for a mobile robot. IEEE Journal of Robotics and Automation RA-2, 14–23 (1986)
6. Brooks, R.A.: A robot that walks; emergent behaviors from a carefully evolved network. A.i. memo 1091. Institute of Technology, Massachusetts (1989)
7. Brooks, R.A., Connell, J.H.: Asynchronous distributed control system for a mobile robot. In: Proceedings of SPIE, Cambridge, MA, pp. 77–84 (1986)
8. Cruse, H.: The control of body position in the stick insect Carausius morosus, when walking over uneven surfaces. Biological Cybernetics 24, 25–33 (1976)
9. Cruse, H.: The function of the legs in the free walking stick insect. Carausius morosus. J. Comp. Physiol. A 112, 235–262 (1976)
10. Cruse, H.: What mechanisms coordinate leg movement in walking arthropods. Trends in Neurosciences 13(1), 15–21 (1990)
11. Cruse, H., Brunn, D., Bartling, C., Dean, J., Dreifert, M., Schmitz, J.: Walking - a complex behavior controlled by simple networks. Adaptive Behavior 3(4), 385–418 (1995)
12. Cruse, H., Kindermann, T., Schumm, M., Dean, J., Schmitz, J.: Walknet - a biologically inspired network to control six-legged walking. Neural Networks 11(7-8), 1435–1447 (1998)
13. Cruse, H., Schilling, M.: From egocentric systems to systems allowing for theory of mind and mutualism. In: Doursat, R. (ed.) Proceedings of the ECAL 2011. MIT Press, Paris (in Press 2011)
14. Dean, J.: A model of leg coordination in the stick insect, *Carausius morosus* II. description of the kinematic model and simulation of normal step patterns. Biological Cybernetics 64, 403–411 (1991)
15. Dürr, V., Schmitz, J., Cruse, H.: Behaviour-based modelling of hexapod locomotion: Linking biology and technical application. Arthropod. Struct. Devel. 33(3), 237–250 (2004)
16. Lewinger, W., Harley, C., Ritzmann, R., Branicky, M., Quinn, R.: Insect-like antennal sensing for climbing and tunneling behavior in a biologically-inspired mobile robot. In: Proceedings of IEEE/ICRA, pp. 4176–4181 (2005)
17. Pearson, K.G.: The control of walking. Scientific American 235, 72–86 (1976)
18. Pearson, K.G., Fourtner, C.R., Wong, R.K.: Nervous control of walking in the cockroach. In: Stein, R.B., Pearson, K.G., Smith, R.S., Bedford, J.B. (eds.) Control of Posture and Locomotion, pp. 495–514. Plenum Press, New York (1973)
19. Pfeiffer, F.: The TUM walking machines. Phil. Trans. R. Soc. A 365, 109–131 (2007)
20. Schmitz, J., Schneider, A., Schilling, M., Cruse, H.: No need for a body model: Positive velocity feedback for the control of an 18-DOF robot walker. Applied Bionics and Biomechanics 5(3), 135–147 (2008)

21. Schneider, A., Cruse, H., Schmitz, J.: Decentralized control of elastic limbs in closed kinematic chains. The International Journal of Robotics Research 25(9), 913–930 (2006)
22. Schneider, A., Cruse, H., Schmitz, J.: Winching up heavy loads with a compliant arm: A new local joint controller. Biological Cybernetics 98(5), 413–426 (2008)
23. Spenneberg, D., Kirchner, F.: The Bio-Inspired SCORPION Robot: Design, Control & Lessons Learned. In: Climbing & Walking Robots, Towards New Applications, pp. 197–218. I-Tech Education and Publishing, Wien (2007)
24. Wendler, G.: Laufen und Stehen der Stabheuschrecke: Sinnesborsten in den Beingelenken als Glieder von Regelkreisen. Zeitschrift für vergleichende Physiologie 48, 198–250 (1964)

Force Controlled Hexapod Walking

Shalutha De Silva and Joaquin Sitte

Abstract. In this paper we describe the dynamic simulation of an 18 degrees of freedom hexapod robot with the objective of developing control algorithms for smooth, efficient and robust walking in irregular terrain. This is to be achieved by using force sensors in addition to the conventional joint angle sensors as proprioceptors. The reaction forces on the feet of the robot provide the necessary information on the robots interaction with the terrain. As a first step we validate the simulator by implementing movement control by joint torques using PID controllers. As an unexpected by-product we find that it is simple to achieve robust walking behaviour on even terrain for a hexapod with the help of PID controllers and by specifying a trajectory of only a few joint configurations.

Keywords: Robot Walking, PID controllers, Hexapod robot, Dynamic Simulation, Force Control.

1 Introduction

Legged locomotion offers vastly superior mobility in natural terrain compared to wheeled locomotion. The current drawback of legged locomotion is that walking robots are enormously more complex than their simple wheeled counterparts in both their mechanical realisation and in the co-ordinated control of the legs. Therefore the construction of fast, smooth and robust walking robots remains an elusive goal. The current popularity of humanoid robotics is stimulating much research and development in biped walking with inspiration taken from human walking. In nature

Shalutha De Silva
Queensland University of Technology, Brisbane, Queensland, Australia
e-mail: `sm.desilva@student.qut.edu.au`

Joaquin Sitte
Queensland University of Technology, Brisbane, Queensland, Australia
e-mail: `j.sitte@qut.edu.au`

the majority of creatures have either four, six or eight legs. Four or six legs offer a useful alternative to biped robots as demonstrated by Boston Dynamics' Big Dog four legged robot [9]. Six legs have the advantage of allowing a statically stable tripod gait. Hexapod walking has been well researched in insects providing knowledge that can inspire hexapod robot walking technology. A noteworthy characteristic of insect walking is the apparent high level of autonomy of the legs [3]. This poses the question whether it is possible for hexapod walking to arise from the co-operation of autonomous legs without explicit communication between the legs. The work reported here is an initial part of the research to answer this question. Such a scheme would provide distributed control of a large number of degrees of freedom and possibly robustness in irregular terrain. For a robot to walk with autonomous legs each leg has to use its sensor signals to generate the right torque at the joints. The autonomous legs we use for our study have joint angle sensors and three component force sensors. The forces acting on a foot of a legged creature depend on the topography, characteristics of the ground and the pose of the creature, including the configuration and actions of the other legs. By sensing the forces on the foot the leg gains information on the pose of the robot as a whole implicitly mediated by the robot's body. In animals, proprioreceptors give information on the position of the limbs and the tension in the muscles.

Joint angle sensors are commonly used for articulated robots, but torque or force sensing is not widely used in robot walking. Joint angles alone do not give sufficient sensory input for generating an appropriate autonomous action leading to smooth and efficient walking gaits. Joint angles do not give any information about the very essential interaction of the creature with the ground. They do not tell if there is contact with the ground, or whether the ground is soft or hard. Force sensors that measure the reaction forces on the foot can provide this information. For example, Hori et al. [5] measures the reaction forces for maintaining stability by re-configuring the distribution of total weight among the legs.

Before building a hexapod robot with force sensors [4] we use simulation for investigating autonomous leg behaviours that result in successful co-operative walking. For this we need a simulator capable of reproducing the dynamics of an articulated body with joints driven by muscles, pneumatics or electric motors.

In most of the simulations of articulated bodies, be it for robots, games or movies, movement is produced using kinematics. Especially in smaller robots the control of joints are achieved using stepper motors. The joint is moved to the desired position in a succession of small discrete steps. Trajectories are prescribed as a dense succession of points. However in reality legs move as a result of the application of torque to each joint. The torque accelerates the joint and therewith the limb, resulting in a velocity that produces a displacement. To control the movement of a leg it is necessary to apply a time varying torque such that the limb follows the right trajectory. To determine the displacement produced by of the application of a torque to a joint it is necessary to solve the corresponding dynamic equations of motion. Dynamic simulation has recently been used by [8] to modify the ankle joint of a biped robot before physically modifying the joint. How dynamic simulation combined with kinematic simulation can be used for virtual prototyping of a hexapod robot is discussed in [6].

In nature central pattern generators form the basis for repetitive leg motion in walking and these patterns are modulated by sensory input for adaptation to irregular terrain. Such mechanisms could be useful for walking robots [2]. For the purpose of validating the dynamic hexapod robot simulator we made the simple choice of using PID controllers for generating the torque required to reach a specified joint angle configuration. This choice and further details of the dynamic simulation are described in Section 2. In Section 3 we describe results of the simulation with PID controllers that already show a level of robustness that is interesting by itself. Section 4 contains our plan for further work and finally we draw conclusions in section 5.

2 Simulation of Hexapod Walking

2.1 Simulation Tools

There are many software packages that provide tools for the simulation and visualisation of rigid body dynamics. After reviewing Matlab Sim Mechanics, Microsoft Robotics Studio, Vortex toolkit, Bullet Physics Library and Open Dynamics Engine, we chose ODE (Open Dynamics Engine) [11]. ODE is an open source library of C language functions used by researchers [7, 10, 12] and game developers. The popular Webots robot simulator, developed by Cyberbotics Ltd., also relies on ODE for accurate physics simulation [7]. We use the ODE library with Microsoft Visual Studio 2008 C++ to build the hexapod robot simulation. ODE also includes the *drawstuff* 3D visualisation library.

2.2 Configuration of the Hexapod

The main consideration for choosing the body shape of the hexapod is the equivalence of leg positions that is achieved with hexagonal symmetry around a vertical axis. For simplicity the body is disk shaped with legs attached at 60 degree intervals as shown in figure 1.

Each leg has 3 sequential segments: coxa, femur and tibia. The coxa attaches to the body, the femur to the coxa and the tibia to the femur, each through a motor powered hinge joint giving each leg three degrees of freedom (DOF). The hexapod has a total of 18 DOF. The body has 0.5 m radius and a thickness of 5 cm. The coxa is 0.1 m long, and the femur and tibia are each 0.25 m long. Each leg has a mass of 0.6 kg and the mass of main body is 6 kg giving a total of 9.6 kg mass to the hexapod. Other values can be easily chosen. Figure 2 shows the numbering of the legs and labelling of the joints.

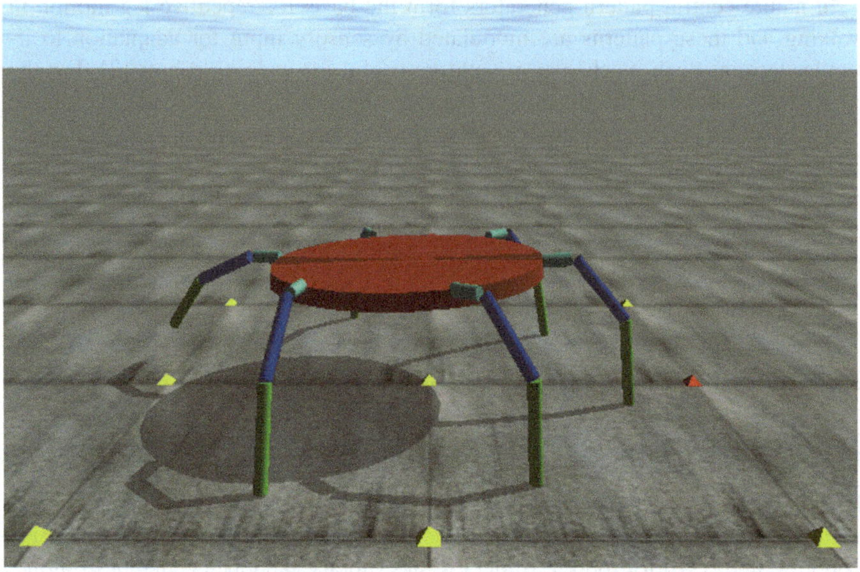

Fig. 1 Hexapod in ODE simulation environment

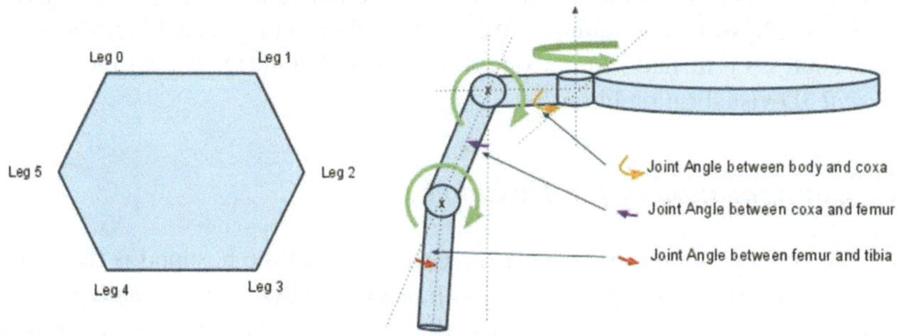

Fig. 2 Leg configuration of the hexapod and the rotation axis for each joint. The coloured arrows indicate the measurement of joint angles.

2.3 Kinematics vs Dynamics

The motion of a real robot is determined by the acting forces and the laws of dynamics. Accordingly the position and speeds of the robot's body parts have to be obtained by solving the dynamic equations of motion. We call this dynamic simulation as opposed to kinematic simulation that is often used in robot simulation. In kinematic simulation the cause of movement in an articulated body are the changes in joint angles.

2.3.1 Kinematics

In kinematic simulation the positions of all the connected bodies over time are computed from previously specified trajectories. These trajectories may not comply with Newton's laws. Kinematic simulation is prevalent in most of the animation movies. To obtain realistic looking movement, trajectories are captured from the motion of real actors and then the animated character is moved along this trajectory. There are two main methods for kinematic simulation of robots:

- Forward kinematics: In forward kinematics the position of the end of the kinematic chain, in our case the foot, is calculated from the given joint angles.
- Inverse kinematics: In inverse kinematics the end position is given and the objective is to calculate the required joint angle for each joint. When the joints are moved to the calculated joint angles, the articulated body is at the desired position.

2.3.2 Dynamics

A change in joint angle, or more correctly, a change in the angular velocity of the joint results from the application of a torque either by a muscle or a motor. The control of motion is exerted by the control of the torque applied to the joints. The resulting motion depends on the mass and the geometry of the kinematic chain and the external reaction forces resulting from contact with other bodies.

2.4 Torques and Reaction Forces

In a real hexapod and in simulation the control algorithm provides us with the values of the torques applied at the joints. In a real robot the reaction forces acting on the feet on the ground could be measured by force sensors. Simulating force sensors requires computing the reaction forces by solving the dynamic equations of motion when the collision of the foot with the ground occurs. When ODE detects a collision it creates a special joint called a *contact joint* for the duration of one time step. ODE obtains the reaction forces by solving the dynamic equations of motion with the constraint of the contact. The ODE engine provides the functionality to read the forces applied on each of the colliding bodies when a contact joint is created.

When a leg of our hexapod needs to be moved, in the current configuration we apply a torque on the corresponding joints of that leg. The important thing here is that we control the torque on each joint using the PID controllers. In fact we use target angle(position) to indirectly control the torque. The PID controller produces a variable torque over time that will take each joint(hence the attached limbs and the legs) from an initial position to a target position. In this way we do not have to specify the torque at every instant(or every time step). We input the target position for the joint and the controller does the task of moving from current to target position and then holding that position for us.

2.5 The PID Controller

A Proportional-Integral-Derivative (PID) controller is a generic feedback control loop mechanism used widely in industrial control systems. A PID controller calculates a control signal $u(t)$ from the difference (error) $e(t)$ between the current value of the variable to be controlled and a target value. The output $u(t)$ of a PID controller is the sum of three terms: proportional (P), integral (I) and derivative (D)

$$u(t) = K_p e(t) + K_i \int e(t)dt + K_d \frac{d}{dt} e(t) \tag{1}$$

The PID controller provides a simple means to calculate the torque to be applied to a joint such that it turns to the target angle irrespective of the load. For our purpose the error will be the difference between the target joint angle $\alpha^\star$ and the joint angle α at time t

$$e(t) = \alpha^\star - \alpha(t) \tag{2}$$

and the output $u(t)$ will be a torque. The constants K_p, K_i and K_d have to be chosen so that the output torque makes the joint reach the target angle as fast as possible and with little overshoot and oscillation around the target angle. One important point is that the error does not have to be small for the PID controller to function well.

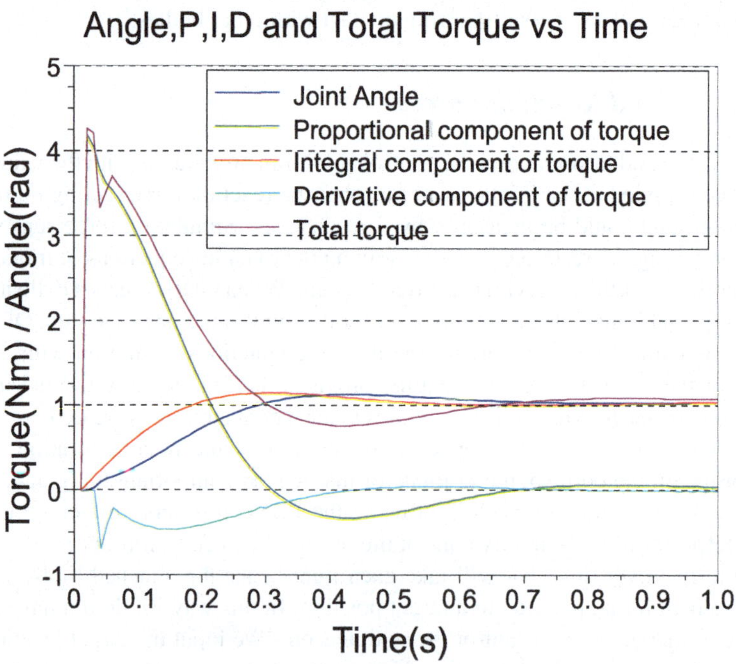

Fig. 3 Joint angle, total torque and magnitude of the PID terms for a target angle of 60 degrees for hip joint 02 of a leg, starting from 0 degrees. The other two joints remain constant at 0 degrees.

Figure 3 shows the joint angle and the torque over time for the joint between coxa and femur as it moves under PID control from being at rest at 0° to being at rest at 60°. The femur comes at rest in the new position in 0.7 seconds. The figure also shows the contribution of each of the PID terms over time.

The proportional term acts as spring pulling the joint to the target position while the integral term builds up the constant torque needed to hold the joint in its target position. The differential term provides the damping to reduce overshoot and oscillation. This example shows a numerical instability at the start arising from the derivative computation is a difference of small numbers which magnifies the initial errors of the integration of the equation of motion. To mitigate this the derivative term is ignored for the first two time steps.

3 Results

In this section we show the results of making the haxapod robot walk in a tripod gait using PID controllers for the joint angles. The fundamental gait for a hexapod is the tripod gait. The tripod gait gets its name from having always three feet on the ground. With three feet on the ground the hexapod is statically stable. Usually the legs with their feet on the ground are the front and rear leg on one side and the middle leg on the opposite side while the remaining feet are off the ground moving forward to a new foothold position. The legs on the ground are in the *stance* phase and their role is to push the body forward. The other three legs in the air are said to be in the *swing* phase. As soon as the legs in the air have set down and found their new footholds they go into the stance phase freeing the legs previously in the stance phase to switch into the swing phase. Walking comes about by the two groups of three legs alternately doing the swing and the stance phases.

The provision of a PID controller to each of the 18 joints already gives the hexapod the ability for quite robust motion. For example by providing a single target joint configuration the hexapod can get up from laying on the floor, for example with all legs stretched out, to a stable standing position. The robot will do this even if a leg temporarily slips on the ground. With the simple PID controllers the hexapod's movement exhibits a robustness that is reminiscent to that of Boston Dynamics' Big Dog robot [9]. The hexapod can be made to walk in a tripod gait on flat ground by only specifying a sequence of a few target joint configurations. For the circular body of our hexapod we divide the six legs into two groups: the even numbered legs 0,2 and 4 in Group 1, and the odd numbered legs 1,3 and 5 in Group 2. The two groups will alternate between the swing and the stance phases. Walking starts by moving the legs into a start configuration. In this configuration the coxa of all legs are parallel and perpendicular to the direction of walking. In the starting position the Hip joint 1 (Joint between body and coxa) of the legs 0,1,3 and 4 are moved by 60 degrees to be parallel with coxa of legs 2 and 5. From there on walking consists of the sequential repetition of the swing and stance phases described below.

3.1 Swing Phase

The swing phase starts at the end of the stance phase of, say Group 1. The feet of the legs of Group 1 are on the ground and are in their most backward position in relation to the body. The feet have to be taken off the ground and moved forward as far as possible to a new foothold. Because the start and the end of the swing phase are on the ground just specifying the end angles as the target positions for the PID controllers will not guarantee that the feet are lifted off the ground. Therefore an additional intermediate configuration with the foot off the ground has to be specified for each leg of the Group. The two chosen configurations are specified in Table 1. The first target position is with the feet off the ground and halfway to their forward relative position. As soon as all 3 legs of the group have reached the second target configuration and thus are again on the ground, the stance phase starts for this group of legs. At this point all the legs of the other group which were on the ground in their stance phase begin their swing phase.

Table 1 Target angles for the legs (leg group 1) moving forward in swing phase (legs are off the ground). The Hip joint 1 angles are relative to the body 's horizontal radial outward direction at the location of the joint.

	Target Position-1(degrees)	Target Position-2(degrees)	Step Size(m)
Hip joint 1-Leg 0	20°	20°	0.068
Hip joint 1-Leg 2	20°	20°	0.068
Hip joint 1-Leg 4	20°	20°	0.068
Hip joint 2-Legs 0,2,4	60°	5°	N/A
Knee joint-Legs 0,2,4	45°	10°	N/A

3.2 Stance Phase

The legs in Group 1 have just completed the swing phase and start the stance phase. The legs must stay on the ground where they touched down and must push the body forward. To do this the legs now need to move backwards relative to the hexapod body, as specified in Table-2. If the legs are moved backwards by turning only the Hip joint 1 then the feet cannot remain in their position on the ground. This is because the foot will follow a circular trajectory centred on Hip joint 1, increasing first their horizontal distance from the body and then returning to their initial position. This can only happen if the feet slide on the ground, which is a wast of energy. To prevent this, hip joint 2 and the knee joint need to change during the stance phase so that the feet remain at the same distance from the body during the whole stance phase, as shown in Figure-4.

Leg 2 (group 1) is taken as an example to illustrated this movement of the legs during the stance phase. At the end of the swing phase the foot of the leg is on the ground at Point A. After completing the stance phase, the foot should be at Point C, relative to the body. Again we need to specify an intermediate configuration to

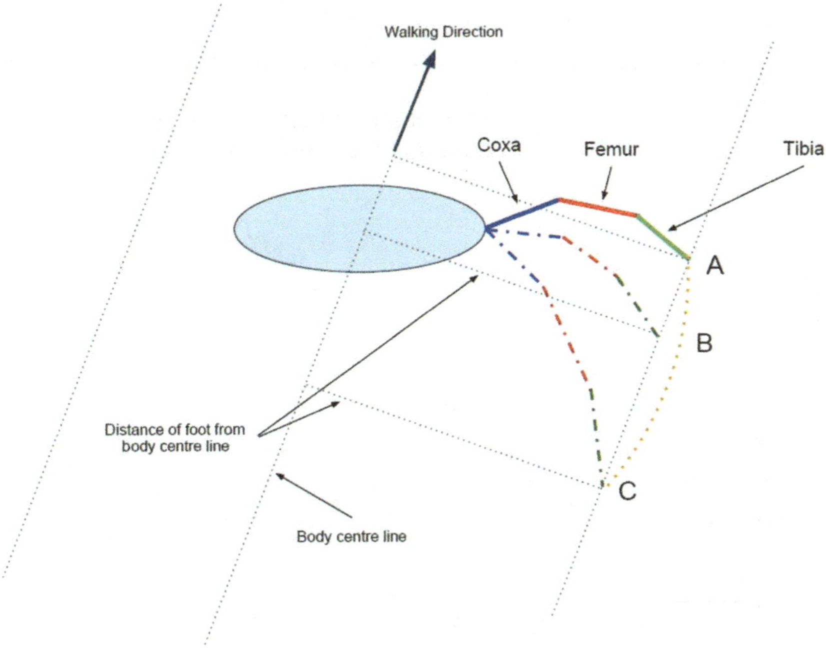

Fig. 4 Stance Phase - Movement of Leg 2 relative to the hexapod body

prevent the foot following a circular path, in relation to the body, we specify an intermediate position for the stance phase (Point B in Figure-4) that has the same distance of the foot from the body as the initial and final target configurations.

Table 2 Target angles for the legs (leg group 2) moving backward relative to the robot body. The Hip joint 1 angles are relative to the body 's horizontal radial outward direction at the location of the joint.

	Target Position(degrees)	Step Size(m)
Hip joint 1-Leg 0	-20°	0.068
Hip joint 1-Leg 2	-20°	0.068
Hip joint 1-Leg 4	-20°	0.068
Hip joint 2-Legs 0,2,4	5°	N/A
Knee joint-Legs 0,2,4	10°	N/A

When Group 1 has completed the swing phase, Group 2 will follow the swing sequence while Group 1 follows the stance sequence, and so forth. The time of the start of the next phase is determined by the moment when the previous configuration has been reached within a specified error tolerance and when the legs in swing phase touches the ground. By following this sequence

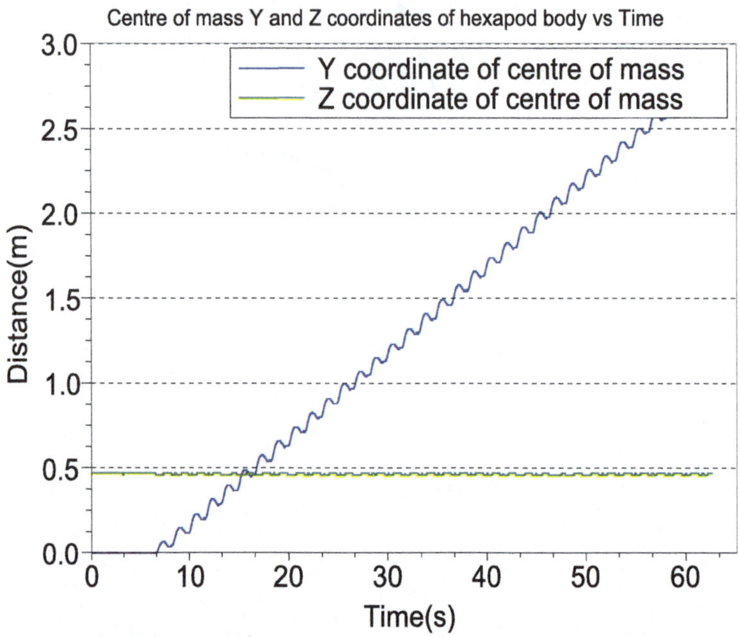

Fig. 5 Y and Z coordinate values of centre of mass of the hexapod body. Y is the directoin of walking and Z is the height from terrain.

- Hexapod does not collapse to the ground
- Keeps fairly a straight direction of walking
- When a leg slips, it recovers from the slippage and tries to get to the target position

The Figure 5 shows the y and z cordinates of the hexapod body over time. This illustrates that the hexapod is walking in y axis direction and keeps the body at a consistent height from terrain during walking. This also gives an idea about the speed of the robot, which is around 2.7m/min.

This shows that a hexapod walking in tripod gait on even ground can be achieved with 18 synchronised PID controllers and by specifying a small number of intermediate joint configurations. A video of the hexapod walking simulation can be watched on YouTube.[1]

4 Future Work

The simple and somewhat unexpected result obtained with the PID controllers is only preliminary. The aim remains to achieve robust walking through the

[1] http://www.youtube.com/user/hexapodamire

co-operation of autonomous legs. We plan to reach this goal in the successive stages described next.

4.1 PID Control with Trajectory Way Points

This is the stage reached so far and described in this paper.

4.2 Evolving Trajectory Way Points

Determine a set of optimal intermediate configurations using evolutionary optimisation. The next step would be to define a function which takes in to account the current position and the reaction forces in order to generate a torque.

4.3 Avoid Specification of Joint Configurations

The goal is to avoid the specification of joint target configuration. Without specifying target configurations PID controllers cannot be used. The PID controllers will be replaced by controllers that computes the instantaneous joint torque from the legs joint angles and the x,y and z components of the reaction forces on the foot as illustrated in figure 6. A total of 18 similar controllers will be attached to each leg joint of the hexapod.

The controller (torque function) will be implemented by a feed-forward neural network with the proper weights found by evolutionary optimisation. Evolving neural

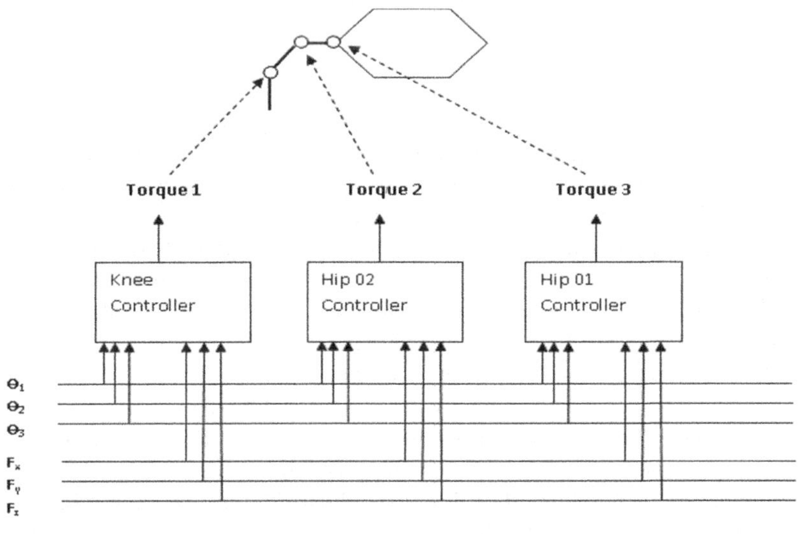

Fig. 6 All 3 controllers of a leg. Inputs are the same but the output is different.

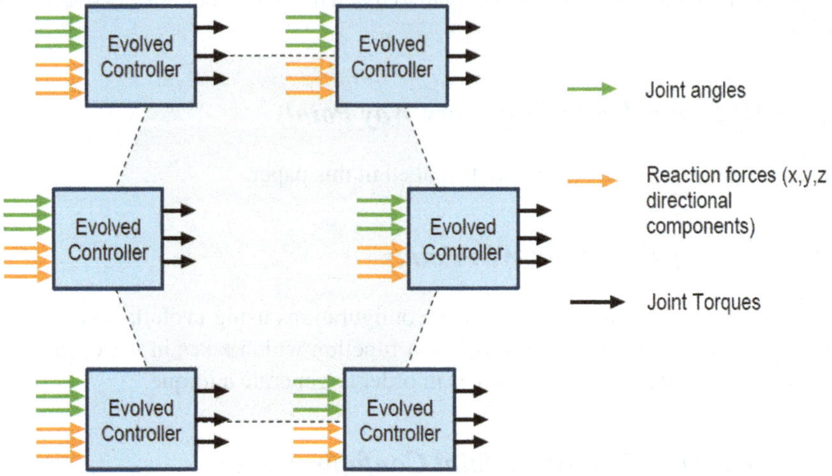

Fig. 7 Control architecture for the hexapod robot

network controllers for a hexapod has been done before [1] but without using force feedback.

Figure 7 shows the complete control architecture for the hexapod robot. The tree joint controllers are shown as a single controller with torque outputs for each joint. When one controller changes its output the effect, transmitted mechanically by the robot body, will be felt by all the other leg controllers as changes in their reaction force inputs.

5 Conclusion

We have shown that by generating the torque at each joint with a PID controller a simulated hexapod robot can achieve a wide range of motions by specifying only a small number of joint configurations between the initial pose and the final pose. For example the robot can stand up from a laying on the ground pose with all legs stretched out by just specifying the standing configuration. Furthermore the hexapod robot can be made to walk with this approach. The walking speed depends on how soon the target positions are reached using the PID controllers. This in turn depends on the weight and size of the limbs and the PID controller constants. The speed can only be controlled indirectly through by scaling the torque output by the PID controllers. The movement generated with PID controllers is remarkably robust as demonstrated by recovery from temporary slippage of one or more feet on the ground. This is reminiscent of the extraordinarily robust walking achieved by the Big Dog four legged robot developed by Boston Dynamics under a DARPA(U.S Defence Advanced Research Project Agency) project [9].

These apparent positive features of movement generation with PID controllers are limited to movement on flat surfaces. Smooth and robust walking on irregular terrain requires force sensing. The dynamic simulation we have implemented and demonstrated allows us to obtain the reaction forces in simulated terrain and provides us with the tool need to progress the work as outlined in section 4.

References

1. Belter, D., Kasinski, A., Skrzypczynski, P.: Evolving feasible gaits for a hexapod robot by reducing the space of possible solutions. In: IEEE/RSJ International Conference on Intelligent Robots and Systems, IROS 2008, pp. 2673–2678. IEEE (2008)
2. Delcomyn, F., Nelson, M.E.: Architectures for a biomimetic hexapod robot. Robotics and Autonomous Systems 30(1-2), 5–16 (2000)
3. Dürr, V., Schmitz, J., Cruse, H.: Behaviour-based modelling of hexapod locomotion: linking biology and technical application. Arthropod Structure & Development 33(3), 237–250 (2004)
4. Gorinevsky, D.M., Formalsky, A.M., Schneider, A.Y.: Force control of robotics systems. CRC Press (1997)
5. Hori, T., Kobayashi, H., Inagaki, K.: Force control for hexapod walking robot with torque observer. In: Proceedings of the IEEE/RSJ/GI International Conference on Intelligent Robots and Systems, IROS 1994. Advanced Robotic Systems and the Real World, vol. 2, pp. 1294–1300. IEEE (2002)
6. Mahapatra, A., Roy, S.S.: Computer Aided Dynamic Simulation of Six-Legged Robot. International Journal of Recent Trends in Engineering 2(2), 146–151 (2009)
7. Michel, O.: Cyberbotics Ltd. WebotsTM: Professional mobile robot simulation. International Journal of Advanced Robotic Systems 1(1), 39–42 (2004)
8. Omer, A.M.M., Ghorbani, R., Lim, H., Takanishi, A.: Semi-passive dynamic walking for biped walking robot using controllable joint stiffness based on dynamic simulation. In: IEEE/ASME International Conference on Advanced Intelligent Mechatronics, AIM 2009, pp. 1600–1605. IEEE (2009)
9. Playter, R., Buehler, M., Raibert, M.: BigDog. In: Proceedings of SPIE, vol. 6230, p. 623020 (2006)
10. Shen, W.M., Krivokon, M., Chiu, H., Everist, J., Rubenstein, M., Venkatesh, J.: Multimode locomotion via SuperBot reconfigurable robots. Autonomous Robots 20(2), 165–177 (2006)
11. Smith, R.: Open dynamics engine (ODE) (2006)
12. Zordan, V.B., Majkowska, A., Chiu, B., Fast, M.: Dynamic response for motion capture animation. ACM Transactions on Graphics (TOG) 24(3), 697–701 (2005)

Author Index

The manufacturer's authorised representative in the EU is Springer Nature Customer Service Centre GmbH, Europaplatz 3, 69115 Heidelberg, Germany. If you have any concerns regarding our products, please contact ProductSafety@springernature.com

Printed and bound by CPI Group (UK) Ltd, Croydon, CR0 4YY
07/01/2026
02030930-0003